Sensors Data Processing Using Machine Learning

Sensors Data Processing Using Machine Learning

Editors

Peter Hockicko
Róbert Hudec
Patrik Kamencay

MDPI Basel • Beijing • Wuhan • Barcelona • Belgrade • Novi Sad • Cluj • Manchester

Editors
Peter Hockicko
University of Žilina
Žilina
Slovakia

Róbert Hudec
University of Žilina
Žilina
Slovakia

Patrik Kamencay
University of Žilina
Žilina
Slovakia

Editorial Office
MDPI
St. Alban-Anlage 66
4052 Basel, Switzerland

This is a reprint of articles from the Special Issue published online in the open access journal *Sensors* (ISSN 1424-8220) (available at: https://www.mdpi.com/journal/sensors/special_issues/Sensors_data_ML).

For citation purposes, cite each article independently as indicated on the article page online and as indicated below:

Lastname, A.A.; Lastname, B.B. Article Title. *Journal Name* **Year**, *Volume Number*, Page Range.

ISBN 978-3-7258-1171-7 (Hbk)
ISBN 978-3-7258-1172-4 (PDF)
doi.org/10.3390/books978-3-7258-1172-4

Contents

About the Editors

Peter Hockicko

Peter Hockicko defended his Ph.D. theses in the field of physics of condensed matter and acoustics in 2008. Since 2013, he worked as a Tutor, a Lecturer, and a Researcher with the Department of Physics, Faculty of Electrical Engineering and Information Technology, University of Žilina, where he has been working as an Associate Professor. His scientific research interest includes relaxation processes in materials.

Róbert Hudec

Róbert Hudec received the M.S. and Ph.D. degrees from the Faculty of Electrical Engineering and Informatics, Technical University of Košice, in 1998 and 2003, respectively. In 2016, he became a Full Professor with the University of Žilina, Slovakia, where he currently leads the Department of Multimedia and Information-Communication Technology. In 2017, he and his collaborators from VUTCH-CHEMITEX Žilina, Slovakia, received the Gold Fatima Award for a smart garment prototype that continuously monitors the EKG signals.

Patrik Kamencay

Patrik Kamencay received the Ph.D. degree in telecommunications from the University of Žilina, Slovakia, in 2012. He is currently an Associate Professor with the Department of Multimedia and Information-Communication Technology, University of Žilina. His research interests include digital processing areas as diverse as image data collection, processing and machine learning methods, and neural networks, with a focus on image data segmentation, classification, 2-D/3-D recognition, and 3-D reconstruction.

sensors

MDPI

Editorial

Sensors Data Processing Using Machine Learning

Patrik Kamencay *, Peter Hockicko and Robert Hudec

Faculty of Electrical Engineering and Information Technology, University of Zilina, 010 26 Zilina, Slovakia; peter.hockicko@feit.uniza.sk (P.H.); robert.hudec@feit.uniza.sk (R.H.)
* Correspondence: patrik.kamencay@feit.uniza.sk

Various sensors utilize computational models to estimate measured variables, and the generated data require processing. Data processing involves transforming data from a given format into a more usable and desirable form, rendering them more meaningful and informative. Machine learning (ML), deep learning (DL), and artificial intelligence (AI) have proven effective for this purpose. The entire process can be automated using machine learning algorithms, mathematical modeling, or various statistical techniques.

The aim of this Special Issue was to compile research on data processing through machine learning and deep learning. It features both original and review articles that address research and development in data processing using machine learning (ML) and deep learning (DL). These areas include solutions designed for smart devices.

The first paper [1] focuses on detecting toxicity in online discussions. The authors used classification models that incorporate machine learning methods to classify short texts on social networking sites into multiple degrees of toxicity. Their models used both classic methods of machine learning, such as naive Bayes and SVM (support vector machine), as well as ensemble methods, such as bagging and RF (random forest). The models were created using text data, which they extracted from social networks in the Slovak language. Finally, an application was created based on machine learning models, which can be used to detect the degree of toxicity of new social network comments, as well as for experimentation with various machine learning methods. The best results were achieved using an SVM—average value of accuracy = 0.89 and F1 = 0.79. This model also outperformed the ensemble learning by the RF and Bagging methods; however, the ensemble learning methods achieved better results than the naïve Bayes method.

The second paper [2] introduces a framework for evaluating segments of physical and digital infrastructure, which is designed to assess their features and readiness for facilitating the deployment of Connected and Automated Vehicles (CAVs). It delves into the equipment and methodology employed to collect and analyze the necessary data for automated scoring of infrastructure segments. The authors illustrate the assessment methodology using two types of data: connectivity and positioning data for evaluating the infrastructure's connectivity and localization performance, and image data for road signage detection through a Convolutional Neural Network (CNN). The data collection and analysis were conducted in both urban and suburban settings. The primary communication challenge in the examined area is latency, particularly in infrastructure segments located at busy intersections or near various points of interest. The study observed lower localization accuracy in dense areas with large buildings and trees, limiting the visibility of localization satellites. To address the challenge of automated traffic sign recognition precision assessment, the authors proposed a CNN that achieved a precision rate of 99.7%.

The authors of [3] introduce an enhanced IoT-based system to assist teachers in managing classroom activities in adherence to COVID-19 restrictions. The system, which comprises three components—an entry Gate node, IoT nodes, and a server—comprehensively monitors the number of individuals in the classroom and their spatial distribution. The Gate node, positioned at the entrance, tracks individuals entering or exiting the room through door crossing detection, while IoT nodes, based on Arduino with NodeMCU

Citation: Kamencay, P.; Hockicko, P.; Hudec, R. Sensors Data Processing Using Machine Learning. *Sensors* **2024**, *24*, 1694. https://doi.org/10.3390/s24051694

Received: 6 February 2024
Accepted: 1 March 2024
Published: 6 March 2024

modules and ultrasonic distance sensors, collect data on seat occupancy. The server, hosted on a Raspberry Pi, allows teachers to connect to it via a web application from a classroom computer or smartphone. The teacher can configure and modify system settings through the graphical user interface (GUI) provided by the web application. A straightforward algorithm assesses the distance between occupied seats, ensuring compliance with imposed restrictions. Notably, this system prioritizes privacy, distinguishing it from camera-based alternatives.

Meanwhile, the authors of [4] suggest undertaking the text classification task using the two previously mentioned models for two languages (English and Brazilian Portuguese) across distinct datasets. According to their findings, DistilBERT exhibits a training time approximately 45% faster for both English and Brazilian Portuguese compared to its larger counterpart. Furthermore, it is around 40% smaller yet maintains approximately 96% of language comprehension skills, particularly for balanced datasets.

In [5], a multi-delay identification method is proposed based on improved time-correlation analysis. Initially, the data undergo gray relational analysis for preprocessing, leading to the construction of a time delay sequence and a data matrix for time correlation. The multi-delay identification problem is subsequently reformulated as an integer optimization problem. The optimization is performed using an enhanced discrete state transition algorithm to acquire multi-delay. Lastly, to assess its performance, the proposed method is compared with the unimproved time delay identification method and the model without an identification method, utilizing a Neodymium (Nd) component content model constructed by a wavelet neural network. The proposed algorithm enhances optimization accuracy, convergence speed, and stability. The effectiveness of the proposed method is further validated by the significant improvement in the performance of the component content model after time delay identification, particularly in the context of the rare earth extraction process.

The analysis presented in [6] focuses on the attention heat map of benchmarks, revealing that prior models placed greater emphasis on individual phrases rather than capturing the holistic semantic information of the entire sentence. Additionally, a strategy was introduced to disperse attention away from opposing sentiment words, preventing one-sided judgments. A two-stream network was devised, incorporating the gradient reversal layer and feature projection layer within the auxiliary network. The gradient reversal layer was employed to invert the gradient of features during training, optimizing parameters based on the reversed gradient in the backpropagation stage. An auxiliary network was utilized to extract backward features, which were then integrated into the main network along with the standard features obtained by the main network. This approach was implemented across three baseline models—TextCNN, BERT, and RoBERTa—using sentiment analysis and sarcasm detection datasets. The outcomes demonstrated a 0.5% enhancement for sentiment analysis datasets and a 2.1% improvement for sarcasm detection datasets.

The authors of [7] investigate the detrimental effects of packet loss on the video quality encoded using different combinations of compression parameters and resolutions. Their research utilizes a dataset comprising 11,200 full HD and ultra HD video sequences encoded in H.264 and H.265 formats across five bit rates, incorporating a simulated packet loss rate (PLR) ranging from 0 to 1%. Objective assessment relied on peak signal-to-noise ratio (PSNR) and Structural Similarity Index (SSIM) metrics, while subjective evaluation employed the widely recognized absolute category rating (ACR). Their results confirmed the anticipated decline in video quality with an increase in packet loss rate, irrespective of compression parameters. The experiments also revealed that the quality of sequences affected by PLR diminishes with higher bit rates. The paper concludes with recommendations for compression parameters suitable for various network conditions.

The primary goal of [8] was to optimize the conventional 3DCNN model and introduce a novel architecture that integrates 3DCNN with Convolutional Long Short-Term Memory (ConvLSTM) layers. Through experiments conducted on the LoDVP Abnormal Activities dataset, UCF50 dataset, and MOD20 dataset, the results highlight the superior performance of the 3DCNN + ConvLSTM fusion in the realm of human activity recognition. The proposed model is suitable for real-time human activity recognition applications and holds potential for further improvement with the incorporation of additional sensor data. To offer a comprehensive comparison, they evaluate the proposed 3DCNN + ConvLSTM architecture across these datasets. The authors achieved 89.12% precision for the LoDVP Abnormal Activities dataset, and for the modified UCF50 dataset (UCF50mini) and MOD20 dataset, they achieved 83.89% and 87.76% precision, respectively. In summary, their study underscores the efficacy of combining 3DCNN and ConvLSTM layers to enhance accuracy in human activity recognition tasks, positioning the proposed model as a promising candidate for real-time applications.

The authors of [9] outline the development of an Internet of Things (IoT)-based connected university system. Though various smart solutions have emerged at the university, their adoption has been limited among users. The IoT-based connected university system addresses this by facilitating the integration of multiple subsystems, allowing end-users to access diverse solutions through a unified interface. Employing a microservices architecture, the system prioritizes robustness, scalability, and universality. Currently, four subsystems are implemented: indoor navigation, parking assistants, smart classrooms/offices, and news aggregation from university life. The paper comprehensively details the principles governing each subsystem and presents the system's implementation as both a web interface and a mobile application. A detailed account of the indoor navigation subsystem using Bluetooth beacons is also provided. The paper includes a thorough presentation of the Bluetooth-based indoor navigation concept, considering diverse node placements. Real-world tests were conducted to assess the feasibility of the navigation module, employing deterministic fingerprinting algorithms for precise estimation of users' device positions.

The research presented in [10] evaluates the usability of several Apple MacBook Pro laptops for basic machine learning research applications, encompassing text-based, vision-based, and tabular data. Four distinct benchmarks were executed, employing four MacBook Pro models—M1, M1 Pro, M2, and M2 Pro. A Swift-script was employed to train and assess four machine learning models utilizing the Create ML framework, in three iterations. The script also recorded performance metrics, particularly time-related outcomes. The findings are presented in tabular form, facilitating a comparative analysis of each device's performance and the influence of their respective hardware architectures.

The research presented in [11] introduces an inventive data augmentation strategy aimed at identifying distinct student behaviors by leveraging focused behavioral attributes. The primary goal is to alleviate the pedagogical workload. The first step is to curate a concise dataset tailored for discerning student learning behaviors, followed by the application of data augmentation techniques to significantly expand its size. Moreover, the architectural prowess of the Extended-efficient Layer Aggregation Networks (E-ELAN) is harnessed to effectively extract a diverse array of learning behavior features. Notably, integrating the Channel-wise Attention Module (CBAM) focal mechanism into the feature detection network enhances the network's ability to detect key cues relevant to student learning behaviors, thereby improving feature identification precision. The methodology concludes with the classification of the extracted features through a dual-pronged conduit: the Feature Pyramid Network (FPN) and the Path Aggregation Network (PAN). Empirical evidence vividly demonstrates the potency of the proposed methodology, yielding a mean average precision (mAP) of 96.7%. This accomplishment surpasses comparable methodologies by a substantial margin of at least 11.9%, conclusively highlighting the method's superior recognition capabilities. This research has significant implications for teaching evaluation systems, reducing the burden on educators while enhancing the objectivity and accuracy of teaching evaluations.

The authors of [12] explore a self-supervised binary classification algorithm designed for defect image classification within ductile cast iron pipe (DCIP) images. Utilizing the CutPaste-Mix data augmentation strategy, they amalgamate defect-free data with enhanced data, feeding them into a deep convolutional neural network. Gaussian Density Estimation is then employed to compute anomaly scores, facilitating the classification of abnormal regions. The proposed approach has been implemented in several real-world scenarios, encompassing equipment installation, data collection, and experimentation. The results showcase the robust performance of the method, which is evident in both the DCIP image dataset and practical field applications, achieving an impressive 99.5 AUC (Area Under Curve). It is a cost-effective method for providing data support for subsequent DCIP surface inspection model training.

In [13], three images—Sentinel-2, GF-1, and Landsat 8—were chosen, and three sample selection methods, namely grouping selection, entropy-based selection, and direct selection, were applied. Subsequently, the selected training samples were utilized to train three supervised classification models—random forest (RF), support-vector machine (SVM), and k-nearest neighbor (KNN). The classification results of the three images were then evaluated. The experimental outcomes indicated similar performances among the three classification models. Notably, the grouping selection method achieved higher classification accuracy using fewer samples compared to the entropy-based method. Furthermore, compared to the direct selection method with an equal number of samples, the grouping selection method exhibited superior performance. Hence, the grouping selection method demonstrated the most favorable outcomes. Additionally, when employing the grouping selection method, the image classification accuracy demonstrated an increase with the augmentation of the number of samples within a specified sample size range.

Author Contributions: Conceptualization, P.K. and R.H.; Methodology, P.K., P.H. and R.H.; Formal analysis, P.K., P.H. and R.H.; Investigation, P.K., P.H. and R.H.; Resources, P.K. and R.H.; Writing—original draft preparation, P.K. and R.H.; Supervision, P.H. and P.K.; Project administration, R.H. and P.K.; Funding acquisition, R.H. All authors have read and agreed to the published version of the manuscript.

Funding: This research was supported by a Slovak Research and Development Agency under contract No. APVV-21-0502: BrainWatch: System for automatic detection of intracranial aneurysms.

Acknowledgments: The Guest Editors wish to thank all the authors who submitted manuscripts for consideration in *Sensors*, and the reviewers for their hard work during the review process. Furthermore, we sincerely thank the editors of *Sensors* for their help and support. We hope that the readers enjoy reading the articles within this Special Issue. Finally, the Guest Editors wish to acknowledge partial support from the Slovak Research and Development Agency under contracts no. PP-COVID-20-0100: DOLORES.AI: The pandemic guard system.

Conflicts of Interest: The authors declare no conflict of interest.

List of Contributions

1. Machová, K.; Mach, M.; Adamišín, K. Machine Learning and Lexicon Approach to Texts Processing in the Detection of Degrees of Toxicity in Online Discussions. *Sensors* **2022**, *22*, 6468. https://doi.org/10.3390/s22176468.
2. Cucor, B.; Petrov, T.; Kamencay, P.; Pourhashem, G.; Dado, M. Physical and Digital Infrastructure Readiness Index for Connected and Automated Vehicles. *Sensors* **2022**, *22*, 7315. https://doi.org/10.3390/s22197315.
3. Matuska, S.; Machaj, J.; Hudec, R.; Kamencay, P. An Improved IoT-Based System for Detecting the Number of People and Their Distribution in a Classroom. *Sensors* **2022**, *22*, 7912. https://doi.org/10.3390/s22207912.
4. Silva Barbon, R.; Akabane, A.T. Towards Transfer Learning Techniques—BERT, DistilBERT, BERTimbau, and DistilBERTimbau for Automatic Text Classification from Different Languages: A Case Study. *Sensors* **2022**, *22*, 8184. https://doi.org/10.3390/s22218184.
5. Lu, R.; Liu, H.; Yang, H.; Zhu, J.; Dai, W. Multi-Delay Identification of Rare Earth Extraction Process Based on Improved Time-Correlation Analysis. *Sensors* **2023**, *23*, 1102. https://doi.org/10.3390/s23031102.
6. Seng, D.; Wu, X. Enhancing the Generalization for Text Classification through Fusion of Backward Features. *Sensors* **2023**, *23*, 1287. https://doi.org/10.3390/s23031287.
7. Bienik, J.; Uhrina, M.; Sevcik, L.; Holesova, A. Impact of Packet Loss Rate on Quality of Compressed High Resolution Videos. *Sensors* **2023**, *23*, 2744. https://doi.org/10.3390/s23052744.

8. Vrskova, R.; Kamencay, P.; Hudec, R.; Sykora, P. A New Deep-Learning Method for Human Activity Recognition. *Sensors* **2023**, *23*, 2816. https://doi.org/10.3390/s23052816.
9. Matuska, S.; Machaj, J.; Hutar, M.; Brida, P. A Development of an IoT-Based Connected University System: Progress Report. *Sensors* **2023**, *23*, 2875. https://doi.org/10.3390/s23062875.
10. Kasperek, D.; Antonowicz, P.; Baranowski, M.; Sokolowska, M.; Podpora, M. Comparison of the Usability of Apple M2 and M1 Processors for Various Machine Learning Tasks. *Sensors* **2023**, *23*, 5424. https://doi.org/10.3390/s23125424.
11. Wang, Z.; Li, L.; Zeng, C.; Yao, J. Student Learning Behavior Recognition Incorporating Data Augmentation with Learning Feature Representation in Smart Classrooms. *Sensors* **2023**, *23*, 8190. https://doi.org/10.3390/s23198190.
12. Zhang, H.; Sun, Q.; Xu, K. A Self-Supervised Model Based on CutPaste-Mix for Ductile Cast Iron Pipe Surface Defect Classification. *Sensors* **2023**, *23*, 8243. https://doi.org/10.3390/s23198243.
13. Zhang, H.; He, J.; Chen, S.; Zhan, Y.; Bai, Y.; Qin, Y. Comparing Three Methods of Selecting Training Samples in Supervised Classification of Multispectral Remote Sensing Images. *Sensors* **2023**, *23*, 8530. https://doi.org/10.3390/s23208530.

sensors

MDPI

Article

Machine Learning and Lexicon Approach to Texts Processing in the Detection of Degrees of Toxicity in Online Discussions

Kristína Machová *, Marián Mach and Kamil Adamišín

Department of Cybernetics and Artificial Intelligence, Faculty of Electrical Engineering and Informatics, Technical University of Košice, Letná 9, 04200 Kosice, Slovakia
* Correspondence: kristina.machova@tuke.sk; Tel.: +421-902-683736

Abstract: This article focuses on the problem of detecting toxicity in online discussions. Toxicity is currently a serious problem when people are largely influenced by opinions on social networks. We offer a solution based on classification models using machine learning methods to classify short texts on social networks into multiple degrees of toxicity. The classification models used both classic methods of machine learning, such as naïve Bayes and SVM (support vector machine) as well ensemble methods, such as bagging and RF (random forest). The models were created using text data, which we extracted from social networks in the Slovak language. The labelling of our dataset of short texts into multiple classes—the degrees of toxicity—was provided automatically by our method based on the lexicon approach to texts processing. This lexicon method required creating a dictionary of toxic words in the Slovak language, which is another contribution of the work. Finally, an application was created based on the learned machine learning models, which can be used to detect the degree of toxicity of new social network comments as well as for experimentation with various machine learning methods. We achieved the best results using an SVM—average value of accuracy = 0.89 and F1 = 0.79. This model also outperformed the ensemble learning by the RF and Bagging methods; however, the ensemble learning methods achieved better results than the naïve Bayes method.

Keywords: web mining; detection of degrees of toxicity; machine learning; lexicon approach; text data processing

Citation: Machová, K.; Mach, M.; Adamišín, K. Machine Learning and Lexicon Approach to Texts Processing in the Detection of Degrees of Toxicity in Online Discussions. *Sensors* **2022**, *22*, 6468. https://doi.org/10.3390/s22176468

Academic Editors: Peter Hockicko, Róbert Hudec and Patrik Kamencay

Received: 28 July 2022
Accepted: 25 August 2022
Published: 27 August 2022

1. Introduction

Social networks today allow us to express publicly our agreement or disagreement with other people, their opinions or social behavior. This freedom is often abused in the online space and that is why we can see social networks that are full of toxic comments dealing with the ongoing pandemic, political or social situations. The increase in textual data on the Internet has stimulated the emergence of new scientific fields that examine short texts in the online space and look for toxic posts, trolls, that try to detect the polarity of comments, and that try to solve other tasks within the processing of textual data.

We focused our research on toxicity detection while distinguishing several degrees of toxicity. Our research was motivated by a serious problem of the spread of toxicity and anti-social behavior in the online space that almost all of us use. All types of antisocial behaviors affect democracy in many countries and contribute to the polarization of a society (Tristan Harris—former expert on the ethics of a design in Google, and cofounder of the Center for Human Technologies). Often, the spreading of toxic posts tries to manipulate the opinions of users looking for answers in the online space. Modern social media companies utilize constant monitoring of users to keep their attention and consequently to ensure the success of advertising. These information technologies can, thus, channel users toward content that causes rudeness, a lack of trust, loneliness and societal polarization, and they can indirectly help electoral manipulation and the spread of populism. For these reasons, we consider it important to help with the automatic monitoring and removal of inappropriate content on social media.

We can divide all toxic posts and comments into types based on the cause of toxicity, for example: cultural–ethnically based, insulting a cultural–ethical group of people; racially based, insulting one race; based on religion; sexually based; physically based—insults that are based on health or physical condition adaptations; personal, insulting a specific person; politically based—insults that are based on a political opinion.

Such online harassment affects both adults and children. Research published in [1] focused on identifying the groups of people who are most at risk. The most influential attributes of online harassment in adults are age and gender and this mostly concerns young men aged 18 to 29 and women aged 18 to 24.

Study [2] analyses hate speech using a web interface with a focus on the most used social networks, such as Twitter, YouTube, and Facebook. Individual social networks try to protect their users. For example, users who are bothered by inappropriate posts can report these posts. Then, the administrators of social networks check the reported posts and delete them if they are really toxic, but this manual regulation is no longer enough today; therefore, automated systems are being developed to detect toxicity. This automatic detection is based on either machine learning methods or a lexicon-based approach.

We have used a lexicon-based approach in the past in our work focused on the detection of suspicious online reviewers [3]. In this work, we used a lexicon-based approach in a sentiment analysis method as an alternative to machine learning methods for the detection of trolls. Here, we instead use a lexicon-based approach in a new method for the automatic labelling of extracted posts in a dataset. We can summarize the contribution of our paper as follows:

- Creation of a dataset by extracting comments from social media in the Slovak language. There is lack of text data for processing in the Slovak language.
- Automatic labelling of extracted data using our lexicon-based method, which represents our original approach to texts labelling.
- Creation of a lexicon of toxic words in the Slovak language as an essential resource needed for the lexicon-based labelling method.
- Comparison of detection models trained by classic machine learning methods (e.g., naïve Bayes, and SVM) and ensemble learning (e.g., Bagging, and RF).
- Creation of an application using the detection models.

1.1. The Toxicity Detection Using Machine Learning

The detection of the degrees of toxicity in online discussions is a multi-classification problem; therefore, using the classification methods of supervised machine learning methods is a natural choice. An example of this approach is the work published in [4], which presents an approach to determine the toxicity of vulgar posts on Twitter using partial learning with a teacher and the logical regression method, where they achieved a true positive rate of 75.1%.

Determining hateful comments in the Italian language was dealt with in the work of [5]. They used SVM (support vector machine) and LSTM (long short-term memory) models trained on a sample of 17,000 comments from Facebook. For the experiment, they used only the data that was annotated by at least three annotators. They divided the comments into three classes: comments without hate, comments with a slight hint of hate, and hateful comments. They further divided the hateful comments into categories based on what the comments were about, namely, religion, physical or mental health, socio-economic status, politics, race, or gender. The result was two documents, one with 3356 comments, divided into 2816 non-hateful comments, 410 mildly hateful comments, and 130 strongly hateful comments. The other document had 3575 comments, divided into 2789 non-hateful comments and 786 hateful comments. Their best results were an accuracy of 80.60% using an SVM and 79.81% using a LSTM for the model training.

A work dedicated to the Arabic language [6] detected the use of offensive words. To create the model, they chose data from Aljazeera.net, which contained 400,000 comments on approximately 10,000 articles. Subsequently, 32,000 offensive comments were selected from them. The selected comments were annotated using CrowdFlower, which divided them into three groups: obscene, offensive, and neutral comments. They achieved their best result of a 0.98 precision using the LOR (log odds ratio) with unigrams.

Since online comments or posts are often informal, unstructured, and poorly written, a problem arises when classical models are trained to detect toxicity. Due to these problems, the authors in [7] proposed to detect the toxicity of posts using the lexical syntactic feature (LSF) architecture, which is used to detect offensive content and to identify potential offensive users in interactive media. As a result, the LSF architecture performed significantly better than the existing methods in detecting toxic posts, achieving a 98.24% accuracy in detecting toxic (attacking) posts and up to a 77.9% accuracy in identifying those users. This method, however, was only able to find 78.86% of toxic contributions.

Additionally, the work of [8] used the SVM, RF and naïve Bayes machine learning methods combined with TF-IDF (term frequency—inverse document frequency) and word embedding representations for cyberbullying and aggressiveness detection in Tweets in the Chilean and Mexican Spanish languages. In this work, all the SVM models obtained better results than the others, with up to 89.2% accuracy and an 89% F1 rate.

Moreover, the work of [9] used neural network methods (e.g., BiGRU—bidirectional gated recurrent unit and BiLSTM—bidirectional long short-term memory) and classic methods of machine learning in combination with TF-IDF and GloVe (global vector) representations for cyberbullying detection. Across all the preprocessing steps, the logistic regression displayed the highest average performance amongst all the machine learning techniques used, followed by SVM, XG Boost (extreme gradient boosting), and naïve Bayes in that order. They achieved the best results using neural networks—with accuracy and F1 scores as high as ~95% and ~98%, respectively.

The machine learning approaches are also commonly used for sentiment analysis. Sentiment analysis is, however, related to toxicity recognition, because toxicity in online spaces usually represents a negative opinion. There are some works which have used machine learning approaches for sentiment analysis, for example, [10] developed an ensemble learning scheme using DT (decision trees), SVM, RF (random forest—of decision trees) and KNN (k-nearest neighbors) for a sentiment analysis of COVID-19 related comments. Additionally, in the work of [11], deep learning models for a sentiment analysis were used in recommender systems. There are also some related works using sentiment analysis based on machine learning.

1.2. Lexicon-Based Approaches to Toxicity Detection

The lexicon-based approach was originally used for the sentiment analysis of texts, but it has also been used in the creation of recommender systems [11]. The detection of toxicity using a lexicon-based approach analyses individual words, phrases and post or comment sentences using lexicons. There are several types of lexicons and they can be divided according to the language they use (the most widespread are English lexicons) and the goal of analysis on what words are used (for example, the recognition of toxic, positive, or negative posts). Lexicons can also be created by merging several other lexicons, translating foreign language lexicons into another language or by adding certain words to the already existing lexicon that can help us in recognition toxicity.

WordNet is the most famous lexicon, which contains a database of English words such as nouns, adjectives, adverbs, and verbs, that are grouped into synonym sets or so-called synsets. WordNet contains 117,000 English synsets and groups them according to their meaning. WordNet also indicates the semantic relationships between words [12].

The Macquarie Semantic Orientation Lexicon (MSOL) is an English dictionary that contains 30,458 positive and 45,942 negative words and phrases. It is a generated lexicon created using a Roget-like thesaurus and Macquarie thesaurus [13].

A very important issue in lexicon building is the correct annotation of words in the lexicon by a measure of toxicity. This annotation can be provided by humans, but more efficient is an annotation using an optimization method. In the paper of [14], PSO (particle swarm optimization) and BBPSO (bare-bones particle swarm optimization) were used. The problem with lexicon-based approaches is that they cannot extract opinions with a specific orientation. In that work, the corpus-based approach was used to solve this problem. For universal words that are not specific to the domain, they look for other sentimental words in a corpus that already have their specific orientation and then adapt this universal lexicon for a new list of sentimental words specific to the given corpus.

2. Materials and Methods

When solving the task of detection, namely, the degree of toxicity of online posts, we decided to use a hybrid method, which consisted of two approaches in two steps. The lexicon-based approach was used in the first step to label the toxic comments in the dataset. The machine learning approach was then used in the second step to train the classification models using machine learning methods.

For training the models, we used our own text data, which we labelled with the degree of toxicity using the lexicon approach. The data was extracted from Facebook and Instagram in the Slovak language. From the machine learning methods, we decided to use classical methods such as naive Bayes, random forest, SVM and bagging, which are usually successful in text classification.

2.1. Data Description

We prepared the dataset by downloading the comments in the Slovak language from Facebook and slightly fewer comments from Instagram. The posts that contained comments came primarily from the news profiles, namely, "*Televízne noviny TV Markíza*" and "*Television TA3*". Several dozen comments also came from the public profiles of specific people. We downloaded the comments in two ways. The first was an extraction using the freely available Export Comments downloader (http://exportcomments.com/ (accessed on 26 July 2022)). To download using this tool, it is necessary to enter the URL address of the post as an input, and then download them in .xlsx or .csv formats. This method of data extraction has its limitations.

The second method we used was to create our own tool for the comment extraction. This was implemented in the Python language, with help from logging into our Facebook account and sending a POST request to "https://m.facebook.com/login.php" (accessed on 26 July 2022) with data stored in the JSON format that contained our login information. After logging into the account, we gave the program the URL address of the post on Facebook that we wanted to download. Subsequently, the page data was retrieved using a GET request. Each comment on the page was stored in the same *<div>* variable and marked with the same class. After searching for comments in the code, we saved them in a .csv file.

The comments downloaded by us had to be pre-processed before further use. The Slovak language is characteristic in its use of common diacritical marks. It is the same with punctuation marks—periods, commas, question marks, exclamation points, colons, and others. Some people on social networks use diacritical marks and punctuation marks, but most people do not use them. When processing the comments, it was necessary to unify the texts and remove all of them. Another pre-processing step was to convert uppercase letters to lowercase letters, since our lexicon only contains words with lowercase letters. The uppercase letters when determining toxicities play almost no role. Some comments also contained an image or a link to another page, which were also unusable for our model training, and deleted.

The resulting dataset contained 3092 labelled texts of posts. This dataset titled, "Toxic_training_data.csv" is available at: https://kristina.machova.website.tuke.sk/ (accessed on 26 July 2022), in the folder RESEARCH at the end of page between "Useful links" (https://kristina.machova.website.tuke.sk/useful/Toxic_training_data.csv (accessed on 26 July 2022)).

2.2. Used Lexicon Approach for Data Labelling

The lexicon-based approach was used in our work to find toxic words from a pre-created lexicon in the extracted comments, and then to label these comments with their degree of toxicity. This labelling is necessary so that we can train detection models to determine the degree of toxicity of unknown comments using supervised learning methods. We classified the comments into four classes, namely:

- *neutral posts*—do not contain any negative words from the lexicon marked by us,
- *weakly toxic posts*—contain fewer toxic words from the lexicon,
- *moderately toxic posts*—contain words from the group of moderately toxic words from the lexicon,
- *very toxic posts*—contain extremely toxic words of the lexicon.

We have created the lexicon for labelling using our domain-independent dictionary created in the past, "lexicon_Small_human.json" (available at: https://kristina.machova.website.tuke.sk/ (accessed on 26 July 2022) in the folder RESEARCH at the end of page between "Useful links") for determining the polarity of posts, which contained negative words (also positive, but we did not take them into account in this work). We edited and supplemented this lexicon of negative words with toxic words of varying degrees of toxicity. We annotated the degrees of the toxicity of words manually. In the lexicon, there are words labelled with the values "1" for the least toxic words, "2" for moderately toxic words and "3" for very toxic offensive words and swear words in the Slovak language. The dictionary is stored in the JSON format and contains a total of 809 toxic and offensive words, of which there are:

- 224 words with a toxicity level "1" (mildly toxic words),
- 243 words with a toxicity level "2" (moderately toxic words),
- 342 words with a toxicity level "3" (very toxic words).

The resulting Slovak language lexicon "Lexicon_of_toxic_words.json" is available at: https://kristina.machova.website.tuke.sk/useful/Lexicon_of_toxic_words.json (accessed on 26 July 2022).

For the pre-processing of data sets, the lexicon-based approach to labelling and use of training methods, we used the Java programming language. After pre-processing, each comment was divided into an array of words—tokens—and stored in a variable. The evaluation of the comment was carried out from individual words in this field, where each word was looked up in the created toxic lexicon and the toxicity value of the word was returned. We used two ways to evaluate the final toxicity of a comment:

1. the label value is the sum of the individual toxic words in the comment,
2. the label value is the value of the most toxic word in the comment.

When learning and testing the models, we found that the second method of determining the final value of a comment as the value of the most toxic word achieves better results. A comment that did not contain any words from the toxic lexicon was labelled as neutral. The values of the labels are in the form of a "float", and we needed to create some differences between the degrees of toxicity of the comments. In our case, they were set as follows:

- the value 0.0 belongs to Neutral,
- the value 1.0 belongs to Low toxic,
- the value 2.0 belongs to Moderately toxic,
- values 3.0 and more belong to Very toxic comments.

The dataset of labelled comments using the lexicon-based approach was subsequently stored in the form of a CSV file with the columns "Comments" (containing comments in the form of text (string)) and "Class" (containing the values Neutral, Low toxic, Moderately toxic or Very toxic in the form of a nominal attribute). An illustration of the data can be seen in Figure 1. To use the Weka tool for model training, we needed to convert the data from CSV to ARFF.

No.	1: Comments Nominal	2: Class Nominal
1	STRACH je svina	Moderaly toxic
2	Musim bohuzial lebo natlaky vyhrazky atd Uz by t...	Low toxic
3	Ani slovko na markize v tv novinach o 17 00 hod ...	Neutral
4	ESTE SA STALE NENAUCILI ZAKONY NEPREK...	Neutral
5	V niektorych okresoch nie	Neutral
6	Biznis je Matelkov biznis	Neutral

Figure 1. Illustration of the dataset of labelled comments extracted from social networks.

2.3. Used Methods for Models Training

We used classical learning as well as ensemble learning for training the detection models. From the classic machine learning approaches, we chose the naïve Bayes classifier as a baseline method, as well as an SVM, because of its performance in related works [5,8–10]. This method is very efficient in text data processing. From the ensemble methods, we selected bagging as a basic approach and random forests, because it could increase the precision of the final classification by vote among a set of de-correlated tree models.

The naïve Bayes classifier (NB) is a probabilistic classifier based on Bayes' theorem and independence assumption between features. The advantage of this algorithm is its simplicity and clarity, as each class is characterized by a probabilistic description. On the contrary, the disadvantage of the method, as mentioned, is the assumption that attributes are independent. NB is often applied as a baseline method; however, it is reported to be outperformed by support vector machines [15].

Support vector machines (SVM) separate the sample space into two or more classes with the widest margin possible. The method was originally a linear classifier, which creates a decision boundary for the separation of examples in the space into two classes "1" and "−1" [16], as can be seen in Figure 2.

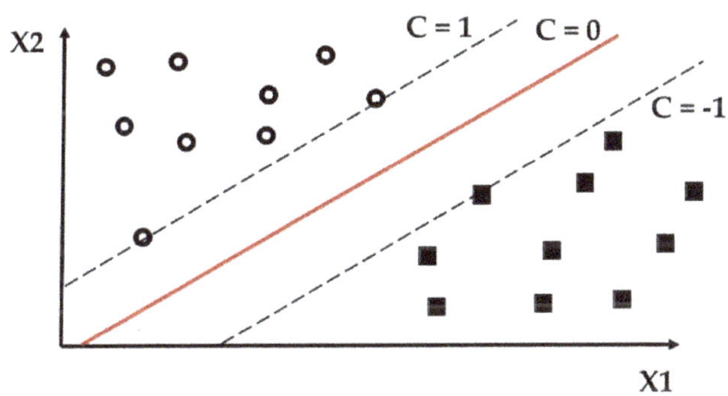

Figure 2. Illustration of the linear SVM.

The linear SVM is suitable only for linear data, therefore, it is not a good selection for text data, but it can relatively efficiently perform a non-linear classification using kernel functions [17]. The trick of kernels functions is illustrated by Figure 3.

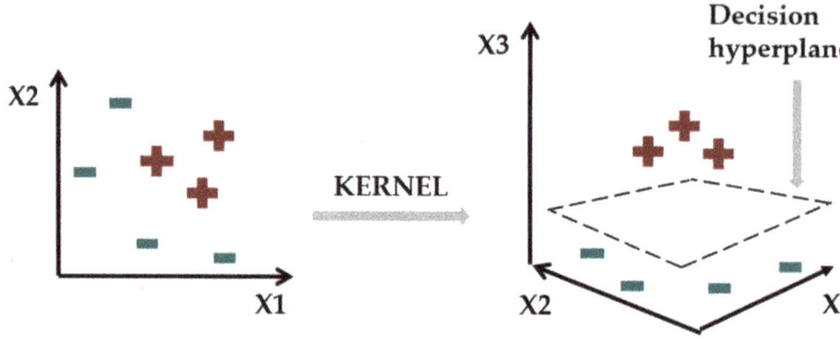

Figure 3. Illustration of transformation of the n-dimensional space into n + 1 dimensional space using a kernel function (there are examples of two classes – the first one consist of red pluses and the second one consist of green minuses).

The objective is to maximize the width of the boundary, which is known as the primal problem of support vector machines [18].

Bagging is a classification method belonging to the group of learning by a set of methods—so called ensemble learning—which helps improve the stability and accuracy of machine learning. The algorithm creates several Bootstrap samples so that each sample works as an independent dataset. After selecting "*m*" samples, a partial classifier is generated for each of them. Then, the results are averaged, and the resulting class is selected by the vote of all partial classifiers [19], which is illustrated in Figure 4.

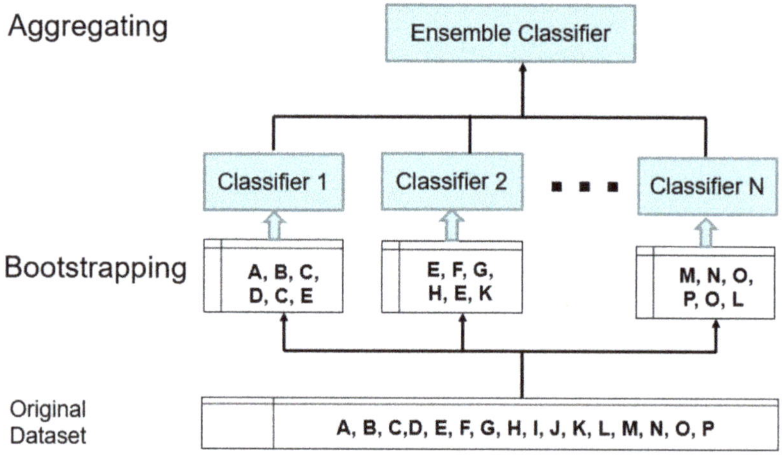

Figure 4. Illustration of two steps of Bagging method—Bootstrapping and Aggregating.

The random forests (RF) method also belongs to the ensemble learning group of methods. It is based on decision trees (DT) and the number of DTs is a parameter. The random forests method tries to minimize the variance by creating de-correlated DTs using a random selection of a subset of attributes. The method determines the result by a vote of the individual generated trees [20].

The complexities of the used and proposed methods are different. The naïve Bayes algorithm has the complexity $O(N \times M)$, where N is the number of attributes – words and M is the number of examples in the dataset. The complexity of the SVM depends on the used optimization. Using quadratic programming, the complexity becomes $O(N^3)$, but the complexity of the sequential minimal optimization depends only on support vectors. The complexity of decision trees is $O(N \times M)$ but for RF it is $O(K \times N \times M)$, where K is the number of generated trees. Bagging also has a similar complexity. The complexity of our lexicon-based labelling method is $O(M \times L \times T)$, where M is the number of labelled texts in the dataset, L is the number of words in the lexicon and T is the number of terms in a unique labelled text.

3. Methodology of Research

Our research was focused at first on the creation of a new lexicon of toxic words in the Slovak language, which was an inevitable resource for our lexicon-based labelling method. The resulting Slovak language lexicon, "Lexicon_of_toxic_words.json" is available at: https://kristina.machova.website.tuke.sk/useful/Lexicon_of_toxic_words.json (accessed on 26 July 2022). The lexicon was created from English lexicons by the translation of negative words to the Slovak language and it was extended by toxic words in the Slovak language. Next, we designed a lexicon-based labelling method. We extracted short texts from social media to create a dataset, which was then labelled using our lexicon-based method. The labelled dataset (available at: https://kristina.machova.website.tuke.sk/useful/Toxic_training_data.csv (accessed on 26 July 2022)) was used for the training of the classic methods (NB, SVM) and ensemble methods (bagging, RF). The models were evaluated using the sensitivity, FP rate, specificity, FN Rate, accuracy, precision and F1 score. Then, the best models were used in the application for toxicity detection. The methodology is illustrated in Figure 5.

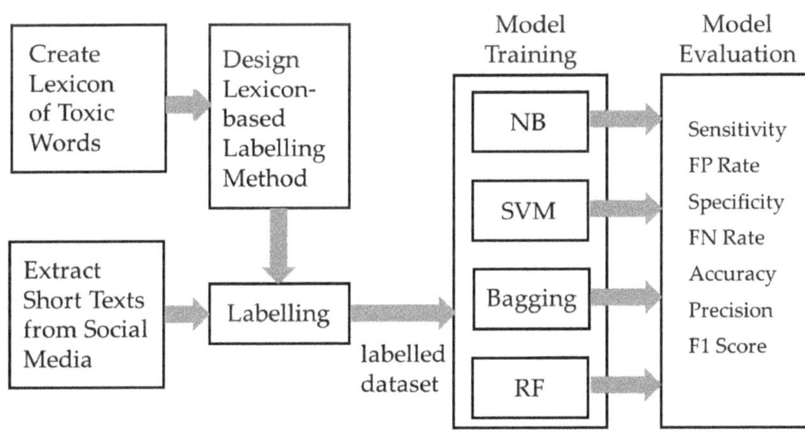

Figure 5. Methodology of research devoted to the creation of the detection models for detection of a degree of toxicity in social media.

We decided to use the freely available Weka tool to train the classification models. This tool can be used in two ways. The first is to import this tool in Java as a library. With its help, it is possible to use all the functions that the tool contains. The second way is to use the graphical interface of the Weka program. We decided to combine these two methods by using the graphical interface of the Weka program to create, test and save the models, and then, to use the created models for a new data classification, we used the library in the Java language. The advantage of such a combination is the ease of setting parameters when training models using a graphical interface and a simple evaluation of model quality using the Java library.

For the model training, the Weka GUI Chooser offers several tools to choose from. We worked with the Explorer tool. The training consisted of several steps:

- selection of training data,
- use of filters for data pre-processing,
- selection of the required model,
- setting of the model,
- selection of the method of testing the model,

- evaluation of the achieved results.

In the process of the model training, we started from the parameters that were predefined in the Weka tool and then changed them until we obtained the most effective model. Manual tuning of the parameters was used. The parameter settings of the best models for each machine learning method are listed in tables below.

The input for training the models was data stored in a CSV file, which contained two columns of data, namely, text data (String) containing comments, and a column with nominal values (Nominal), containing the toxicity value of the given comment. The total number of 3091 comments were divided into following categories:

- Moderately toxic—776 data,
- Low toxic—757 data,
- Neutral—779 data,
- Very toxic—779 data.

We used the following measures of effectivity of the models: Sensitivity, False Positive Rate, Specificity, False Negative Rate, Accuracy, Precision and F1 Score. They were calculated according to the formulas in Figure 6 where:

- TP is the number of true positive classifications,
- FP is the number of false positive classifications,
- TN is the number of true negative classifications,
- FN is the number of false negative classifications.

Sensitivity	$TPR = \frac{TP}{P} = \frac{TP}{TP+FN}$
False Positive Rate (FPR)	$FPR = \frac{FP}{N} = \frac{FP}{FP+TN}$
Specificity	$SPC = \frac{TN}{N} = \frac{TN}{TN+FP}$
False Negative Rate (FNR)	$FNR = \frac{FN}{P} = \frac{FN}{FN+TP}$
Accuracy	$ACC = \frac{TP+TN}{TP+FP+TN+FN}$
Precision	$PPV = \frac{TP}{TP+FP}$
F1 Score	$F1 = \frac{2TP}{2TP+FP+FN}$

Figure 6. Formulas of the used measures of the effectivity of models.

3.1. Models Training Using Classic Methods of Machine Learning

We have focused at first on classic methods such as naïve Bayes multinomial (NBM) and support vector machines (SVM). These methods generate models that can be intuitively well explained. Naïve Bayes offers conditional probabilities, which represent the measure of belonging attributes to classes for recognition. Support vector machines are often used because they form a mathematical model of a hyperplane dividing two or more classes in a space. The model contains information about the importance of particular attributes in it. The baseline method for the experiments was the NBM.

With the naïve Bayes method, there are several ways we can create models. The first one is to build a model using the function "*NaiveBayesMultinomial-Text*". In this case, the "*FilteredClassifier*" function is not used, but the data is processed directly without a filtered classifier. The training inputs are full sentences from the comments along with the labelling obtained using the lexicon approach. The settings which we achieved the best results with are listed in Table 1. The NBM—Text model was validated using a 10-fold cross-validation. This means, that 9/10 of the data were used for training a model, which was consequently validated on an unseen 1/10 of the data, which were not used for training. The total number of data entries was 3091, of which 2086 were correctly categorized, which represents 67.49%, while 1005 data were categorized incorrectly, which represents 32.51%. Detailed validation results are shown in Table 2. These results are not bad if we consider that the random selection for four classes has a precision of 0.25.

Table 1. The settings of the Naïve Bayes Multinomial Text model training.

Settings	Values
lowercaseTokens	True
minWordFrequency	2.0
norm	2.0
numDecimalPlaces	3
stemmer	LovinsStemmer

Table 2. Effectivity of Naïve Bayes Multinomial Text model for recognition of four degrees of toxicity in online discussions.

Measures	Very Toxic	Moderately Toxic	Low Toxic	Neutral
Sensitivity	0.671	0.663	0.742	0.655
FP Rate	0.078	0.061	0.094	0.200
Specificity	**0.896**	**0.918**	**0.874**	**0.773**
FN Rate	0.329	0.367	0.258	0.345
Accuracy	**0.827**	**0.830**	**0.834**	**0.741**
Precision	0.743	0.776	0.719	0.525
F1 Score	0.705	0.697	0.730	0.583

The second way to train a naïve Bayes classifier is using the function "*NaiveBayes-Multinomial*" in combination with the "*FilteredClassifier*" function. In this case, text input is transformed into the vectors of words. We used "*StringToWordVector*" as a filter. The settings with which we achieved the best results with this model are listed in Table 3 and the validation results are shown in Table 4.

Table 3. The settings of the Naïve Bayes Multinomial model training.

Settings	Values
IDFTransform	True
TFTransform	True
lowCaseToken	True
stemmer	SnowballStemmer
wordsToKeep	2000

Table 4. Effectivity of Naïve Bayes Multinomial model for recognition of four degrees of toxicity in online discussions.

Measures	Very Toxic	Moderately Toxic	Low Toxic	Neutral
Sensitivity	0.702	0.693	0.651	0.401
FP Rate	0.135	0.172	0.139	0.072
Specificity	**0.812**	**0.773**	**0.812**	**0.904**
FN Rate	0.298	0.307	0.349	0.600
Accuracy	**0.777**	**0.748**	**0.763**	**0.749**
Precision	0.637	0.575	0.603	0.651
F1 Score	0.668	0.629	0.626	0.496

The total number of records during the model training was 3091, of which 1890 were correctly categorized, which represents 61.15%, while 1201 data were incorrectly categorized, which represents 38.85%. The data was validated using a 10-fold cross-validation.

Both naïve Bayes models achieved the best results in specificity and accuracy. The naïve Bayes multinomial text was better in the detection of three degrees of toxicity—very, moderately and low toxic. The neutral comments were better classified by the naïve Bayes multinomial using a vector representation of the text.

The second classic method of machine learning—SVM—was used for training the other models. This method is usually very successful in text processing. We trained two SVM models. For the first one, the function *FilteredClassifier* was used with the filter *StringToWordVector*, which transforms the text of comments into the vectors of words. The settings of training and the filter settings for the SVM1 model are presented Table 5. The validation results are shown in Table 6.

Table 5. The settings of the SVM1 model training.

Settings	Values
IDFTransform	True
TFTransform	True
Debug	True
loweCaseTokens	True
stemmer	NullStemmer
wordsToKeep	3000
batchSize	200
kernel	PolyKernel
numDecimalPlaces	3

Table 6. Effectivity of SVM1 model for recognition of four degrees of toxicity in online discussions.

Measures	Very Toxic	Moderately Toxic	Low Toxic	Neutral
Sensitivity	0.628	0.723	0.712	0.621
FP Rate	0.093	0.112	0.102	0.132
Specificity	**0.880**	**0.854**	**0.866**	**0.839**
FN Rate	0.372	0.277	0.288	0.379
Accuracy	**0.804**	**0.814**	**0.820**	**0.775**
Precision	0.694	0.684	0.695	0.613
F1 Score	0.659	0.703	0.703	0.617

The total number of data entries during the training was 3091 comments, of which 2139 were correctly classified, representing 67.1%, while 952 were incorrectly classified, representing 33.9%. We obtained this result by a 10-fold cross-validation.

The next SVM2 model, with a slightly better result, differed from the previous SVM1 model only in the settings of the filter. Namely, the stemmer attribute had the value *LovinsStemmer*. All other filter and model settings remained the same as for the SVM1. In the following Table 7, we can see an improvement of the results of the SVM2 model in all indicators compared to the previous SVM1 model. The total number of records was 3091, of which there were 2473 correctly classified comments into all classes, which represents 80%, while a total of 618 were incorrectly classified data, which represents 20%. The model was evaluated using a 10-fold cross-validation.

Table 7. Effectivity of SVM2 model for recognition of four degrees of toxicity in online discussions.

Measures	Very Toxic	Moderately Toxic	Low Toxic	Neutral
Sensitivity	0.711	0.791	0.828	0.870
FP Rate	0.044	0.053	0.051	0.119
Specificity	**0.950**	**0.938**	**0.939**	**0.867**
FN Rate	0.289	0.209	0.172	0.128
Accuracy	**0.884**	**0.897**	**0.908**	**0.868**
Precision	0.846	0.834	0.839	0.714
F1 Score	0.773	0.812	0.834	0.785

In comparison with the baseline NBM models, the SVM models achieved better results. In addition, a small change in the settings (stemmer = LovinsStemmer) produced better results in the specificity above 90 percent.

3.2. Models Training Using Ensemble Learning

In several works, ensemble learning provided better and more reliable results for classification than single machine learning methods. We decided at first to use a basic method of ensemble learning, namely, bagging. Using the bagging method, we generated a set of models. Similar to the previous methods, we used the *FilteredClassifier* function, where we applied a filter with the following settings presented in Table 8. The test results of this model are presented in Table 9.

Table 8. The settings of filters in the ensemble learning using Bagging method.

Settings	Values
IDFTransform	True
TFTransform	True
lowCaseToken	True
stemmer	LovinsStemmer
wordsToKeep	1000

Table 9. Effectivity of Bagging method for recognition of four degrees of toxicity in online discussions.

Measures	Very Toxic	Moderately Toxic	Low Toxic	Neutral
Sensitivity	0.574	0.719	0.799	0.901
FP Rate	0.052	0.055	0.052	0.184
Specificity	**0.948**	**0.932**	**0.933**	**0.791**
FN Rate	0.426	0.281	0.201	0.378
Accuracy	**0.842**	**0.870**	**0.894**	**0.821**
Precision	0.813	0.813	0.832	0.622
F1 Score	0.673	0.763	0.815	0.736

The bagging method was validated using a 10-fold cross-validation. The total number of training data was 3091, of which there were correctly classified 2312 comments, which represents 74.80%, while 779 comments were incorrectly classified, which represents 25.2%.

The results presented in Table 9 did not confirm the tendency of composite classification by an ensemble of models to give better results than models learned by single machine learning methods. The results of the bagging method were worse than the SVM2 model, although better than the SVM1 model; therefore, we trained and tested one more model using the ensemble learning, namely, a random forest (RF) of decision trees, which is the newest and most successful from the ensemble methods. The "*FilteredClassifier*" function was also used for training the RF. In this function, the RF method was specified as the classifier, and "*StringToWordVector*" was set as the filter. The "*LovinsStemmer*" function was set for the stemmer attribute. We can see the random forest setting in Table 10. The test results of this model are presented in Table 11. They are unfortunately not better than the results of the bagging method.

Table 10. The settings of the ensemble learning using Random Forest method.

Settings	Values
breakTiesRandomly	True
debug	True
numDecimalPlaces	3
numIterations	200
seed	10

Table 11. Effectivity of Random Forest model for recognition of four degrees of toxicity in online discussions.

Measures	Very Toxic	Moderately Toxic	Low Toxic	Neutral
Sensitivity	0.525	0.713	0.699	0.875
FP Rate	0.031	0.059	0.070	0.237
Specificity	**0.961**	**0.923**	**0.910**	**0.732**
FN Rate	0.475	0.287	0.301	0.125
Accuracy	**0.831**	**0.858**	**0.848**	**0.771**
Precision	0.850	0.803	0.765	0.555
F1 Score	0.650	0.755	0.730	0.680

The RF model was validated using a 10-fold cross-validation. The total number of comments was also 3091, of which 2173 were correctly classified, which represents 70.3%, while 918 comments were incorrectly classified, which is 29.7%. This model achieved the best results among all the models we trained using the random forest method.

3.3. *Application for Recognition of Degrees of Toxicity*

We used the most precise models for recognition of the degrees of toxicity of online comments in the application for analyzing newly extracted posts (they were not used for the model training). For an easier analysis of new data with the option of selecting the type of model, we created an application in Java that includes an annotation using the lexicon-based method, and also allows the use of all the best models. The user can choose which model they want to use to analyze new comments. After starting the program, the first thing that appears is a window with a menu that offers us a choice of several options:

- vocabulary approach—a method based on the lexicon for the searching and annotation of toxic words,
- SVM model—the use of the SVM model to analyze the toxicity of new comments,
- random forest model—analysis of online comments using the random forest model,
- naive Bayes model—using the naive Bayes model to analyze online comments,
- bagging model—analysis using the bagging model,
- info—contains final results for comparing the quality of models,
- exit—ending the program.

After selecting one of these options, a new window will open (see Figure 7), which is used to analyze the text of a comment. In the upper part, we can select the file which contains the text to be analyzed. The program allows us to choose one of two file types, namely, csv or arff.

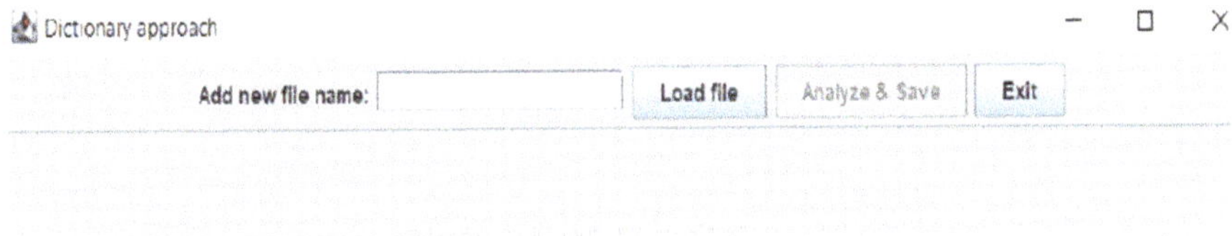

Figure 7. Window for a text data loading and an analysis starting.

Next, the user can start the analysis of the selected document of comments and save the results. The analysis produces a table containing the analyzed comments from the selected file together with their toxicity values, as illustrated in Figure 8. A file with this data will also be saved as a csv.

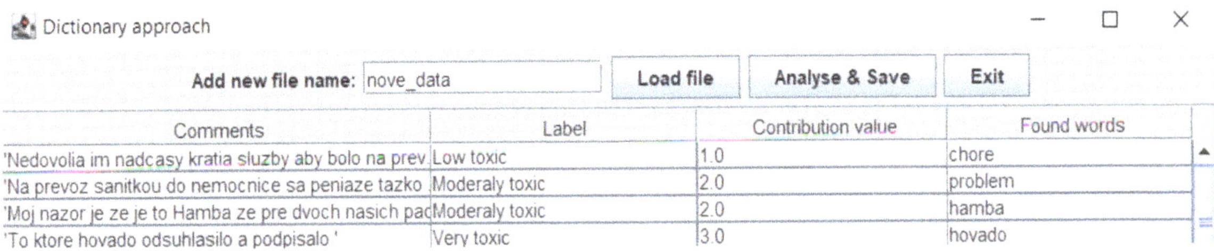

Figure 8. The illustration of results of the lexicon-based analysis of comments from the selected and loaded file.

4. Discussion and Conclusions

The most accurate model for the degree of toxicity recognition of online comments is the SVM 2 with an average accuracy value of 0.890, which exceeded all other models at each efficiency measure. The second most accurate model is the model created using ensemble learning bagging with an average accuracy value of 0.857, which is only slightly less than the SVM 2. The third most accurate model is the random forest model with an average accuracy value of 0.827. The SVM 1 and naive Bayes multinomial text models have very comparable results with differences in accuracy up to the third decimal place. The least successful is the naive Bayes multinomial model with an average accuracy of 0.759. The average results of all the models are illustrated in Figure 9. Our experiments did not confirm the tendency of composite classification by an ensemble of models to give better results than models learned by single machine learning methods.

Table 12 shows a comparison of the predictive performance of our best model to the best models from previous studies focused on the same or similar problem. We can see from this table of best results, that our SVM model achieved better results than the SVM models of previous studies, but the deep neural network model based on BiLSTM was more effective than our best model. On the other hand, the effectiveness of different approaches as presented in Table 12 can be understood as indicative only. The reason for this is that the experiments, the results of which are collected in the table, were performed by different authors employing different textual datasets (with datasets differing not only in content but in their used language as well). Nevertheless, the table identifies the BiLSTM method as a possible candidate for the next research on detecting toxicity in online discussions in the Slovak language.

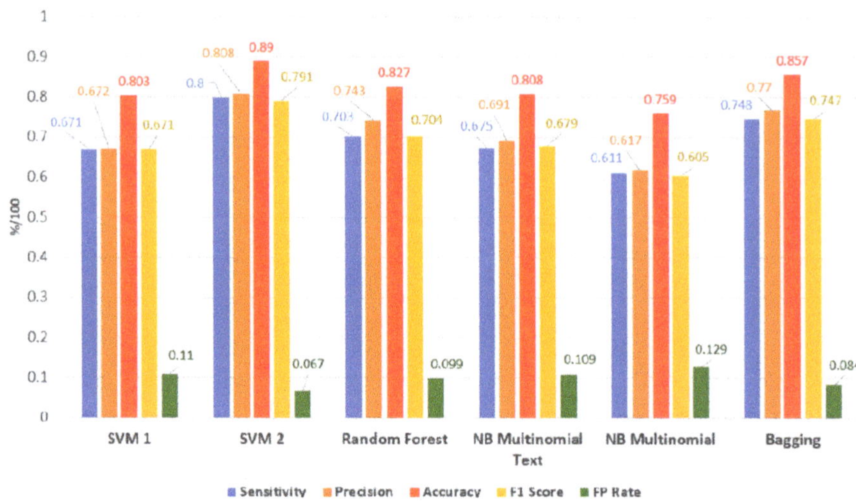

Figure 9. The average results of all models in Sensitivity, Precision, Accuracy, F1 Score and FP Rate measures.

Table 12. Comparison of results in Accuracy of our best model to the best models from previous studies focused on the same respectively similar problem.

Studies	Methods	Type of Detection	Best Results
[5]	SVM, LSTM	Hate posts	Acc (SVM) = 0.699
[7]	LSF *	Offensive posts	Acc (LSF) = 0.778
[8]	SVM, NB, RF	Aggressive language	Acc (SVM) = 0.892
[9]	SVM, LR **, BiGRU, BiLSTM	Cyberbullying	Acc (BiLSTM) = 0.948
Our	NB, SVM, Bagging, RF	Toxic posts	Acc (SVM) = 0.908

* Lexical Syntactic Features, ** Logistic Regression, Acc—Accuracy.

This article dealt with the detection of the degree of toxicity of posts in online discussions on social networks. It describes the procedure for training models using both single and ensemble machine learning methods and their comparison. The article also deals with the creation of a lexicon of toxic words and the use of this lexicon to label short texts in a dataset by the degree of toxicity of texts. We can say that the novelty of the article is mainly based on the lexicon approach for the Slovak language and in the comparison of the effectivity of the detection models trained using classic and ensemble learning. An application for the extraction of comments from social media was designed to help with acquiring the text data for the generation of the detection models. We have used the extracted and labelled comments as a dataset to train the classification models. Additionally, an application was created to detect the degree of toxicity in new comments, based on the most successful models.

In future, we could focus on the neural ensemble models [21], since the ensemble models based on classical machine learning methods did not achieve the results we hoped for. In particular, an ensemble could use methods, such as BiLSTM and the more accurate BiGRU according to [22]; we could also extend the text processing to include the processing of images available in online comments, for example, emoticons. Future research could also focus on using various ensemble strategies [23] to increase the detection performance of a set of models, for example, by using the random forest voting strategy.

Another possibility for future research is using new optimization methods in the SVM method, such as ant colony optimization or particle swarm optimization [24], the dynamic feature weight selection [25], and feature extraction integrating principal component analysis and local binary patterns [26]. Using a kernel extreme learning machine could also bring improved results [26].

Author Contributions: Conceptualization, K.M.; methodology, K.M.; software, K.A.; validation, K.A.; formal analysis, K.M.; investigation, K.M.; resources, K.M.; data curation, K.A.; writing—original draft preparation, K.M.; writing—review and editing, K.M.; visualization, M.M.; supervision, K.M.; project administration, M.M.; funding acquisition, K.M. All authors have read and agreed to the published version of the manuscript.

Funding: This work was partially supported by the Slovak Grant Agency of the Ministry of Education and Academy of Science of the Slovak Republic under grant no. 1/0685/21.

Data Availability Statement: Machine learning models were trained on the dataset "Toxic_training_data.csv" available at: https://kristina.machova.website.tuke.sk/useful/Toxic_training_data.csv (accessed on 26 July 2022). The lexicon of the Slovak language "Lexicon_of_toxic_words.json" is also available at: https://kristina.machova.website.tuke.sk/useful/Lexicon_of_toxic_words.json (accessed on 26 July 2022).

Conflicts of Interest: The authors declare no conflict of interest. The funders had no role in the design of the study; in the collection, analyses, or interpretation of data; in the writing of the manuscript, or in the decision to publish the results.

References

1. Veletsianos, G.; Houlden, S.; Hodson, J.; Gosse, C. Women scholars' experiences with online harassment and abuse: Self-protection, resistance, acceptance, and self-blame. *New Media Soc.* **2018**, *20*, 4689–4708. [CrossRef]
2. Vrysis, L.; Vryzas, N.; Kotsakis, R.; Saridou, T.; Matsiola, M.; Veglis, A.; Arcila-Calderón, C.; Dimoulas, C. A Web Interface for Analyzing Hate Speech. *Future Internet* **2021**, *13*, 80. [CrossRef]
3. Machova, K.; Mach, M.; Vasilko, M. Comparison of Machine Learning and Sentiment Analysis in Detection of Suspicious Online Reviewers on Different Type of Data. *Sensors* **2022**, *22*, 155. [CrossRef] [PubMed]
4. Xiang, G.; Fan, B.; Wang, L.; Hong, J.; Rose, C. Detecting Offensive Tweets via Topical Feature Discovery over a Large Scale Twitter Corpus. In Proceedings of the 21st ACM International Conference on Information and Knowledge Management-CIKM'12, Maui, HI, USA, 29 October 2012; Association for Computing Machinery: New York, NY, USA, 2012; pp. 1980–1984, ISBN 9781450311564. [CrossRef]
5. Del Vigna, F.; Cimino, A.; Dell'Orletta, F.; Petrocchi, M.; Tesconi, M. Hate me, hate me not: Hate speech detection on Facebook. In Proceedings of the First Italian Conference on Cybersecurity (ITASEC17), Venice, Italy, 17–20 January 2017; pp. 86–95.
6. Mubarak, H.; Darwish, K.; Magdy, W. Abusive Language Detection on Arabic Social Media. In Proceedings of the First Workshop on Abusive Language Online, Vancouver, BC, Canada, 4 August 2017; Association for Computational Linguistics: New York, NY, USA, 2017; pp. 52–56. [CrossRef]
7. Chen, Y.; Zhou, Y.; Zhu, S.; Xu, H. Detecting Offensive Language in Social Media to Protect Adolescent Online Safety. In Proceedings of the 2012 ASE/IEEE International Conference on Social Computing and 2012 ASE/IEEE International Conference on Privacy, Security, Risk and Trust, Amsterdam, The Netherlands, 3–5 September 2012; IEEE Computer Society: Washington, DC, USA, 2012; pp. 71–80, ISBN 9780769548487. [CrossRef]
8. Lepe-Faúndez, M.; Segura-Navarrete, A.; Vidal-Castro, C.; Martínez-Araneda, C.; Rubio-Manzano, C. Detecting Aggressiveness in Tweets: A Hybrid Model for Detecting Cyberbullying in the Spanish Language. *Appl. Sci.* **2021**, *11*, 10706. [CrossRef]
9. Raj, C.; Agarwal, A.; Bharathy, G.; Narayan, B.; Prasad, M. Cyberbullying Detection: Hybrid Models Based on Machine Learning and Natural Language Processing Techniques. *Electronics* **2021**, *10*, 2810. [CrossRef]
10. Kandasamy, V.; Trojovský, P.; Al Machot, F.; Kyamakya, K.; Bacanin, N.; Askar, S.; Abouhawwash, M. Sentimental Analysis of COVID-19 Related Messages in Social Networks by Involving an N-Gram Stacked Autoencoder Integrated in an Ensemble Learning Scheme. *Sensors* **2021**, *21*, 7582. [CrossRef]
11. Dang, C.N.; Moreno-García, M.N.; De la Prieta, F. An Approach to Integrating Sentiment Analysis into Recommender Systems. *Sensors* **2021**, *21*, 5666. [CrossRef]
12. WordNet—A Lexical Database for English. 2022. Available online: https://wordnet.princeton.edu/ (accessed on 26 July 2022).
13. Mohammad, S.; Dunne, C.; Dorr, B. Generating High-Coverage Semantic Orientation Lexicons from Overtly Marked Words and a Thesaurus. In Proceedings of the 2009 Conference on Empirical Methods in Natural Language Processing, Singapore, 6–7 August 2009; Association for Computational Linguistics: Stroudsburg, PA, USA, 2009; pp. 599–608.
14. Machova, K.; Mikula, M.; Gao, X.; Mach, M. Lexicon-based Sentiment Analysis Using the Particle Swarm Optimization. *Electronics* **2020**, *9*, 1317. [CrossRef]
15. Hastie, T.; Tibshirani, R.; Friedman, J.H. *Elements of Statistical Learning: Data Mining, Inference, and Prediction*; Stanford University: Stanford, CA, USA, 2009; ISBN 9780387848570.

16. Widodo, A.; Yang, B.-S. Support vector machine in machine condition monitoring and fault diagnosis. *Mech. Syst. Signal Process.* **2007**, *21*, 2560–2574. [CrossRef]
17. Siddhartha, S. Kernel Trick in SVM. 2022. Available online: https:/mediumcom/analytics-vidhya/how-to-classify-non-linear-data-to-lineardata-bb2df1a6b781 (accessed on 26 July 2022).
18. James, G.; Witten, D.; Hastie, T.; Tibshirani, R. *An Introduction to Statistical Learning-with Applications in R. Springer Texts in Statistics*; Springer: Berlin/Heidelberg, Germany, 2014; pp. 1–445, ISBN 978-1-4614-7137-0. [CrossRef]
19. Rocca, J.J.; Rocca, B. Ensemble methods: Bagging, Boosting and Stacking. *Towards Data Sci.* **2019**. Available online: https://towardsdatascience.com/ensemble-methods-bagging-boosting-and-stacking-c9214a10a205 (accessed on 26 July 2022).
20. Kevin Zhou, S. *Medical Image Recognition, Segmentation and Parsing, Machine Learning and Multiple Object Approaches*; MICCAI Society Book Series; Elsevier: Amsterdam, The Netherlands, 2016; ISBN 978-0-12-802581-9.
21. Hrúz, M.; Gruber, I.; Kanis, J.; Boháček, M.; Hlaváč, M.; Krňoul, Z. One Model is Not Enough: Ensembles for Isolated Sign Language Recognition. *Sensors* **2022**, *22*, 5043. [CrossRef] [PubMed]
22. Tng, S.S.; Le, N.Q.K.; Yeh, H.-Y.; Chua, M.C.H. Improved Prediction Model of Protein Lysine Crotonylation Sites Using Bidirectional Recurrent Neural Networks. *J. Proteome Res.* **2021**, *21*, 265–273. [CrossRef] [PubMed]
23. Ben Atitallah, S.; Driss, M.; Almomani, I. A Novel Detection and Multi-Classification Approach for IoT-Malware Using Random Forest Voting of Fine-Tuning Convolutional Neural Networks. *Sensors* **2022**, *22*, 4302. [CrossRef] [PubMed]
24. Zhou, X.; Ma, H.; Gu, J.; Chen, H.; Deng, W. Parameter adaptation-based ant colony optimization with dynamic hybrid mechanism. *Eng. Appl. Artif. Intell.* **2022**, *114*, 105139. [CrossRef]
25. An, Z.; Wang, X.; Li, B.; Xiang, Z.; Zhang, B. Robust Visual Tracking for UAVs with Dynamic Feature Weight Selection. *Appl. Intell.* **2022**, 1–14. [CrossRef]
26. Chen, H.; Miao, F.; Chen, Y.; Xiong, Y.; Chen, T. A Hyperspectral Image Classification Method Using Multifeature Vectors and Optimized KELM. *IEEE J. Sel. Top. Appl. Earth Obs. Remote Sens.* **2021**, *14*, 2781–2795. [CrossRef]

sensors

MDPI

Article

Physical and Digital Infrastructure Readiness Index for Connected and Automated Vehicles

Boris Cucor [1], Tibor Petrov [2], Patrik Kamencay [1,*], Ghadir Pourhashem [2] and Milan Dado [1]

1 Faculty of Electrical Engineering and Information Technology, University of Zilina, 010 26 Zilina, Slovakia
2 Department of International Research Projects—ERAdiate+, University of Zilina, 010 26 Zilina, Slovakia
* Correspondence: patrik.kamencay@uniza.sk

Abstract: In this paper, we present an assessment framework that can be used to score segments of physical and digital infrastructure based on their features and readiness to expedite the deployment of Connected and Automated Vehicles (CAVs). We discuss the equipment and methodology applied for the collection and analysis of required data to score the infrastructure segments in an automated way. Moreover, we demonstrate how the proposed framework can be applied using data collected on a public transport route in the city of Zilina, Slovakia. We use two types of data to demonstrate the methodology of the assessment-connectivity and positioning data to assess the connectivity and localization performance provided by the infrastructure and image data for road signage detection using a Convolutional Neural Network (CNN). The core of the research is a dataset that can be used for further research work. We collected and analyzed data in two settings—an urban and suburban area. Despite the fact that the connectivity and positioning data were collected in different days and times, we found highly underserved areas along the investigated route. The main problem from the point of view of communication in the investigated area is the latency, which is an issue associated with infrastructure segments mainly located at intersections with heavy traffic or near various points of interest. The low accuracy of localization has been observed mainly in dense areas with large buildings and trees, which decrease the number of visible localization satellites. To address the problem of automated assessment of the traffic sign recognition precision, we proposed a CNN that achieved 99.7% precision.

Keywords: cooperative, connected and automated mobility; infrastructure readiness assessment; connectivity data; positioning data; convolutional neural network

Citation: Cucor, B.; Petrov, T.; Kamencay, P.; Pourhashem, G.; Dado, M. Physical and Digital Infrastructure Readiness Index for Connected and Automated Vehicles. *Sensors* **2022**, *22*, 7315. https://doi.org/10.3390/s22197315

Academic Editors: Burak Kantarci and Mengchu Zhou

Received: 26 July 2022
Accepted: 22 September 2022
Published: 27 September 2022

1. Introduction

Connected and Automated Vehicles (CAVs) are expected to bring tremendous social, economic and environmental benefits, including increased road safety, addressing of road congestion and decreased environmental impact due to less wasted fuel thanks to improved vehicle management [1,2]. However, even the strongest supporters of the idea that vehicles should be as independent of the infrastructure as possible already accept the fact that automated driving can safely work only in specific Operational Design Domains (ODDs) [3]. Therefore, the physical and digital infrastructure already plays a role in the design and functioning of CAVs.

The preparation of infrastructure for automated driving is a multifaceted challenge, including components such as connectivity, provision of localization and mapping services, machine-readable road signage, and CAV-friendly road geometry and is a time-consuming, costly task requiring thorough planning. As highlighted in reference [4], automated driving functions relying on the infrastructure are often regarded from the perspective of a chicken-and-egg situation—infrastructure investments are postponed in an expectation that vehicle manufacturers take the lead and implement the related applications and vice versa. While Automated Driving Systems (ADS) are still under development, some of their basic requirements on the physical and digital infrastructure are already clear. Therefore, a complex assessment framework to help infrastructure providers evaluate the state of the infrastructure in terms of its readiness to support automated driving can be beneficial to help identify the essential areas of intervention and plan the investments and the timeline of infrastructure upgrades.

Artificial Intelligence (AI) is one of the key enabling technologies for the deployment of CAVs. Being an essential component responsible for a CAVs perception of the environment and decision making, application of AI in the context of CAVs has been a thoroughly investigated topic by both academia and industry. Interested readers are referred to the works of Ma et al. [5] and Li et al. [6] for thorough surveys on the state-of-the-art of implementing AI in CAVs.

Cybersecurity is another crucial aspect of CAVs operation that needs to be carefully considered. Ge et al. [7] proposed an algorithm to address the problem of resilient and safe platooning control of CAVs that are under denial-of-service attacks disrupting the V2V communication. Khan et al. [8] developed a conceptual system dynamics model to analyze the cybersecurity of CAVs. The model integrates six critical areas and their corresponding parameters that either enable or mitigate attacks on CAVs operation.

Several studies have already explored the key requirements on infrastructure for automated driving and assessed their impact on the performance of CAV operation. Carreras et al. [9] proposed a classification scheme to classify and harmonize the capabilities of a road infrastructure to support CAVs. Similarly to SAE levels of Driving Automation [10], the authors propose five levels of infrastructure support ranging from no support up to support sufficient to facilitate cooperative driving.

Mackenzie et al. [11] performed an assessment of line markings at multiple sites along Australia's Great Southern Highway using two vehicles equipped with a lane departure warning system and two cameras. The authors conclude that the failure to accurately detect line marking crossing events can be most often attributed to the absence of a marked line, vehicle travel speed being lower than the speed recommended for system operation by the manufacturer, bad condition of lane markings and the low line marking retro-reflectivity and/or daylight brightness.

Magyari et al. [12] conducted a study on sight distances at unsignalized intersections, comparing the minimum required sight distances between automated and human-driven vehicles. The authors demonstrated that automated vehicles require 10–40 m shorter sight distances than conventional vehicles.

Liu et al. [13] identified infrastructure aspects that should be considered to be upgraded based on the gap between their current state and future requirements of CAVs. The authors concluded that the main infrastructure intervention areas that currently require attention should be traffic signs and road markings, communications, pavement structure and road surface, parking lots, service stations, safe harbor areas, roundabouts, bridges, drainage and geotechnical aspects.

Nitsche et al. [14] conducted a study about the requirements on road transport infrastructure for highly automated vehicles focusing on automation Levels 2–4. The methods used in the study consist of a literature review and an online survey with 54 multidisciplinary experts. The study identified the factors with the largest impact on the performance of ADS in three categories: lane assistance systems, collision avoidance systems and speed control systems. The authors also argued that the complex urban environments, temporary work zones and poor visibility due to bad weather conditions are the major infrastructure challenges for automated driving systems.

Madadi et al. [15] carried out an assessment of road network readiness based on a workshop including experts who judged images of specific infrastructure segments. Each of the two rounds of judgements was followed by a group discussion and a summary. The authors conclude that the experts identified many similar issues for different instances of roads and intersections.

Although there are numerous approaches already presented in the literature that aim at assessing the readiness of infrastructure for automated driving, those approaches are either limited to the assessment of a single performance indicator, do not provide measurable performance indicators at all, or rely on a per-segment evaluation performed by experts. Evaluation by a group of experts is both time-consuming and expensive for evaluating larger infrastructure networks. Moreover, it also brings the challenge of personal perception and subjective evaluation of readiness, as can be seen from often contradictory human assessments of the same infrastructure segment.

To allow large-scale, automated assessment of the physical and digital infrastructure readiness for automated driving, key aspects of the infrastructure contributing to the safe and reliable operation of CAVs and their minimum requirements in terms of measurable performance indicators have to be identified. Furthermore, a common methodology for the collection of required data, their processing and evaluation is essential to be developed, which, to the best of our knowledge, has not been presented in the available literature yet.

It is noteworthy to mention that the proposed assessment framework does not cover all the requirements of automated driving, as many of them are still under development. However, the framework provides a tool to infrastructure providers that can quickly and cheaply assess the extent to which the already known fundamental requirements are met and support them with identifying segments and specific interventions needed for increasing their readiness for automated mobility. The framework presented herein aspires to contribute toward solving the above-mentioned chicken-and-egg loop by providing a simple yet robust basis to identify the areas where infrastructure investments are necessary, regardless of the further developments in automated driving functions down the line.

The scientific and practical contribution of this paper can be summarized as follows:

- We propose a novel assessment framework to help infrastructure operators evaluate the readiness of physical and digital infrastructure for automated driving based on a set of indicators derived from the literature review on infrastructure requirements for CAV;
- We propose a data-collection setup and data processing methodology for collecting and evaluating data on the infrastructure necessary for the application of the assessment framework;
- We demonstrate the data collection and processing approach as well as the experimental results on a part of infrastructure in the city of Žilina, Slovakia.

The rest of the paper is organized as follows. The framework for assessment of physical and digital infrastructure for CAVs is described in Section 2. Section 3 contains a description of the data collection for connectivity, positioning and image data. The applied methods and methodology are described in Section 4. The achieved experimental results are presented in Section 5. Finally, Section 6 concludes by summarizing the results of this study, its contribution and further suggestions for future research.

2. Framework for Assessment of Physical and Digital Infrastructure Readiness for CAVs

In this section, we present the framework that we developed to assess the infrastructure readiness for Cooperative, Connected and Automated Mobility (CCAM). The framework is based on an extensive literature review of the requirements of CCAM, as well as on current industry best practices, anticipating the future requirements of various components of automated driving.

The framework aims to assess the infrastructure in five key areas crucial for CCAM implementation—connectivity, localization, machine-readable signage, and maps and object detection. For each area, a set of indicators is presented. A scoring grid mapping a score to an indicator value is assigned to each indicator. It is worth noting that the number of indicators selected for assessment of each area, as well as their corresponding assessment grids, may be subject to change and will be continually revised as new requirements for CCAM emerge or the currently identified ones are further clarified and quantified. The indicators for the current framework version has been selected in regard to the data that are either available now, or can be collected with a currently available technology. It should be noted that currently, it is not practical to provide an overarching index for each of the assessment areas since the level of impact of individual indicators on the performance of CAVs is not entirely clear yet. Therefore, the infrastructure assessment should be performed on a per-indicator basis.

The result of an assessment of an infrastructure segment is a numerical index, which represents the readiness of the evaluated road segment for CAVs deployment in the context of the corresponding indicator. This index can be useful for objective evaluation of the current status of infrastructure readiness and for planning and prioritizing future infrastructure upgrades, expansion and investments to increase its readiness to CAVs. The framework is presented in Tables 1–5.

Table 1. The assessment framework—connectivity area.

Indicator	Value	Score
Communication latency	x < 1 ms	1
	1 ms ≤ x < 50 ms	0.75
	50 ms ≤ x < 100 ms	0.5
	x ≥ 100 ms	0
Message loss	x < 0.001%	1
	0.001% ≤ x < 10%	0.5
	x ≥ 10%	0
Bitrate per vehicle	x ≥ 1 Gbit/s	1
	24 Mbit/s ≤ x < 1 Gbit/s	0.75
	8.5 Mbit/s ≤ x < 24 Mbit/s	0.5
	300 kbit/s ≤ x < 8.5 Mbit/s	0.25
	x < 300 kbit/s	0

Table 2. The assessment framework—localization area.

Indicator	Value	Score
Average number of satellites	x > 20	1
	15 ≤ x < 20	0.75
	10 ≤ x < 15	0.5
	5 ≤ x < 10	0.25
	x ≤ 5	0
Number of using satellites	x ≥ 4	1
	x = 3	0.75
	x = 2	0.5
	x = 1	0.25
	x = 0	0
GNSS lateral localization error	x < 0.1 m	1
	0.1 m ≤ x ≤ 0.2 m	0.5
	x > 0.2 m	0

Table 3. The assessment framework—object detection distance area.

Indicator	Value	Score
Intersection sight distance	V_d = 62.5 km/h	
	x ≥ 183 m	1
	148 m ≤ x < 183 m	0.75
	113 m ≤ x < 148 m	0.25
	x < 113 m	0
	V_d = 112.5 km/h	
	x ≥ 329 m	1
	266 m ≤ x < 329 m	0.75
	203 m ≤ x < 266 m	0.25
	x < 203 m	0
Infrastructure for remote sensor sharing available	Yes	1
	No	0

Table 4. The assessment framework—quality of maps area.

Indicator	Value	Score
Quality of maps	Static and dynamic infrastructure is available on the map and available to the CAV. Based on the information on the map the vehicle can perceive microscopic traffic situations in real time.	1
	Digital map with detailed lane information and static road signs is available. Traffic lights, short-term road works and variable message signs have to be recognized by AVs.	0.75
	Digital map is available but vehicle has to recognize lane geometry and/or road signs.	0.25
	No digital map is available. The vehicle has to recognize road geometry and traffic signs on its own.	0

Table 5. The assessment framework—machine-readable signage area.

Indicator	Value	Score
Precision of horizontal signage detection	$x \geq 99\%$	1
	$90\% \leq x < 99\%$	0.5
	$80\% \leq x < 90\%$	0.25
	$x < 80\%$	0
Precision of vertical signage detection	$x \geq 98\%$	1
	$x < 98\%$	0

2.1. Connectivity of Vehicles

It is already a well-accepted fact that connectivity and V2X communication will play a crucial role in the automation of road transport and addressing its safety challenges. A good example is a work by Zadobrischi et al. [16], who developed a system for analysis and management of dangerous situations that can detect a wide range of potentially hazardous conditions, including driver's psychosomatic conditions, as well as attributes of nearby vehicles and pedestrians. The system communicates with various traffic safety elements through V2X radio frequency (RF) or Visible Light Communication (VLC). In order to share the detected hazards beyond the nearby connected vehicles in direct RF or VLC reach, a connectivity infrastructure based on either Dedicated Short-Range Communications (DSRC) Roadside Units (RSUs), or cellular networks has to be in place.

We would like to point out that the assessment framework for the connectivity area is technology agnostic, i.e., communication infrastructure based on any current (e.g., DSRC, 4G- or 5G-based Cellular-V2X, VLC), or future communication technology can be considered for V2X communication as long as it meets the corresponding performance indicators.

The values of indicators used to evaluate the connectivity area are based on the already identified requirements for V2X communications as well as on projected bandwidth needs for AVs. While the amount of data collected by SAE Level 5 AV's sensors is expected to be huge [17], it is important to note that the majority of these raw measurements will be processed and utilized locally.

The lowest possible boundary of the bit rate has been set to 300 kbit/s. This bit rate corresponds to a connected vehicle broadcasting Cooperative Awareness Messages (CAM) with a maximum length of 1500 bytes while receiving CAMs from one neighboring vehicle at the same time with a message generation frequency of 10 Hz. If the infrastructure is not able to support this level of service, then no deployment of Cooperative ITS is possible, and therefore, the score of such an infrastructure segment would be equal to zero. On the contrary, if the 300 kbit/s bit rate is available for each vehicle at the given road segment, a score of 0.25 is awarded, indicating that at least a basic CAM service is available.

To achieve a score of 0.5, the infrastructure segment has to allow at least bi-directional sharing of sensory information on top of the basic CAM service. A sharing of footage from one automotive-grade camera with a resolution of 1280 × 1080 capturing a fairly complex scene, including buildings and vegetation, at 30 frames per second encoded by H.265 codec, including one 128 kbit/s audio track has been assumed. The resulting bit rate necessary to facilitate such traffic is 8.5 Mbit/s.

An infrastructure segment with a score of 0.75 is able to facilitate at least a bi-directional sensor sharing of one camera and a LIDAR sensor as well as a basic CAM service. From collected data, we empirically estimated the average bit rate of a 16-ray automotive LIDAR sensor to 7.66 Mbit/s. Therefore, the resulting minimum bit rate per vehicle necessary to achieve a score of 0.75 is 24 Mbit/s.

It is widely accepted that highly automated vehicles will utilize communication links with bit rates beyond 1 Gbit/s for extensive sensor sharing and operational data exchange. Therefore an infrastructure segment providing this level of service is awarded a score of 1.

To achieve the highest score in the message loss indicator, the communication infrastructure has to demonstrate at least 99.99% availability of service. On the contrary, if the average packet loss is above 10%, it might mean a steady information loss from more than one communicating vehicle, which, depending on the specific message content, might be unacceptable. Therefore, such a segment receives a score of zero.

The last evaluated indicator within the connectivity area is communication latency. The indicator values corresponding to the scores were derived from the networking and connectivity requirements of V2X communication services presented in [18].

2.2. Localization of Vehicles

Precise localization is a fundamental element of automated driving. Global Navigation Satellite Systems (GNSS) have become common tools to determine the precise location of vehicles and other road participants. Within the localization area of the framework, we evaluate four indicators (see Table 2) that provide insight into the availability and precision of the localization achievable by the GNSS at the given infrastructure segment.

Most GNSS techniques work with as few as five satellites. However, the redundancy is important for a number of reasons. First, the large number of satellites increases GNSS availability by providing service even if local obstructions block a significant part of the sky—a situation very common, especially in urban environments. Second, as demonstrated in [19], the performance of a three-constellation system, which sees only satellites more than 32 degrees above the horizon is equivalent to a single-constellation system in an open-sky scenario. Third, GNSS systems are developed independently, which allows performing cross-checking between constellations, enabling integrity guarantees [20]. Therefore, the average number of satellites as well as the number of available constellations a GNSS receiver can see when driving along the evaluated road segment are important indicators impacting the availability and performance of the localization service.

The values of the GNSS lateral localization error indicator corresponding to individual scores were derived using a methodology described in [21]. The lateral localization error below 0.1 m means that the vehicle is capable of determining its lane on a local road reliably (assuming a lane width of 3 m and a curvature of 20 m). If the lateral localization error is below 0.2 m, the vehicle is capable of determining its lane reliably on a highway (assuming a lane width of 3.6 m and speed of up to 137 km/h). If the lateral localization error is above 0.2 m, the vehicle might not be capable of determining its lane reliably.

It is worth noting here that the precision of the map also has to be factored in when evaluating the lateral localization precision. We assumed an equal error budget for the GNSS and the map.

2.3. Object Detection

The distance and reliability of object detection play a crucial role in automated driving as it is one of the key inputs into the CAV's decision-making. Currently deployed CAVs use a multitude of sensors for object detection and advanced algorithms to classify those objects and assign them meaning. Most commonly used technologies use Light Detection and Ranging (LIDAR), Radio Detection and Ranging (RADAR) and camera sensors.

These sensors are usually embedded in the vehicle, but the infrastructure too can support the object detection either by transmitting information from sensors deployed along the road to the CAVs, hence extending their sensing capability beyond the onboard sensors' line-of-sight, or by its geometry that accounts for the limitations of CAV sensing technologies.

An important attribute of the road infrastructure affecting the ability of CAV to detect other vehicles in time and prevent potentially dangerous situations, especially in an urban setting, is Intersection Sight Distance (SD). According to the methodology presented in the study [12], the required intersection sight distance along a major road SD for an automated combination truck can be computed as:

$$\mathrm{SD} = 0.278\ V_{\mathrm{d}} t_{\mathrm{c}} - \frac{V_{\mathrm{d}}(t_{\mathrm{R1}} - t_{\mathrm{R2}})}{3.6}, \tag{1}$$

where t_c is an acceptable time gap to enter the major road, V_d is the design speed of the major road in km/h, t_{R1} is the reaction time of a conventional vehicle in seconds, t_{R2} is the reaction time of automated vehicle in seconds.

The critical gaps for various vehicle types (passenger car, single-unit truck, combination truck) and maneuver types (left turn, right turn, crossing) are provided in the American Association of State Highway and Transportation Officials Green Book [22]. From the road design point of view, we consider the worst-case scenario of a combination truck trying to perform a left turn at a STOP-controlled intersection. In such a scenario, the value of t_c is 11.5 s.

Dixit [23] estimated the value of an automated vehicle's reaction time to 0.8 s, while Guzek [24] provided the human driver's reaction time on the brake pedal in the range between 1.2 and 2.2 s. We assume the difference in human-driven and automated vehicle's reaction times of 1 s, a reasonable assumption commonly used in many studies, e.g., Schoettle [25]. However, it is worth noting that some studies, e.g., Rossi [26], point out that in the case when a driver has to take over the driving tasks from a Level 4 automated vehicle, the reaction time of the driver is, in fact, much larger than in the case of a manually driven vehicle. We do not consider this phenomenon for sight distance calculations.

In Table 3, we present the values of SD calculated for a passenger car, single-unit truck and combined truck for speeds of 50 and 90 km/h, which are the usual speed limits for urban roads and rural roads in Europe, respectively. It is worth noting here that the speed limit is usually set in the range of 80–90% of the road's design speed. Therefore, V_d of 62.5 and 112.5 km/h was considered for computing the indicator values for urban and rural roads, respectively. A highway scenario has not been considered as the maneuvers performed on the highway are different from the ones performed on urban and rural roads, i.e., no left turns or crossings are allowed there.

The infrastructure segment that satisfies the SD criteria for the combination truck is assigned a score of 1 in the framework. An intersection that satisfies the criteria for a single-unit truck is assigned a score of 0.75. An intersection that satisfies the minimum SD criteria for a passenger car is assigned a score of only 0.25 since its ability to provide safe maneuvering space to any larger vehicle type than a passenger car might be severely limited.

The indicator "Infrastructure for remote sensor sharing available" refers to the ability of the infrastructure to sense the traffic situation at the assessed road segment and share its sensor data with the CAVs heading to that segment before they are able to detect the situation with their onboard sensors. An example might be a pedestrian crossing equipped with a radar sensor and Infrastructure-to-Vehicle communication capability to share information about the presence of pedestrians in an area where the CAV's sensor detection range might be limited due to obstacles or road geometry.

2.4. Quality of Maps

Just as conventional vehicles, CAVs use outdoor structured roads whose basic attributes such as location and geometry are a priori known. These static road data can be pre-created and provided to the vehicle. In combination with GNSS, inertial navigation and odometry allow the vehicle to perform high-precision (centimeter level) positioning in real-time, reducing the complexity and cost of the CAV's systems significantly [27]. Once the vehicle establishes its precise position on the road, it can use the a priori information from the map to make decisions about maneuvers and navigation, some of which would not be possible relying only on the sensor-based road model recognition methods [28]. Therefore, high-precision maps are considered one of the core enabling technologies for automated driving.

Obviously, the available maps come with different levels of precision and provide different richness of additional information about the infrastructure, ranging from the provision of basic static data on road geometry to highly dynamic high-definition maps updated in real-time and reflecting the current traffic situation.

Table 4 presents the indicator for the Quality of maps framework area with suggested map attributes and corresponding scores.

It is worth noting here that for the purpose of assessment, the score of a road segment without a fully updated map should correspond to the real state at the time of the data collection, i.e., if the traffic signs on the map are not up-to-date, the segment should be scored as there were no traffic signs on the map available at all.

2.5. Machine-Readable Signage

Road segments where either no high-definition map is available or where a mixed traffic of conventional and CAV traffic is expected, CAVs need to detect and recognize road signage using their own sensors. Numerous studies, e.g., [29,30], conclude that the features of road markings that are key for their recognition by human drivers, such as retroreflectivity and contrast, are also important in the case of marking detection by CAVs.

Table 5 presents the assessment framework and indicators proposed for the Machine-readable signage area of infrastructure assessment.

We consider two indicators within this assessment area—precision of horizontal signage detection and precision of vertical signage detection by an automated detection system. We detail the CNN used to evaluate the precision of vertical signage detection indicator from the collected sample data in Section 4.2. In this article, we will consider only vertical marking (vertical signs). Road traffic participants will also be included in the neural network training process.

Waykole et al., in [31], conducted an extensive literature review on lane detection and tracking algorithms for advanced driver assistance systems. The authors conclude that the lane detection and tracking efficiency rate under dry and light rain conditions is near 99% in most scenarios. Therefore, we adopt this value of precision for the infrastructure segment to be scored by a score of 1. To achieve a score of 0.5, a segment of infrastructure has to provide road markings clear enough to allow precision of detection in the range between 90% and 99%, which is equivalent to a precision of a lane detection and tracking system operating during the night at isolated highways. When the precision of horizontal signage detection is between 80% and 90%, the infrastructure markings only provide a performance equivalent to a vanishing point detection system operating on unstructured roads. Such an infrastructure segment is awarded a score of 0.25.

Due to the high variation in detection results in different testing environments, the evaluation of the precision of horizontal signage detection is a complex problem on its own, requiring an extensive definition of test scenarios, which is out of the scope of this paper. Therefore, we refer the interested reader to the relevant works summarized in [31] and relevant automotive standards such as [32] for further details on measurement methodology and test settings.

3. Data Collection and Processing

In this section, we interpret the technical parameters of used data systems and describe the parameters of positioning and the connectivity dataset. We also describe the system for the collection of image data.

3.1. Connectivity and Positioning Data

To collect the connectivity and positioning data, a Mikrotik LtAP LTE6 wireless access point and a Single-Board Computer (SBC) were used. LtAP LTE6 is a compact wireless access point with built-in GPS. LTE connectivity was enabled by the Mikrotik R11e-LTE6 LTE modem connected to miniPCIE slot integrated in the Mikrotik LtAP LTE6 access point, as shown in Figure 1. The used LTE modem belongs to the LTE CAT6 category and provides a maximum download speed of 300 Mb/s and an upload speed of 50 Mb/s. The collection system block diagram is shown in Figure 1.

For LTE transmission, an external 3dB wideband monopole LTE antenna with a resonant frequency from 698–2690 MHz and an input impedance of 50 Ω was used. For GPS reception, a Mikrotik ACGPSA external 26 dB, 50 Ω antenna with a resonant frequency of 1575.4 MHz was used. To ensure effective communication, we installed both antennas on the top of the testing vehicle's roof, as illustrated in Figure 2.

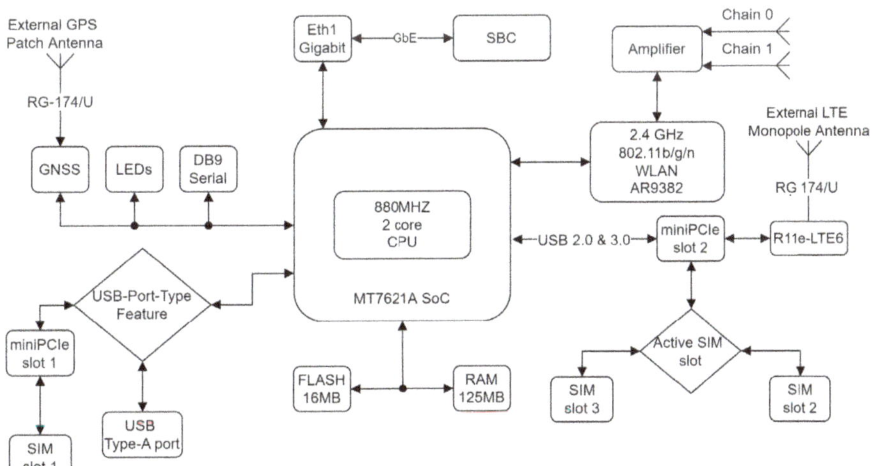

Figure 1. Collection system block diagram [33].

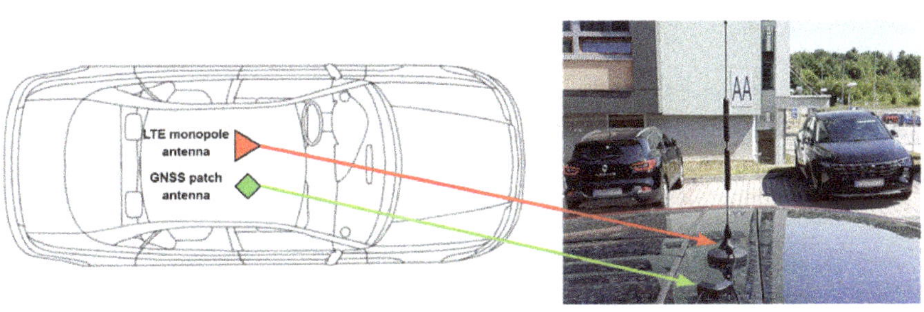

Figure 2. Antenna placement view.

3.1.1. Positioning Data

The used device supports GPS, GLONASS, BeiDou and Galileo GNSS standards. The following telemetry of GNSS was collected as is shown in Figure 3—GPS coordinates, number of satellites used, Dilution of Precision (DOP) and fix quality.

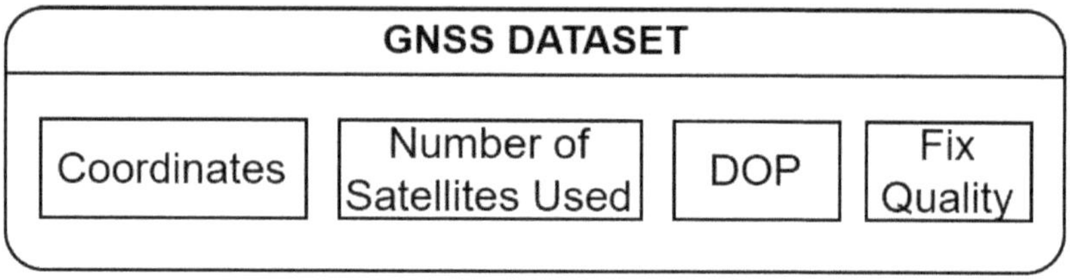

Figure 3. Block scheme of the position dataset.

The Dilution of Precision (DOP) is an important factor in determining positional errors in a GPS system. It is the collection of satellites' geometry constellation from which signals are actually received. Basically, four satellites are the minimum required value to determine a complete positional fix in three dimensions. DOP is calculated using geometrical correlations between the position of the GPS receiver and the positions of the GPS satellites. The exact locations of these satellites relative to the receiver have an effect on the positional error. If the GPS receiver communicates with satellites spread throughout the sky, the calculated position will be more accurate, and the DOP value will be low. However, when satellites are close to each other, the calculated position will be less accurate, and the DOP value will be high [34,35]. In the following table, the DOP value rating is shown (Table 6).

Table 6. DOP value rating [34].

DOP Value	Rating
<1	Ideal
1–2	Excellent
2–5	Good
5–10	Moderate
10–20	Fair
>20	Poor

The following metrics are used to describe DOP. Position Dilution of Precision (PDOP), Horizontal Dilution of Precision (HDOP), Vertical Dilution of Precision (VDOP) and Time Dilution of Precision (TDOP).

PDOP describes the number of satellites used that are spread in the sky. The more satellites directly above GPS receiver are used, the lower the PDOP value is. The effect of DOP on the horizontal position is described by HDOP. The HDOP and horizontal position (latitude and longitude) are better when more GPS satellites are used. The effect of DOP on the vertical (altitude) position is referred to as VDOP. The time difference between the GPS satellites' and the GPS receiver's internal clocks are represented by TDOP. A low TDOP value represents more accurate time synchronization. Because DOP metrics are derived from convergence, they are not independent. For example, a high TDOP value represents worse clock synchronization, and it has an effect on positional error [34,36,37].

The type of signal or technique used by the GPS receiver to establish its location is represented by the GPS fix status telemetry. The number of GPS satellites and techniques used by the GPS receiver are used to determine the GPS fix type technique. In general, the fix quality rating is given by numbers ranging from one to five, as Table 7 shows. The fix quality number represents the type of GPS technique that was used to determine location. Each technique has a different accuracy. The used GPS technique can be GPSFix, Differential GPS (DGPS), Precise Positioning System (PPSFix), Fixed Real Time Kinematic (RTK Fixed) or Float Real Time Kinematic (RTK Float). The GPSFix describes a basic GPS technique or Standard Positioning Service (SPS). SPS is a standard service provided to any user worldwide, without qualification or restrictions. Based on US security interests, the accuracy of this service is determined by the US Department of Defense [34,37]. Unlike GPSFix, the DGPS technique utilizes a network of ground stations used to broadcast the divergence between indicated position by GPS satellites and the real known position. PPSFix stands for most precise localization technique provided by GPS. Only the Federal Government and military have access to this service, which is encrypted. The RTK Fixed technique is used to optimize position accuracy calculated by DGPS. The technique is based on carrier phase measurement of the GPS, GLONASS, Galileo. Thanks to high accuracy, this technique is used for geodetic measurement purposes. On the other hand, RTK Float is a similar technique as RTK Fixed but with low accuracy. The accuracy is decreased by skipping the phase initialization process, which increases the speed of position calculation [34,38,39].

Table 7. GPS FIX status enumeration and technique accuracy [34,38,39].

FIX Quality	Technique	Accuracy (m)
1	GPSFix	15
2	DGPS	0.1
3	PPSFix	<0.03
4	RTK Fixed	0.01–0.02
5	RTK Float	0.75–0.2

3.1.2. Connectivity Data

The following telemetry of LTE communication was collected, as is shown in Figure 4. Communication latency, bandwidth, Signal Interference Noise Ratio (*SINR*), Received Signal Strength Indicator (*RSSI*), Reference Signal Received Quality (*RSRQ*), Reference Signal Received Power (*RSRP*), E-UTRA Absolute Radio Frequency Channel Number (*EARFCN*), Cell Identification (Cell ID), Channel Quality Indicator (*CQI*) and Rank Indicator (Ri).

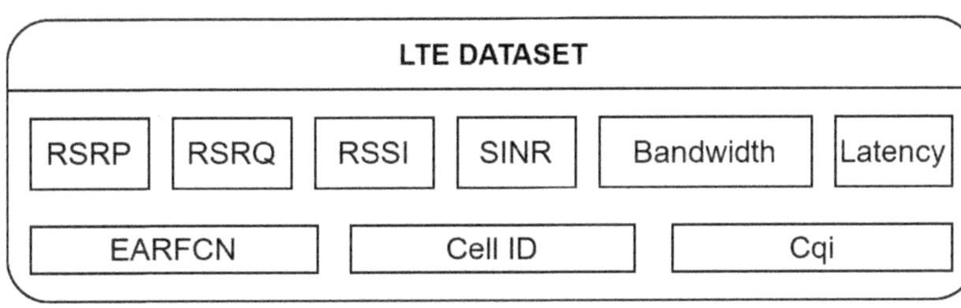

Figure 4. Block scheme of the connectivity dataset.

Our priority was to emulate the Cellular-V2X (C-V2X), as there is currently no DSRC- and VLC-enabled communication infrastructure available along the investigated route. We set up SBC to send communication packets periodically to a virtual server. We chose a packet length of 300 bytes and period of 100 ms since these values are commonly used to represent a transmission of CAM [40]. The virtual server re-sent the packet back to SBC, as is shown in Figure 5.

Each sent and received packet was marked with a time stamp by Network Time Protocol. The two-way latency communication (Δt) was calculated depending on packet transmit time (t_{tx}) and packet received time (t_{rx}), as shown in equation:

$$\Delta t[ms] = t_{rx} - t_{tx}\,. \tag{2}$$

Signal Interference Noise Ratio (*SINR*) represents the signal quality based on the strength of the wanted signal compared to the unwanted interference and noise. The *SINR* is a metric used in cellular networks to determine if a particular frequency resource is acceptable for maintaining a communication link. The network employs *SINR* to track radio link and handover failures. In systems that employ multiple access technologies based on frequency division, the scheduler can take *SINR* into account while allocating frequency resources [41]. It is a signal quality metric that is established by the User Equipment (UE) manufacturer instead of the 3GPP specifications. The basic *SINR* mathematical expression is shown in Equation (3):

$$SINR[dB] = \frac{S}{I + N}, \quad (3)$$

where S stands for the strength of the usable signals. I stands for interference power of signals or channel interference signals from other cells. N stands for background noise, which is proportional to measurement bandwidths and receiver noise coefficients. Table 8 shows the standard *SINR* values and signal quality category.

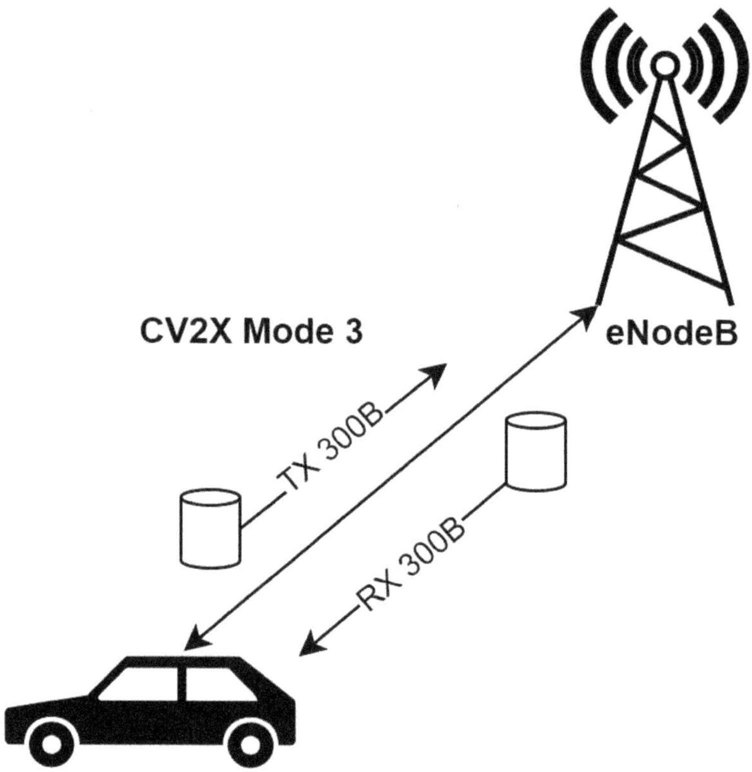

Figure 5. Emulated CV2X Mode 3 communication scheme.

Table 8. Performance indicator standards for *SINR* [42].

Range (dB)	Category
10 to 30	Excellent
3 to 10	Good
0 to −3	Fair
−20 to −3	Poor

Higher *SINR* values can affect the spectral efficiency as it enables the receiver to decode a higher Modulation Coding Scheme (MCS). To provide the best possible User Experience, the network operator attempts to optimize *SINR* at all locations, either by transmitting at a greater power or by avoiding interference and noise [43].

SINR optimization can aid in achieving higher cell capacity by allowing higher QAM modulation, which results in greater peak data rates, fewer missed calls, and an overall better quality of user experience [44].

Received Signal Strength Indicator (*RSSI*) is an LTE metric that states how much overall wideband power measured in symbols have been received, including all interference and thermal noise. UE does not send *RSSI* values to eNodeB. It may be easily calculated using

RSRQ and *RSRP*, which are instead reported by UE. The value is measured in dBm. *RSSI* is defined as [45]:

$$RSSI[dBm] = S + I + N \ , \tag{4}$$

where *S*, *I* and *N* are the same parameters as in the *SINR* equation.

Reference Signal Received Power (*RSRP*) and Refence Signal Received Quality (*RSRQ*) are two main key signal level and quality indicators for current LTE networks. When a UE goes from cell to cell in a cellular network and conducts handover, it performs a measurement of the reference signal strength and quality of serving and neighbor cells for successful execution of the handover process. In essence, it is the power of the received signal from eNodeB by UE [42]. Based on *RSRP*, it is possible to compare the strengths of signals from individual cells in LTE networks. The measurement process is shown in Figure 6.

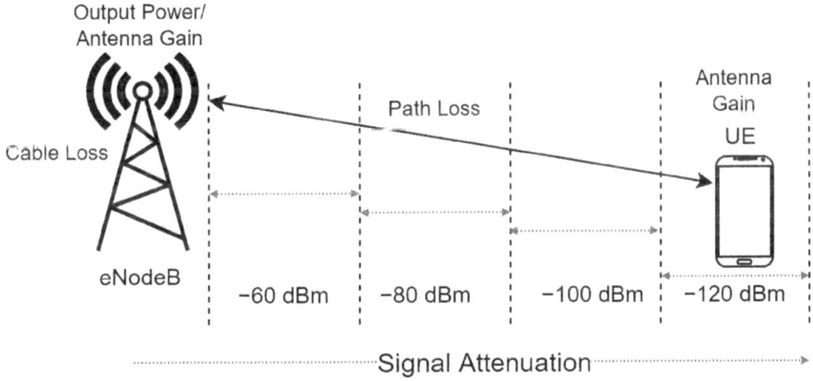

Figure 6. *RSRP* measurement.

Figure 6 shows eNodeB and UE, which receive the reference signal from the eNodeB. The closer the UE is to the eNodeB location, the stronger the received signal. The reporting range of *RSRP* is defined from −140 to −44 dBm with 1 dB resolution [42]. Table 9 shows the standard *RSRP* values and signal quality category.

Table 9. Performance indicator standards for *RSRP* [46].

Range (dBm)	Category
−80 to −44	Excellent
−90 to −80	Good
−100 to −90	Fair
−110 to −100	Poor
−140 to −110	Very Poor

The *RSRP* calculation is shown in Equation (5), where *N* stands for Number of PRBs (Physical Resource Blocks) [42,47,48].

$$RSRP[dBm] = RSSI - 10 * log(12 * N). \tag{5}$$

RSRQ telemetry parameter is the proportion of *RSRP* to wideband power. *RSRQ* represents signal quality received by the UE. The signal, noise, and interference received by the UE also have an effect on the *RSRQ* [40,42]. The following equation [42] is used for *RSRQ* calculation, where *N* stands for Number of Physical Resource Blocks (PRBs).

$$RSRQ[dB] = N * RSRP/RSSI, \tag{6}$$

The reporting range of *RSRQ* is defined from −3 to −20 dB. Table 10 shows the standard *RSRQ* values and signal quality category.

Table 10. Performance indicator standards for *RSRQ* [42].

Range (dB)	Category
−10 to −3	Excellent
−12 to −10	Good
−14 to −12	Fair
−17 to −14	Poor
−20 to −17	Very Poor

Instead of reporting raw carrier frequencies in MHz, LTE base stations use the ETSI E-UTRA Absolute Radio Frequency Channel Number (EARFCN) industry standard to report channel numbers. In LTE technology, EARFCN determines the carrier frequency in the uplink and downlink, the range of which is from 0 to 65,535. The equations below express the relationship between EARFCN and its uplink/downlink carrier frequency [49].

$$F_{\text{downlink}} = F_{\text{DL-low}} + 0.1(N_{\text{DL}} - N_{\text{offs-DL}}), \tag{7}$$

$$F_{\text{uplink}} = F_{\text{UL-low}} + 0.1(N_{\text{UL}} - N_{\text{offs-UL}}), \tag{8}$$

where N_{DL} stands for downlink EARFCN, N_{UL} for uplink EARFCN, $N_{\text{offs-UL}}$ offset used to calculate uplink EARFCN, $N_{\text{offs-DL}}$ offset used to calculate downlink EARFCN. The values $F_{\text{UL-low}}, F_{\text{DL-low}}, N_{\text{DL}}, N_{\text{UL}}, N_{\text{offs-DL}}, N_{\text{offs-UL}}$ are given in [49] by ETSI.

For unique identification of LTE components, the identification numbers are used. As shown in Figure 7, we have three main key identifiers in the LTE cell. The E-UTRAN Cell Identifier (ECI) represents the identity of a cell within a Public Land Mobile Network Identifier (PLMN). ECI consists of 28 bits where the first 20 bits represent the eNodeB ID number and the last 8 bits are stated for cell ID. The sector ID identifies a particular antenna in cell sectors [50,51].

The code rate and modulation are defined by MCS in the LTE. MCS specifies the maximum number of usable bits that can be transferred per Resource Element (RE), and it is affected by radio channel quality. Table 11 shows the CQI-MCS mapping for LTE rel. 12 and beyond. The better channel quality is represented by a higher MCS, and the more useful data can be transmitted. In other words, MCS depends on error probability. In LTE, a Turbo encoder with a 1/3 coding rate is employed. The actual ratio of usable bits to total transmitted bits (useful bits + parity bits) is dependent on the quality of the radio link. The range of coding rates is 0.0762 to 0.9258.

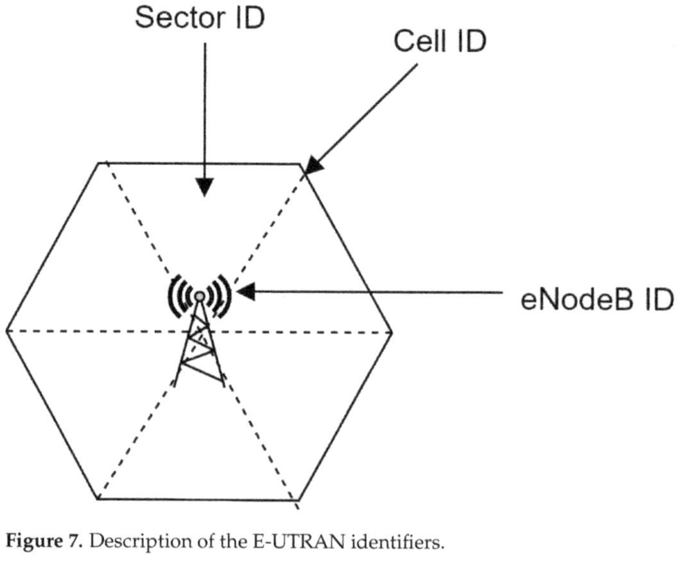

Figure 7. Description of the E-UTRAN identifiers.

Table 11. CQI-MCS mapping for LTE rel.12 and beyond [52].

CQI	Modulation	Code Rate	Bits per RE
1	QPSK	0.0762	0.1524
2	QPSK	0.1885	0.377
3	QPSK	0.4385	0.877
4	QPSK	0.3691	1.4764
5	QPSK	0.4785	1.914
6	QPSK	0.6016	2.4064
7	16QAM	0.4551	2.7306
8	16QAM	0.5537	3.3222
9	16QAM	0.6504	3.9024
10	64QAM	0.7539	4.5234
11	64QAM	0.8525	5.115
12	64QAM	0.6943	5.5544
13	64QAM	0.7783	6.2264
14	64QAM	0.8634	6.9072
15	64QAM	0.9258	7.4064

Radio link quality is estimated based on the Channel Quality Indicator (*CQI*). The *CQI* parameter is reported by UE to the eNodeB. The *CQI* measurement is based on the Cell Reference Signal (CRS) [52]. Better radio condition is represented by higher *CQI* and the higher coding rate, as is shown in the table below. Bits per RE column should be multiplied by the number of data streams to obtain a final value in case of MIMO usage [52].

3.1.3. Collection of Image Data

For image data collection, the OmniVision OV10640 camera system (OmniVision, Santa Clara, CA, USA) was used. This sensor uses a proprietary technology, which delivers an image with a very high dynamic range (HDR). Furthermore, the sensor is encapsulated in a compact package, which can be easily deployed for a wide range of automotive applications (see Table 12). A total of four cameras were used on the bus (one on the windshield recorded the area in front of the bus, one on the rear window recorded the area behind the bus, and one on each side recorded the area on the sides of the bus).

Table 12. Specifications of OmniVision OV10640 camera system.

Camera System Parameter	Specification
Resolution	1280 (H) × 1080 (V)
Mega Pixels	1.3 MP
Supply Voltage	1.7 to 3.47 V
Frame Rate	60 fps
Pixel Size	4.2 μm × 4.2 μm
Dynamic Range	120 dB
Sensitivity	8.4 V/lux-s
SNR	41.5 dB

Three cameras were used in the collection of image data. Two cameras were placed on the sides of the bus and one in the middle of the bus above the windshield (see Figure 8). The cameras located on the sides of the bus had a standard horizontal field of view (52 degrees). The middle camera capturing objects in front of the bus was a fisheye (horizontal field of view of 194 degrees).

Furthermore, the sensor is capable of sampling the recorded scene simultaneously instead of sequentially, which helps to minimize the distortion caused by motion.

The image dataset contains classes representing traffic signs and also classes representing road users, as is shown in Figure 9. Table 13 shows all the classes that our image dataset contains. The first column contains the number of classes, and the second column shows the specification of the given class.

Figure 8. Image data collection.

Figure 9. An example of an image dataset.

Table 13. The classes in the image dataset.

Number of Class	Class Specification	The Overall Number of Data	Training Dataset	Testing Dataset	Validation Dataset
1	Ahead only	1670	1003	500	167
2	Turn left ahead	1670	1003	500	167
3	Turn right ahead	1670	1003	500	167
4	The one-way traffic	1670	1003	500	167
5	The stop sign	1670	1003	500	167
6	Give away	1670	1003	500	167
7	The priority road	1670	1003	500	167
8	The pedestrians	1670	1003	500	167
9	The cyclists	1670	1003	500	167
10	The motorbikes	1670	1003	500	167
11	The scooters	1670	1003	500	167
12	Road closed	1670	1003	500	167
13	Passing prohibited	1670	1003	500	167
14	No entry	1670	1003	500	167
15	Speed limit	1670	1003	500	167
16	No right turn sign	1670	1003	500	167
17	No left turn sign	1670	1003	500	167
18	Two-way traffic ahead	1670	1003	500	167
19	The passenger cars	1670	1003	500	167
20	The Vans/trucks	1670	1003	500	167

4. Methodology

In this section, we describe the processing of connectivity and positioning data and processing of image data using the proposed CNN.

4.1. Processing of Connectivity and Positioning Data

The data collection process was repeated four times on different days and at different time. Data processing was divided into three parts, as illustrated in Figure 10. In the first part, data pre-processing, the data were prepared for processing. Data were collected as text files (.txt), and it was necessary to convert them to Comma-Separated Values (CSV) and separate them. The conversion process and data separation was performed by a python script.

The pre-processed data served as an input to the processing stage. In this stage, the latency data were averaged because latency was measured every 100 ms, and other data were collected every 1000 ms. After the averaging process, all data were synchronized on the basis of a time stamp that was obtained via Network Time Protocol (NTP) during the data collection. A weighting coefficient was assigned to examine parameters on the basis of which it is possible to represent the quality of the digital infrastructure parameters. In the data post-processing stage, data were concentrated, evaluated, and visualized.

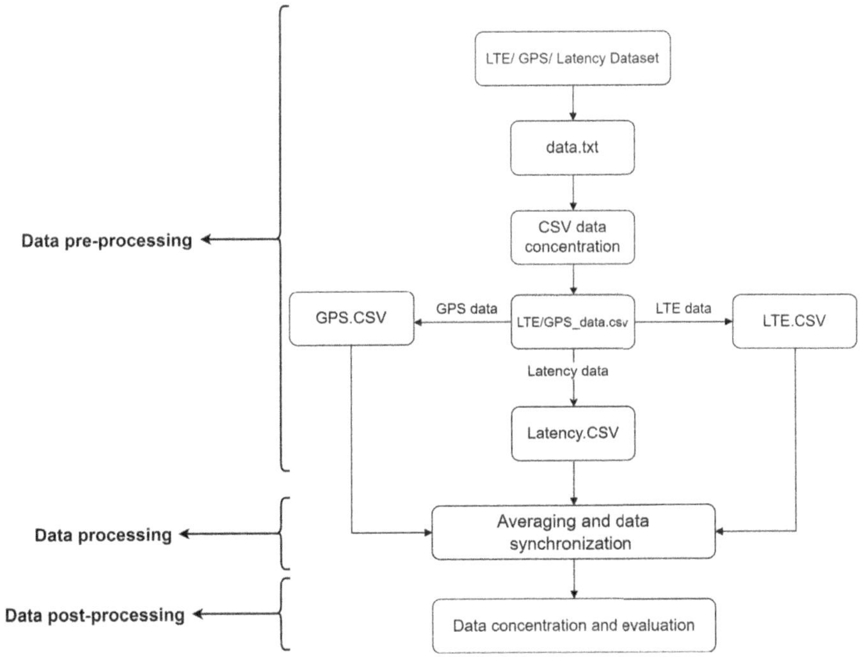

Figure 10. Data processing block diagram.

4.2. Processing of Image Data Using CNN

For traffic sign recognition, we proposed CNN, which is detailed in Figure 11 and Table 14. We selected CNN since, depending on the used hardware, it has a potential to process data in real-time. Hence, it can serve as a basis for the future development of an automated infrastructure readiness assessment system operating in real-time.

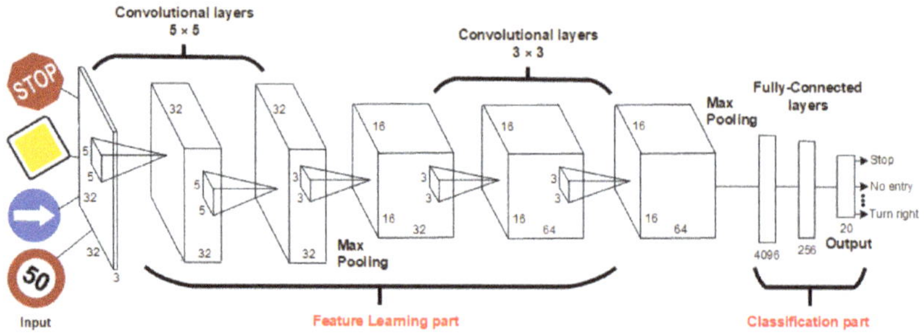

Figure 11. Proposed architecture of CNN.

Table 14. CNN layers.

Layers	Description of Layers
Conv2D_1	32 filters with dimensions 5 × 5, the output is a feature map with dimensions 32 × 32 × 32
Conv2D_2	32 filters with dimensions 5 × 5, the output is a feature map with dimensions 32 × 32 × 32
MaxPooling_1	Filter size 2 × 2, the output is a feature map with dimensions 32 × 16 × 16
Dropout_1	50% neuron shutdown, the output is a 32 × 16 × 16 feature map
Conv2D_3	64 filters with dimensions 3 × 3, the output is a feature map with dimensions 64 × 16 × 16
Conv2D_4	64 filters with dimensions 3 × 3, the output is a feature map with dimensions 64 × 16 × 16
MaxPooling_2	Filter size 2 × 2, the output is a feature map with dimensions 32 × 8 × 8
Dropout_2	50% neuron shutdown, the output is a 32 × 8 × 8 feature map
Flatten_1	4096 neurons
Dense_1	256 neurons
Dropout_3	50% neuron shutdown
Dense_2	20 neurons

The CNN is divided into two parts, the feature learning part (convolutional part) and the classification part. The convolution part is used for data processing. The classification part serves to transform the format of the processed data and to classify the output. Figure 11 shows the block diagram of the proposed CNN. This proposed CNN consists of 12 layers (four 2-D convolutional layers, two MaxPooling layers, three layers for turning off neurons (Dropout) and 3 fully connected layers.

As discussed in [53], the image data that are corrupted by various noises impact the resulting performance of the proposed neural network. For this reason, the noisy image data are recovered with the pre-processing step (using various filters). This step improves the overall performance of the proposed neural network.

The first and input layer is the convolution layer. The input is represented by images with dimensions of 32 × 32 pixels. At the input of the layer, we will therefore have 32 × 32 neurons (1024 arranged in a square matrix). Each pixel in the image is represented by an 8-bit number, ranging from 0–255 for each color. Sometimes it also uses a black and white image, which is represented by one channel in the same range, where 0 represents white and 255 black. In our case, each pixel is represented by three values from the RGB palette. Together, these values form three two-dimensional matrices, which together form the image volume. In this layer, we use 32 filters with dimensions of 5 × 5. We also use the padding parameter set to "Same", which will cause the output feature maps to be the same size as the input image. The output from this layer will be 32 (32 × 32 feature maps for each input image).

The second layer is the convolutional layer, which includes 32 feature maps with dimensions of 32 × 32. It contains 32 filters with a window size of 5 × 5. It also contains a parameter that maintains the same dimensions of the output as the input. The output of the layer will be 32 × 32 × 32 feature maps for each image. The third layer is a merging layer (MaxPooling layer) with a filter size of 2 × 2 and a maximum value criterion. This causes the dimensions of the 32 × 32 × 32 input features to be halved in the output. The number of feature maps remains the same. In the end, we obtain 32 feature maps with dimensions 16 × 16. The fourth layer is dropout with a parameter of 0.5, which means that 50% of the random neurons at the input will be turned off. This layer preserves the previous dimensions of the output. The fifth and sixth layers are convolutional, which includes 32 feature maps with dimensions 16 × 16 and 64 filters with dimensions 3 × 3, respectively. The seventh layer is a merging layer with a filter size of 2 × 2 and a maximum value criterion. The dimensions of the output will be twice as small as the input, i.e., 64 × 8 × 8 with 64 feature maps. The eighth layer is a dropout with a parameter value of 0.5, which means that 50% of the random neurons on the current layer will be turned off. The ninth layer is flattened, which transforms feature maps into fully connected layers. This layer will contain 8 × 8 × 64 neurons, which is a total of 4096 neurons and, therefore, also 4096 outputs. The tenth layer is Dense, which represents the classic fully connected layer. It contains 256 neurons. The last layer is the Dense layer, which classifies the output from the network. It contains 20 neurons (20 classes). Each neuron represents a given class. In this layer, we use the sigmoid activation function, which classifies us with the probability that a given neuron is activated, thus determining to which class it belongs.

5. Experimental Results

In this section, the performance evaluation of the proposed method based on the created dataset is discussed.

5.1. Results for Connectivity and Positioning Data

Analyzed connectivity data are interpreted on the maps with the heat map route, which shows the analyzed route in the city. Figure 12 shows the communication latency on the analyzed route. The blue color represents the latency values less than or equal to 9 ms, and the red color represents values greater than or equal to 800 ms.

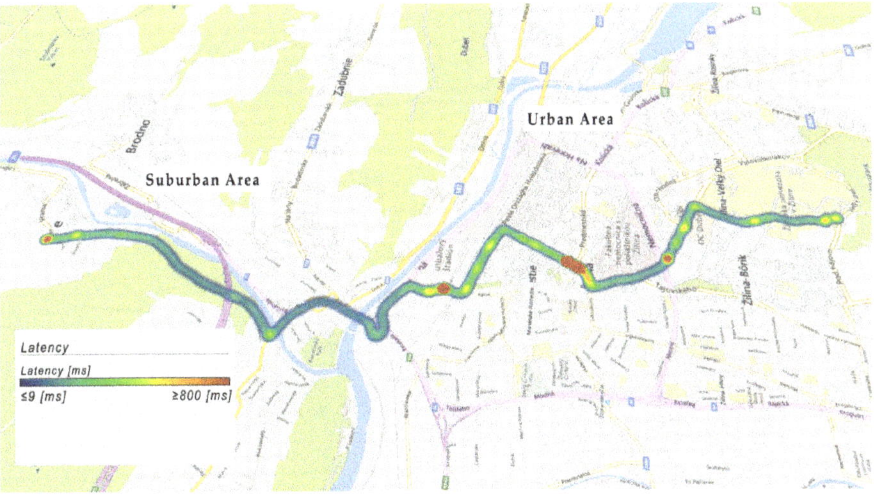

Figure 12. Example of latency data collection.

As we can see, the latency value is higher in urban areas than in suburban areas. The average latency value in urban areas was 83 ms, and in a suburban area, 42 ms. We found spots where the latency was higher than 800 ms in an urban area. These places are interpreted by red color, and they are located mainly at intersections with heavy traffic or near points of interest. The latency value in these places reached the value of 1500 ms. For the deployment of CCAM, it is necessary to support telecommunication infrastructure in these red areas. Please note that during the data collection campaign, no considerable traffic jams or road congestions occurred along the investigated route. In the case of extremely congested traffic, the communication performance is expected to drop even further.

The number of GNSS satellites used, i.e., the number of visible satellites, ranged from 14 to 22. The maximum number of visible satellites was reached in sparsely-built areas with a clear vision of the sky. In Figure 13, the blue color represents the usage of less than or equal to 10 satellites. The red color represents the usage of more than or equal to 20 satellites. The number of visible satellites was lower in dense urban areas and in the suburban area too.

The results for the Signal Interference Noise Ratio are shown in Figure 14. This map interprets the coverage quality of the 4G telecommunication infrastructure. The blue color represents *SINR* values less than or equal to −15 dB, and the red color represents *SINR* values greater than or equal to 32 dB. As shown in Figure 14, the better coverage is in the suburban area compared to the urban area. In the urban area, there were spots in which the *SINR* value was −15 dB, which is also related to poor connectivity parameters.

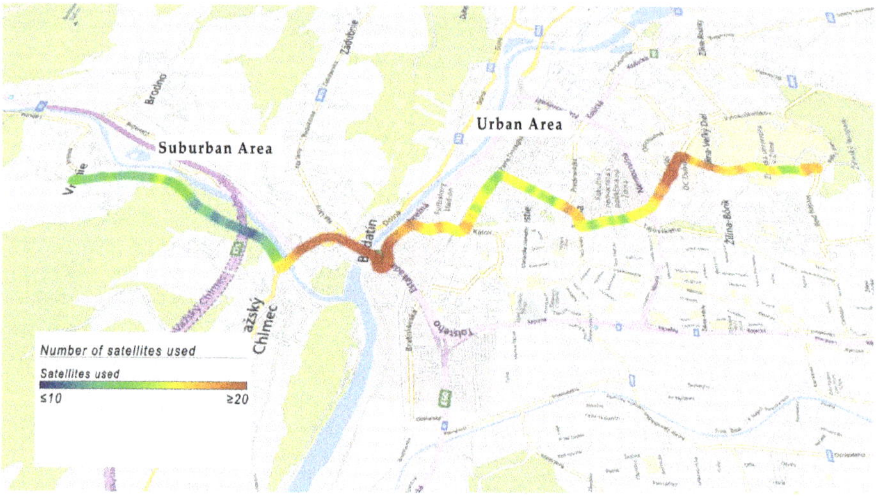

Figure 13. Example of satellite numbers used.

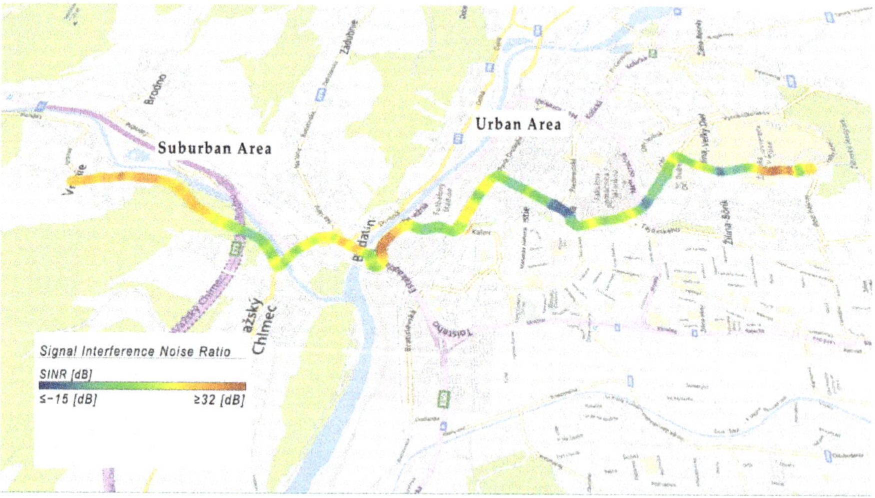

Figure 14. Example of Signal Interference Noise Ratio.

As can be seen in Figure 15, *RSRP* values less than or equal to −120 dBm are represented by blue. *RSRP* values higher or equal to −40 dBm are represented by red. The received power of the reference signal is lower in the suburban area compared to the urban area. It is caused by the fact that there are many eNodeBs in the city, which ensure handover and thus provide higher power of the reference signal. On the other hand, in the suburban area, there is a low number of eNodeBs, which reduces the received power of the reference signal.

The received quality of the reference signal ranges from −15 to −5 (dB). As can be seen in Figure 16, in an urban area at an intersection with heavy traffic, the received quality of reference signal is lower than in other areas. The blue color represents an *RSRQ* value lower than or equal to −15 dB, and the red color represents an *RSRQ* value greater than or equal to −5 dB.

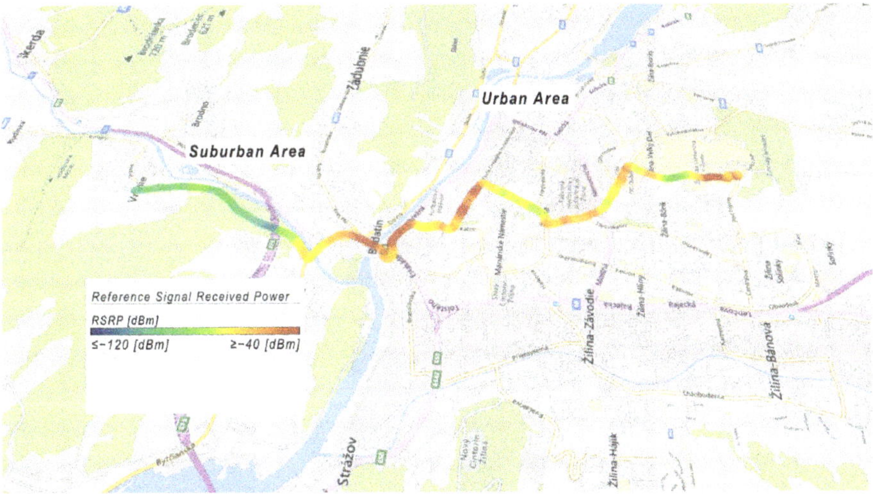

Figure 15. Example of Reference Signal Received Power.

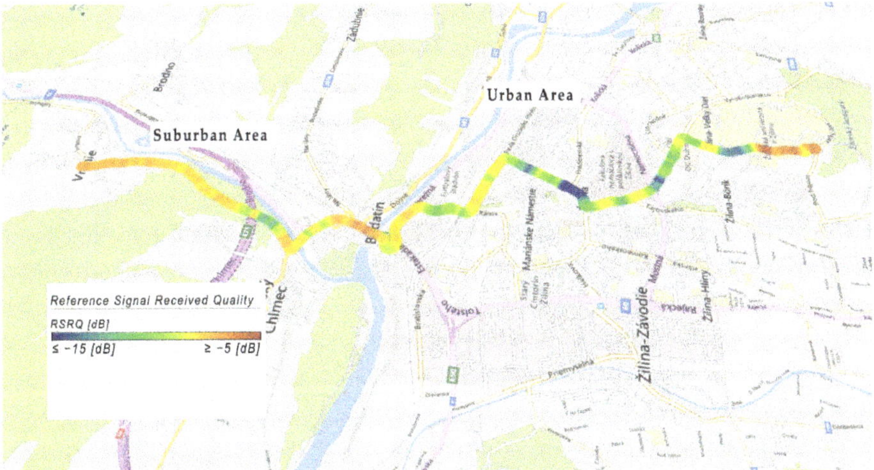

Figure 16. Example of Reference Signal Received Quality.

Fades in received reference signal power are caused by larger communication distance and resulting lower signal power from the individual cells in the 4G network, which is spread over the area. Near the eNodeBs, the quality of the received reference signal was −5 (dB), which represents a better condition of the connection.

5.2. Results for Image Data

The inputs to the proposed CNN were image data of size 32 × 32 × 3. The model was trained on 30 epochs. In the training process, it is important to find the point where the network gives us the best results. If we exceed this threshold, the network learns too much detail, which means that the success on the validation or test model decreases (overtraining of neural network). On the other hand, if we stop learning too early, the network will be untrained. For this reason, we use checkpoints and also dropout layers, which improve our model. In the figure below, can be seen how we split the data into train, validation, and testing. In our work, we use this proportion, but it can be changed as per the requirement. To build CNN for traffic sign classification, the Keras deep learning framework was used.

Each class contains 1670 images. The dataset was divided into training, testing and validation parts in the ratio 60:30:10, as is shown in Figure 17. This means that 60% of images were used for training data, 10% for validation data and 30% for test data. The size of each animal image was 32 × 32 pixels.

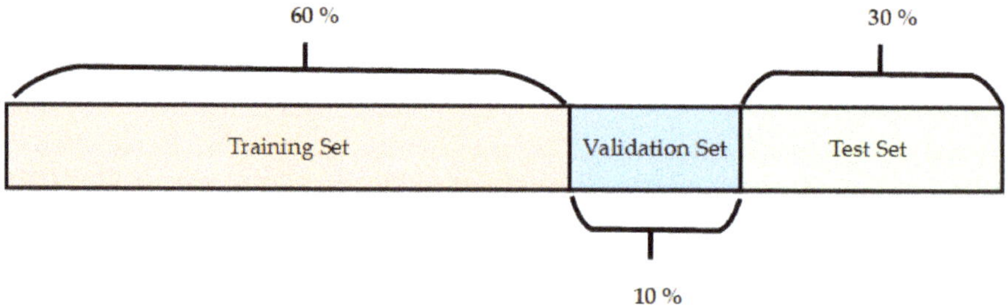

Figure 17. Division of the image dataset (training data, test data and validation data).

The training set and the validation set were used, respectively, to train and optimize the model. The test set was used to check how the model performs on unseen data. As can be seen in Figure 18, precision of 99.7% on our training set (blue line) was obtained. Please note that this precision is very similar to the results presented by other works, such as [54,55].

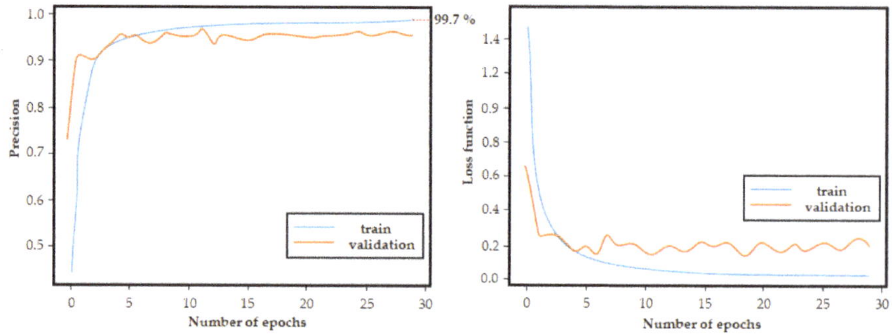

Figure 18. Training and validation precision.

Figure 19 demonstrates the confusion matrix. The rows of the confusion matrix represent the actual class, while the columns of the confusion matrix represent the predicted class. The values along the main diagonal represent images that correctly classified images to be the same class. The correctly classified images across all classes are used to define the classification accuracy. In other words, it is the ratio of the sum of the correctly labeled images to the total number of images in the test dataset. In the case of traffic sign classification, the precision was greater than 99%. On the other hand, in the case of the classification of traffic participants, the precision was 98.4% for pedestrians, 85.5% for cyclists, 86.4% for motorbikes and 96.4% for scooters (see Table 15).

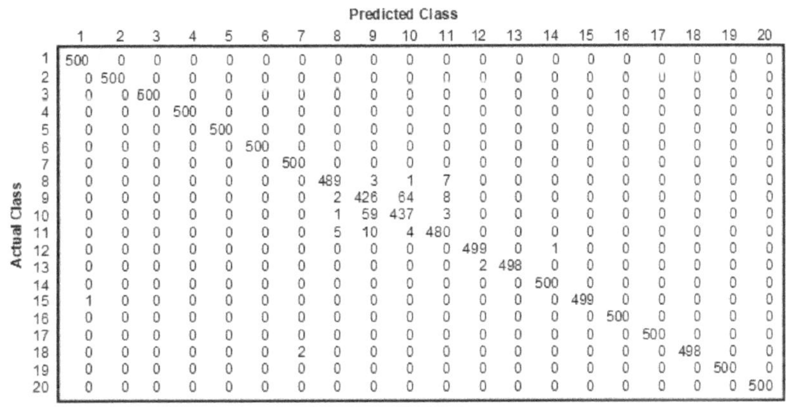

Figure 19. Confusion matrix.

Table 15. Results of image classification.

Number of Class	Precision (%)	Recall (%)	F1 Score (%)
1	99.8	100	99.9
2	100	100	100
3	100	100	100
4	100	100	100
5	100	100	100
6	100	100	100
7	99.6	100	99.8
8	98.4	97.8	98.1
9	85.5	85.2	85.3
10	86.4	87.4	86.9
11	96.4	96.2	96.3
12	99.6	99.8	99.7
13	100	99.6	99.8
14	99.6	100	99.8
15	100	99.8	99.9
16	100	100	100
17	100	100	100
18	100	99.6	99.8
19	100	100	100
20	100	100	100

6. Discussion and Conclusions

The result of our research is a framework that serves to assess the state of physical and digital infrastructure readiness for CCAM. The core of the research is a dataset that can be employed for further research on the topic. Our results can serve as a basis for more effective planning of infrastructure development from the point of view of readiness for CCAM. The main goal of the connectivity and positioning data metering is research on the performance of new generation networks and localization systems for CCAM readiness. As part of this research, we mapped and analyzed the urban and suburban areas. Despite the fact that the analysis and mapping were carried out in different time frames and days, we found underdimensioned areas on the investigated route. The main problem of data communication analysis is the latency. As we described in Section 5.1, we found the critical places in urban areas from the point of view of latency. These places are mainly located at intersections with heavy traffic or near points of interest. A possible solution to high latency is to add microcells to critical places. On the one hand, adding microcells can not only lower the latency, but it can also increase the performance of telecommunication networks and increase coverage. On the other hand, this solution comes with increased infrastructure costs. The best solution for accelerating the implementation of CCAM is the deployment of 5G networks. Currently, 5G networks are not widely deployed, and they are mainly located in metropolitan areas and capital cities in some designated locations. Analysis of the performance of the 5G network from the point of view of CCAM readiness will be our future work. The issue of localization is dense areas with large buildings and trees. It affects the number of visible satellites, which also has an impact on the accuracy of localization. A possible solution is to use 5G networks in conjunction with GNSS. This can improve the better localization accuracy in dense urban areas. Using GPS, digital maps and neural networks, the vehicle can recognize the direction of travel, speed, lane detection and traffic signs. By combining neural network, GPS data and digital maps, it is possible to create a reliable system that could reliably recognize traffic signs.

The problem of traffic sign recognition in order to create an automated system was solved using the proposed neural network. The proposed system opens up new possibilities for further research in our future work. The automation of the traffic sign recognition system is becoming increasingly necessary for its use in road traffic. The example of traffic signs classification using CNN is shown in Figure 20. When the "Speed limit" symbol is shown to the camera system, the trained model identifies it and classifies the traffic sign name as "Speed limit". Classifications and predictions are made in very less time (almost real-time), which benefits drivers. Although the classification of traffic signs has many advantages, there are also some difficulties. For instance, if the traffic sign is covered by trees or any billboard on the side of the road, then it can cause inaccurate traffic sign detection and classification. It may also happen that the vehicle is cruising so fast that the system does not have enough time to correctly recognize the traffic sign. These situations can be very dangerous and lead to traffic accidents.

Figure 20. Example of traffic sign detection and classification.

Various environmental constraints, including lighting, traffic sign distance (the sign is too far away), or shadow, can significantly affect the accurate detection and classification of traffic signs. Therefore, further research in this area is needed.

One of the directions that require further research is traffic light detection, which the industry continues to develop at an ever-increasing pace. Notable examples in this area include recent developments made by several major automotive industry players, whose vehicles already include systems based on either DSRC or image processing for traffic light color recognition and driver alerting.

In our future work, we plan to improve the precision of the proposed neural network for the recognition of traffic signs. We also plan to create a line detection system for automated vehicle driving. The processing of collected data for the infrastructure assessment was not completed in real-time. In our future work, we plan to introduce elements of automatization to the assessment process with the ultimate goal of developing a fully automated system for infrastructure readiness assessment. The research presented in this manuscript is an initial step towards this goal.

Another option for further research is to design a fully automatic road sign recognition system that will work in real-time. This system will use the camera system on the vehicle to detect and recognize traffic signs in real-time. In the event that this system is integrated together with the GPS system, it is also possible to provide the driver with additional practical information about the current restrictions within the current traffic situation on the given road. Based on the comparison of data from GPS and the sign recognition system, this system could warn the driver in the case of disregarding traffic signs. It is worth noting here that the detection and correct classification of live objects is an extremely important aspect of CAV operation. While being beyond the scope of this paper, we aim to incorporate this aspect and address its challenges in our future work as well.

Author Contributions: Conceptualization, T.P., B.C. and P.K.; methodology, T.P.; software, B.C. and P.K.; validation, T.P., P.K. and M.D.; formal analysis, T.P., B.C., P.K. and G.P.; investigation, B.C. and P.K.; resources, M.D.; data curation, T.P.; writing—original draft preparation, T.P., B.C. and P.K; visualization, T.P., P.K. and B.C.; supervision, M.D. and P.K.; project administration, M.D. and P.K.; funding acquisition, M.D., T.P. and P.K. All authors have read and agreed to the published version of the manuscript.

Funding: This publication was realized with the support of Operational Program Integrated Infrastructure 2014–2020 of the project: Innovative Solutions for Propulsion, Power and Safety Components of Transport Vehicles, code ITMS 313011V334, co-financed by the European Regional Development Fund.

Institutional Review Board Statement: Not applicable.

Informed Consent Statement: Not applicable.

Data Availability Statement: The data presented in this study are available on request from the corresponding author.

Acknowledgments: This work has been supported by the Slovak Research and Development Agency under the project No. PP-COVID-20-0100: DOLORES.AI: The pandemic guard system.

Conflicts of Interest: The authors declare no conflict of interest.

Abbreviations

The following abbreviations are used in this manuscript:

ADS	Automated Driving Systems
CAM	Cooperative Awareness Messages
CCAM	Cooperative, Connected and Automated Mobility
CAVs	Connected and Automated Vehicles
CNN	Convolutional Neural Network
CRS	Cell Reference Signal
CQI	Channel Quality Indicator
CV2X	Cellular Vehicle-to-everything
DGPS	Differential Global Positioning System
DOP	Dilution of Precision
EARFCN	E-UTRA Absolute Radio Frequency Channel Number
GNSS	Global Navigation Satellite Systems
GPS	Global Positioning System
HDOP	Horizontal Dilution of Precision
LTE	Long-Term Evolution
ODDs	Operational Design Domains
PDOP	Position Dilution of Precision
RF	Radio Frequency
RI	Rank Indicator
RSRP	Reference Signal Received Power

RSRQ	Reference Signal Received Quality
RSSI	Received Signal Strength Indicator
RTK	Real Time Kinematic
SBC	Single-Board Computer
SD	Sight Distance
SINR	Signal Interference Noise Ratio
SPS	Standard Positioning Service
TDOP	Time Dilution of Precision
UE	User Equipment
VDOP	Vertical Dilution of Precision
VLC	Visible Light Communication
V2I	Vehicle-to-Infrastructure
V2X	Vehicle-to-everything

References

1. Kopelias, P.; Demiridi, E.; Vogiatzis, K.; Skabardonis, A.; Zafiropoulou, V. Connected & autonomous vehicles—Environmental impacts—A review. *Sci. Total Environ.* **2020**, *712*, 135237. [CrossRef] [PubMed]
2. Wadud, Z.; MacKenzie, D.; Leiby, P. Help or hindrance? The travel, energy and carbon impacts of highly automated vehicles. *Transp. Res. Part Policy Pract.* **2016**, *86*, 1–18. [CrossRef]
3. Alessandrini, A.; Domenichini, L.; Branzi, V. Chapter 7-Infrastructures to accommodate automated driving. In *The Role of Infrastructure for a Safe Transition to Automated Driving*; Alessandrini, A., Domenichini, L., Branzi, V., Eds.; Elsevier: Amsterdam, The Netherlands, 2021; pp. 237–366. [CrossRef]
4. van Schijndel-de Nooij, M.; Krosse, B.; van den Broek, T.; Maas, S.; van Nunen, E.; Zwijnenberg, H.; Schieben, A.; Mosebach, H.; Ford, N.; McDonald, M.; et al. *Definition of Necessary Vehicle and Infrastructure Systems for Automated Driving*; European Commission: Brussels, Belgium, 2011.
5. Ma, Y.; Wang, Z.; Yang, H.; Yang, L. Artificial intelligence applications in the development of autonomous vehicles: A survey. *IEEE/CAA J. Autom. Sin.* **2020**, *7*, 315–329. [CrossRef]
6. Li, J.; Cheng, H.; Guo, H.; Qiu, S. Survey on artificial intelligence for vehicles. *Autom. Innov.* **2018**, *1*, 2–14. [CrossRef]
7. Ge, X.; Han, Q.L.; Wu, Q.; Zhang, X.M. Resilient and safe platooning control of connected automated vehicles against intermittent denial-of-service attacks. *IEEE/CAA J. Autom. Sin.* **2022**, *6*, 1–18. [CrossRef]
8. Khalid Khan, S.; Shiwakoti, N.; Stasinopoulos, P. A conceptual system dynamics model for cybersecurity assessment of connected and autonomous vehicles. *Accid. Anal. Prev.* **2022**, *165*, 106515. [CrossRef] [PubMed]
9. Carreras, A.; Daura, X.; Erhart, J.; Ruehrup, S. Road infrastructure support levels for automated driving. In Proceedings of the 25th ITS World Congress, Copenhagen, Denmark, 17–21 September 2018.
10. SAE International. *J3016: Taxonomy and Definitions for Terms Related to Driving Automation Systems for On-Road Motor Vehicles*; SAE International: Hoboken, NJ, USA, 2021.
11. Mackenzie, J.; Dutschke, J.; van den Berg, A.; Kumar, M.; Meuleners, L. *Automated Vehicles and the Readiness of Western Australian Roads*; Road Safety Commission: Easth Perth, WA, USA, 2018.
12. Magyari, Z.; Koren, C.; Kieć, M.; Borsos, A. Sight distances at unsignalized intersections: A comparison of guidelines and requirements for human drivers and autonomous vehicles. *Arch. Transp.* **2021**, *59*, 8–19. [CrossRef]
13. Liu, Y.; Tight, M.; Sun, Q.; Kang, R. A systematic review: Road infrastructure requirement for Connected and Autonomous Vehicles (CAVs). *J. Phys. Conf. Ser.* **2019**, *1187*, 73. [CrossRef]
14. Nitsche, P.; Mocanu, I.; Reinthaler, M. Requirements on tomorrow's road infrastructure for highly automated driving. In Proceedings of the 2014 International Conference on Connected Vehicles and Expo (ICCVE), Vienna, Austria, 3–7 November 2014; pp. 939–940. [CrossRef]
15. Madadi, B.; van Nes, R.; Snelder, M.; van Arem, B. *Image-Based Assessment of Road Network Readiness for Automated Driving: A Judgement Game*; Delft University of Technology: Delft, The Netherlands, 2018.
16. Zadobrischi, E.; Dimian, M. Inter-Urban Analysis of Pedestrian and Drivers through a Vehicular Network Based on Hybrid Communications Embedded in a Portable Car System and Advanced Image Processing Technologies. *Remote Sens.* **2021**, *13*, 1234. [CrossRef]
17. Just One Autonomous Car Will Use 4000 GB of Data/Day. Available online: https://www.networkworld.com/article/3147892/one-autonomous-car-will-use-4000-gb-of-dataday.html (accessed on 5 July 2022).
18. Mueck, M.; Karls, I. *Networking Vehicles to Everything*; DeG Press: Amsterdam, The Netherlands, 2018.
19. Heng, L.; Walter, T.; Enge, P.; Gao, G.X. GNSS Multipath and Jamming Mitigation Using High-Mask-Angle Antennas and Multiple Constellations. *IEEE Trans. Intell. Transp. Syst.* **2015**, *16*, 741–750. [CrossRef]
20. Joubert, N.; Reid, T.G.R.; Noble, F. Developments in Modern GNSS and Its Impact on Autonomous Vehicle Architectures. In Proceedings of the 2020 IEEE Intelligent Vehicles Symposium (IV), Las Vegas, NV, USA, 13–19 November 2020; pp. 2029–2036. [CrossRef]

21. Reid, T.; Houts, S.; Cammarata, R.; Mills, G.; Agarwal, S.; Vora, A.; Pandey, G. Localization Requirements for Autonomous Vehicles. *SAE Intl. J CAV* **2019**, *2*, 173–190. [CrossRef]
22. AASHTO. *A Policy on Geometric Design of Highways and Streets (The Green Book)*; American Association of State Highway and Transportation Officials: Washington, DC, USA, 2021.
23. Dixit, V.V.; Chand, S.; Nair, D.J. Autonomous vehicles: Disengagements, accidents and reaction times. *PLoS ONE* **2016**, *11*, e0168054. [CrossRef]
24. Guzek, M.; Lozia, Z.; Zdanowicz, P.; Jurecki, R.; Stańczyk, T.; Pieniążek, W. Assessment of Driver's Reaction Times in Diverisified Research Environments. *Arch. Transp.* **2012**, *24*, 149–164. [CrossRef]
25. Schoettle, B. *Sensor Fusion: A Comparison of Sensing Capabilities of Human Drivers and Highly Automated Vehicles*; The University of Michigan: Michigan, MI, USA, 2017.
26. Rossi, R.; Gastaldi, M.; Orsini, F.; Cet, G.D.; Meneguzzer, C. A comparative simulator study of reaction times to yellow traffic light under manual and automated driving. *Transp. Res. Procedia* **2021**, *52*, 276–283. [CrossRef]
27. Liu, J.; Xiao, J.; Cao, H.; Deng, J. The Status and Challenges of High Precision Map for Automated Driving. In *Proceedings of the China Satellite Navigation Conference (CSNC) 2019 Proceedings*; Sun, J., Yang, C., Yang, Y., Eds.; Springer: Singapore, 2019; pp. 266–276.
28. Aeberhard, M.; Rauch, S.; Bahram, M.; Tanzmeister, G.; Thomas, J.; Pilat, Y.; Homm, F.; Huber, W.; Kaempchen, N. Experience, Results and Lessons Learned from Automated Driving on Germany's Highways. *IEEE Intell. Transp. Syst. Mag.* **2015**, *7*, 42–57. [CrossRef]
29. Burghardt, T.E.; Popp, R.; Helmreich, B.; Reiter, T.; Böhm, G.; Pitterle, G.; Artmann, M. Visibility of various road markings for machine vision. *Case Stud. Constr. Mater.* **2021**, *15*, e00579. [CrossRef]
30. Matowicki, M.; Přibyl, O.; Přibyl, P. Analysis of possibility to utilize road marking for the needs of autonomous vehicles. In Proceedings of the 2016 Smart Cities Symposium Prague (SCSP), Prague, Czech Republic, 26–27 May 2016; pp. 1–6. [CrossRef]
31. Waykole, S.; Shiwakoti, N.; Stasinopoulos, P. Review on Lane Detection and Tracking Algorithms of Advanced Driver Assistance System. *Sustainability* **2021**, *13*, 1417. [CrossRef]
32. *ISO 17361*; Intelligent Transport Systems—Lane Departure Warning Systems—Performance Requirements and Test Procedures. ISO: Geneva, Switzerland, 2017.
33. Mikrotik Specifications. Available online: https://mikrotik.com/product/ltap_lte6_kit#fndtn-specifications (accessed on 5 July 2022).
34. Office of the Department of Defense, United States of America, G.N. *Global Positioning System Standard Positioning Service Performance Standard*; Office of the Department of Defense: New York, NY, USA, 2020.
35. Isik, O.K.; Hong, J.; Petrunin, I.; Tsourdos, A. Integrity Analysis for GPS-Based Navigation of UAVs in Urban Environment. *Robotics* **2020**, *9*, 66. [CrossRef]
36. Pattanayak, B.; Moharana, L. Analyzing the effect of dilution of precision on the performance of GPS system. In Proceedings of the 2021 IEEE 1st Odisha International Conference on Electrical Power Engineering, Communication and Computing Technology (ODICON), Bhubaneswar, India, 8–9 January 2021.
37. Li, B.; Zhao, K.; Shen, X. Dilution of Precision in Positioning Systems Using Both Angle of Arrival and Time of Arrival Measurements. *IEEE Access* **2020**, *8*, 192506–192516. [CrossRef]
38. ESRI.ArcGIS.Mobile.Gps Namespace. Available online: https://help.arcgis.com/en/arcgismobile/10.0/apis/arcgismobile/api/ESRI.ArcGIS.Mobile~ESRI.ArcGIS.Mobile.Gps_namespace.html (accessed on 5 July 2022).
39. NovAtel. *An Introduction to GNSS*; NovAtel: New York, NY, USA, 2015.
40. ETSI. *LTE: E-UTRA Physical Layer Measurements (3GPP TS 36.214 Version 14.3.0 Release 14)*; ETSI: Antipolis, France, 2017.
41. Bastidas-Puga, E.R.; Galaviz, G.; Andrade, Á.G. Evaluation of SINR prediction in cellular networks. *IETE Technol. Rev.* **2018**, *35*, 476–482. [CrossRef]
42. Putra, G.M.; Budiman, E.; Malewa, Y.; Cahyadi, D.; Taruk, M.; Hairah, U. 4G LTE Experience: Reference Signal Received Power, Noise Ratio and Quality. In Proceedings of the 2021 3rd East Indonesia Conference on Computer and Information Technology (EIConCIT), Surabaya, Indonesia, 9–11 April 2021; pp. 139–144. [CrossRef]
43. Heinz Mellein, M.M. *Assessing a MIMO Channel*; Rohde & Schwarz: Munich, Germany, 2017.
44. 5G NR SINR Measurement and Its Mapping. Available online: https://www.techplayon.com/5g-nr-sinr-measurement-and-its-mapping/ (accessed on 5 July 2022).
45. RSRQ to SINR Relation. Available online: https://arimas.com/2016/04/24/164-rsrq-to-sinr/ (accessed on 5 July 2022).
46. Pramono, S.; Alvionita, L.; Ariyanto, M.D.; Sulistyo, M.E. Optimization of 4G LTE (long term evolution) network coverage area in sub urban. In Proceedings of the 5th International Conference on Industrial, Mechanical, Electrical and Chemical Engineering 2019 (ICIMECE 2019), Surakarta, Indonesia, 17–18 September 2019; AIP Publishing: Melville, NY, USA, 2020.
47. Zhang, X. *LTE Optimization Engineering Handbook*; Wiley–IEEE, John Wiley & Sons: Nashville, TN, USA, 2017.
48. RSRP and RSRQ Measurement in LTE. Available online: https://www.cablefree.net/wirelesstechnology/4glte/rsrp-rsrq-measurement-lte/ (accessed on 5 July 2022).
49. ETSI. *LTE–Evolved Universal Terrestrial Radio Access (E-UTRA). User Equipment (UE) Conformance Specification. Radio Transmission and Reception. Part 1: Conformance testing (3GPP TS 36.521-1 Version 11.2.0 Release 11)*; ETSI: Antipolis, France, 2013.

50. Berthold, K.P.H. Sector Antenna Number Identification for LTE. Available online: https://people.csail.mit.edu/bkph/cellular_repeater_numerology.shtml (accessed on 5 July 2022).
51. LTE Quick Reference. Available online: https://www.sharetechnote.com/html/Handbook_LTE_IDs_in_LTE.html (accessed on 5 July 2022).
52. LTE Modulation and Coding Scheme (MCS). Available online: http://anisimoff.org/eng/lte_mcs.html (accessed on 5 July 2022).
53. Mercorelli, P. A Fault Detection and Data Reconciliation Algorithm in Technical Processes with the Help of Haar Wavelets Packets. *Algorithms* **2017**, *10*, 13. [CrossRef]
54. Zeng, Y.; Xu, X.; Shen, D.; Fang, Y.; Xiao, Z. Traffic Sign Recognition Using Kernel Extreme Learning Machines With Deep Perceptual Features. *IEEE Trans. Intell. Transp. Syst.* **2017**, *18*, 1647–1653. [CrossRef]
55. Zhu, Y.; Yan, W.Q. Traffic sign recognition based on deep learning. *Multimed. Tools Appl.* **2022**, *81*, 17779–17791. [CrossRef]

sensors

MDPI

Article

An Improved IoT-Based System for Detecting the Number of People and Their Distribution in a Classroom †

Slavomir Matuska *, Juraj Machaj, Robert Hudec and Patrik Kamencay

Faculty of Electrical Engineering and Information Technology, University of Zilina, 010 26 Zilina, Slovakia
* Correspondence: slavomir.matuska@uniza.sk; Tel.: +421-41-513-2215

† This paper is an extended version of the published conference paper "IoT Based System for Detecting the Number of People and their Distribution in Classroom. In Proceedings of the 2022 ELEKTRO (ELEKTRO), Krakow, Poland, 23–26 May 2022".

Abstract: This paper presents an improved IoT-based system designed to help teachers handle lessons in the classroom in line with COVID-19 restrictions. The system counts the number of people in the classroom as well as their distribution within the classroom. The proposed IoT system consists of three parts: a Gate node, IoT nodes, and server. The Gate node, installed at the door, can provide information about the number of persons entering or leaving the room using door crossing detection. The Arduino-based module NodeMCU was used as an IoT node and sets of ultrasonic distance sensors were used to obtain information about seat occupancy. The system server runs locally on a Raspberry Pi and the teacher can connect to it using a web application from the computer in the classroom or a smartphone. The teacher is able to set up and change the settings of the system through its GUI. A simple algorithm was designed to check the distance between occupied seats and evaluate the accordance with imposed restrictions. This system can provide high privacy, unlike camera-based systems.

Keywords: IoT-based system; IoT nodes; Raspberry Pi; Arduino-based module; COVID-19

Citation: Matuska, S.; Machaj, J.; Hudec, R.; Kamencay, P. An Improved IoT-Based System for Detecting the Number of People and Their Distribution in a Classroom. *Sensors* **2022**, *22*, 7912. https://doi.org/10.3390/s22207912

Academic Editor: Jari Nurmi

Received: 22 September 2022
Accepted: 14 October 2022
Published: 18 October 2022

1. Introduction

Following the COVID-19 outbreak in 2019, we have been facing different and difficult challenges in all aspects of our lives. One of them is without a doubt the continuous full-time educational process. Online education has its advantages, however, it cannot replace full-time education and student skills gained from face-to-face experience and practice, especially when it comes to education in technology and engineering. Various countries have different approaches that are enabling full-time education and access to facilities for students. Rules to reduce the maximum number of people in a classroom alongside social distancing rules have been introduced widely. These two mentioned restrictions were our primary motivation to propose an improved smart IoT-based system for detecting the number of people and their distribution in the indoor environment (e.g., classroom, or any type of room).

There have been a multitude of studies published in the area of detecting and counting people in indoor spaces. Most of the proposed solutions are based on image processing from cameras installed in the area. For example, Myint and Sein [1] proposed a robust camera-based system which is able to estimate the number of people entering and exiting a room. Their solution is based on Raspberry Pi and software using a pre-trained VGG-16 CNN model and an SVM classifier based on TensorFlow and the Keras library.

Another camera-based system for counting people using Raspberry Pi was proposed by Rantelobo et al. in [2]. This system can distinguish between people entering or leaving a room by performing image processing using background subtraction, morphological transformation, and calculating the contour area of the image. The main advantage of this solution is that the system can run on cheap hardware such as Raspberry Pi. Similar solutions relying on cameras and computer vision algorithms have been presented in [3,4].

Moreover, Hou et al. [5] presented a solution for social distancing detection based on a deep learning model. The primary goal was distance evaluation between individuals to mitigate the impact of the COVID-19 pandemic and reduce the virus transmission rate in indoor spaces. The detection tool was developed to alert people to maintain a safe distance from each other by processing a video feed from cameras used to monitor the environment. A similar system was proposed by Sharma in [6]. This system helps people to ensure proper social distancing in crowded places and highlights the violations of these norms in real time. The proposed system is based on image processing. There are many more published papers (e.g., [7,8]) solving the problem of social distancing using feeds from cameras and computer vision [9,10].

The work presented in [11] proposed a large Convolutional Neural Network (CNN) trained using a single-step model and You Only Look Once version 3 (YOLOv3) on Google Colaboratory to process the images within a database and accurately locate people within the images. The trained neural network was able to successfully generate test data, achieving a mean average precision of 78.3% and a final average loss of 0.6 while confidently detecting the people within the images. Yet another work presented in [12] uses YOLO v3 and Single Shot multi-box Detector (SSD) to detect and count people. The authors analyzed both methods and their comparison of the achieved results showed that the precision, recall, and F1 measure achieved for SSD were higher than for YOLO v3. The main issue related to camera-based systems involves privacy concerns, as data collected by cameras can be misused for face recognition, thus revealing the identity of the individuals in the area [13].

On the other hand, there are multiple works for counting people or measuring social distance which do not require the installation of camera systems. Among such systems, a smart social distancing monitoring system based on Bluetooth and GPS was described in [14]. In this system, an application can offer a solution for monitoring public spaces and reminding users to maintain distance. The work presented in [15] is based on an ultra-wideband radar sensor for a people counting algorithm. The proposed algorithm can operate in real time and is able to achieve a mean absolute error of less than one person. The system in [16] relies on Wi-Fi probing requests to count people in a crowd by taking advantage of people's smartphones.

Another way of counting people or measuring social distance could be using Internet of Things (IoT) technology. The IoT can be described as a network of physical objects (things) that are equipped with sensors, software, and other technologies to connect and exchange data with other devices or systems via the internet or a local network. IoT represents interaction between the physical and the digital world in its simplest form. An IoT object in the world can be a simple sensor equipped with a communication interface or a smart self-driving car equipped with state-of-the-art technology. The advantages of using IoT technology in real-world applications are almost unlimited, and use cases can be found in such disparate areas as Industry 4.0 [17,18], smart agriculture [19,20], smart cities [21,22], smart transportation [23,24], smart homes [25,26], eHealth [27], and wearables [28]. With the massive adoption of IoT technology, it is finding applications in many areas [29,30]. For example, the authors of [29] stated that their platform, based on a combination of IoT and fog cloud, can be used in systematic and intelligent COVID-19 prevention and control. The system involves five use cases, including COVID-19 Symptom Diagnosis, Quarantine Monitoring, Contact Tracing and Social Distancing, COVID-19 Outbreak Forecasting, and SARS-CoV-2 Mutation Tracking [31]. Another IoT-based COVID-19 and Other Infectious Disease Contact Tracing Model was described by the authors of [32]. They presented an RFID-based proof-of-concept for their model and leveraged blockchain-based trust-oriented decentralization for on-chain data logging and retrieval.

The wearable proximity sensing system presented in [33] is based on an oscillating magnetic field that overcomes many of the weaknesses of the current state-of-the-art Bluetooth-based proximity detection. The authors proposed, implemented, and evaluated their system and demonstrated that the proposed magnetic field-based system is much more reliable than previously proposed Bluetooth-based approaches. Another possible solution for monitoring people in indoor environments was introduced by Perra et al. [34]. The proposed device implements a novel real-time pattern recognition algorithm for processing data sensed by a low-cost infrared (IR) array sensor. The device can perform local processing of infrared array sensor data, and in this way is able to monitor occupancy in any space of a building while maintaining people's privacy. A seat-occupancy detection system based on Low-Cost mm-Wave Radar at 60 GHz was presented in [35]. Detection is based on Pulsed Coherent Radar in the unlicensed 60 GHz ISM band. The system can detect the presence of people occupying the seats by measuring small movements of the body, such as breathing. The solution for counting the people in the classroom proposed by Zhang et al. [36] is similar to the work presented in this paper. In their case, the authors used two E18-D80NK photoelectric sensors to count people in a classroom and an hc-sr501 infra-red sensor for detection of seat occupancy. However, the authors presented only the basic principles and hardware design of the system. The solution proposed in this paper is based on a slightly different technology with lower energy consumption. Moreover, it provides a complex solution with a user-friendly GUI and advanced functionalities, e.g., management of the rules and functions supporting deployment of the system.

The remainder of this paper is organized as follows. Section 2 is devoted to a description of the system concept. The system's implementation is described in detail in Section 3, including the implemented methods, software, and hardware design. In Section 4, the achieved results are presented and discussed. Section 5 provides a comparison with other systems proposed for occupancy detection, and Section 6 concludes the paper.

2. System Concept

The main goal of the proposed system is to deliver counting of incoming students when they are entering the classroom as well as detection of the distribution of the students among the seats. The proposed system is portable and can be easily deployed and managed by the teacher. The main purpose of the proposed system is to help teachers with the management of their classes while respecting implemented COVID-19 restrictions. The concept of the proposed system is presented in Figure 1.

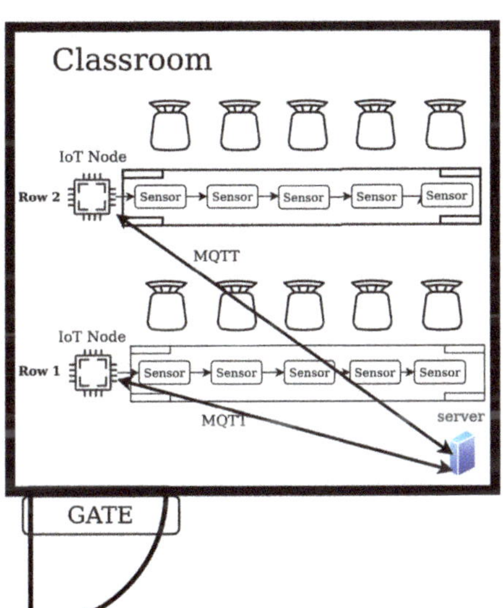

Figure 1. The system concept proposal.

The number of IoT nodes and sensor devices connected to the system is variable, however, there is one server per classroom. The teacher can access the system's Graphical User Interface (GUI) from a mobile device or personal computer in order to set up the system as well as to check whether all restrictions are being obeyed. Each seat available in the classroom is equipped with a single HC-SR04 Ultrasonic Distance Sensor. Up to sixteen distance sensors can be connected to a single IoT node, which is responsible for the evaluation of the seat occupancy on a single row. Each IoT node collects data from the connected sensors and sends data via MQTT protocol to the main server. The IoT nodes are based on the NodeMCU Arduino board. The entrance to the classroom is equipped with a Gate node. The purpose of the Gate node is to count the students entering or leaving the classroom. The Gate node consists of a single NodeMCU board with two HC-SR04 sensors used for door crossing detection by persons entering or leaving the classroom. The HC-SR04 is a low-cost sensor which can provide distance measurements between 2 and 400 cm with non-contact measurements and ranging accuracy up to 3 mm. The sensor accuracy is sufficient for the purposes of occupancy detection. Each sensor module includes an ultrasonic transmitter, a receiver, and a control circuit. The sensor's principle is as follows:

- To trigger measurement output, the pin has to be activated for at least 10 µs;
- The Module automatically sends eight 40 kHz signals and detects pulse signal reflections;
- The distance is calculated as follows:

$$s_t = t * 343/2, \quad (1)$$

where t is the pulse trigger duration and the constant 343 is the speed of sound in m/s.

The sensor implemented in the system operates at 5 volts and consumes 15 mA, while its dimensions are $45 * 20 * 15$ mm. For the sake of comparison, the Infrared Proximity Sensor E18-D80NK used for people counting and HC-SR501 PIR sensor for detecting the seat occupancy in [36] consume 25–100 mA and 65 mA, respectively. Moreover, it is important to note that the infrared proximity sensor has a shorter sensing range compared to the ultrasonic sensor implemented in the proposed solution. On the other hand, the ultrasonic distance sensor has its own disadvantages, such as sensitivity to variations in the ambient temperature and difficulties when reading reflections from soft, curved, thin, and small objects. The server is based on the Raspberry Pi model 4B+ and is responsible for managing the whole system. The proposed solution consists of a Message Queuing Telemetry Transport (MQTT) broker for communication, Node-RED for logic, Mongo database for data storage, and a React-based web application that serves as the GUI. The system is deployed using Docker containers and the docker-compose tool ensures container orchestration.

Cloud Technologies Versus Self-Hosted Solutions

There are two main categories of technology that can be implemented on the server side:

- Cloud-based technology;
- Self-hosted on-premises solutions.

Both of these have their advantages and disadvantages. Cloud technologies are relatively new platforms, however, they can offer a lot of already built-in features which are ready to use without a long setup. On the other hand, users may argue that when their data are not stored on their hardware, they do not have full control over the hardware, nor over the data. The primary goal of cloud technology for IoT is to provide universal functionality for application development as well as ubiquitous access to the data. Therefore, the user of the IoT platform can focus only on the functionality of its product and its value and does not need to care about the hardware itself. There are four categories of IoT platforms:

- IoT cloud platforms;
- IoT connectivity platforms;
- IoT device platforms;
- Analytics platforms.

The well-known IoT platforms in 2022 are represented by the IBM Watson IoT Platform, Particle, AWS IoT Core, Google Cloud IoT Core, Azure IoT Central, and many others. While many of the features provided by these platforms are free of charge, users must pay in order to receive the most out of each of these platforms.

The alternative is self-hosting. In this case, the user has to install and configure services according to their project's needs. In the proposed system, a self-hosted solution based on the Raspberry Pi was chosen. The primary reason for this decision was that there may not be internet access available in every classroom and the Raspberry Pi can create its Wi-Fi network. Therefore, there is no need to add an extra router in order to provide connectivity for the system. With this in mind, the proposed system can be deployed in any classroom without any significant limitations.

3. System Implementation

The first step in system implementation is the design of the communication flow diagram. The flowchart of the system functionalities is presented in Figure 2. Individual users can visit the GUI either from a personal computer in the classroom or through their smartphone. The communication between the GUI and the rest of the system was created using the HTTP protocol. WebSocket-based communication was implemented in order to obtain real-time information when seat occupancy changes. The main communication node of the system is the MQTT broker.

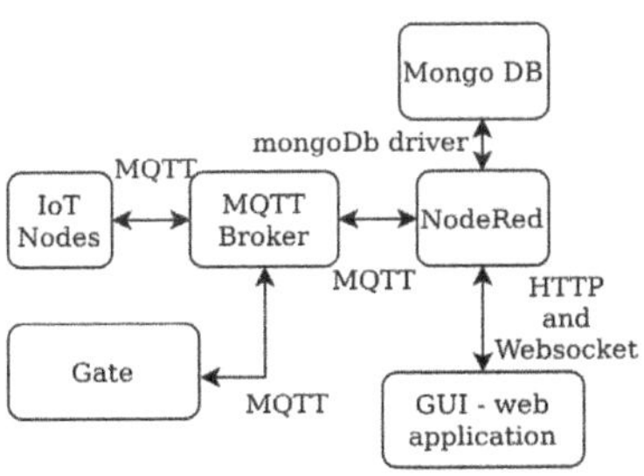

Figure 2. The main system data flowchart.

All communication between nodes and Node-RED applications is handled by this broker. All connected IoT nodes send their data to the broker. The Node-RED application is responsible for data processing and storing the results in the Mongo database server using the Mongo DB driver. At the same time, it provides a RESTful application programming interface (API) and WebSocket endpoint that serves the data for the GUI. The hardware of the system consists of three main blocks:

- Gate node;
- IoT nodes for detecting seat occupancy;
- Self-hosted server.

3.1. The Gate Node

The primary task of the Gate node is to detect people who cross the door, i.e., those entering or leaving the classroom. The Gate node consists of one NodeMCU board equipped with two HC-SR04 distance sensors. The principal functionality of the Gate node is depicted in Figure 3.

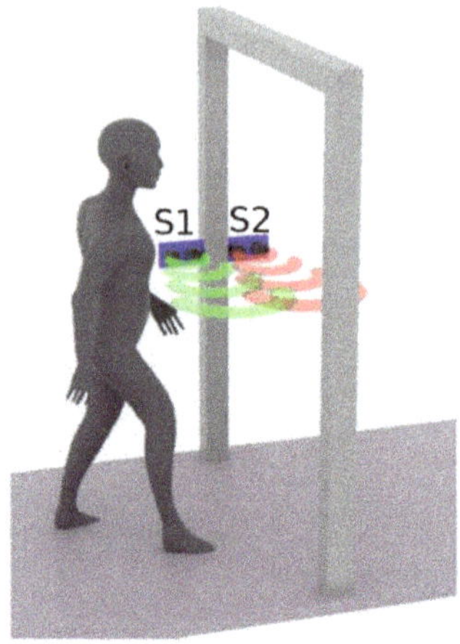

Figure 3. Principal functionality of Gate node.

As mentioned earlier, from the sensing point of view the gate consists of two distance sensors, which are enough to detect when a person is crossing the door of the classroom. However, there is one limitation to this approach, being that only one person at a time can cross the door of the classroom. The flowchart of the software implemented in the Gate node is shown in Figure 4.

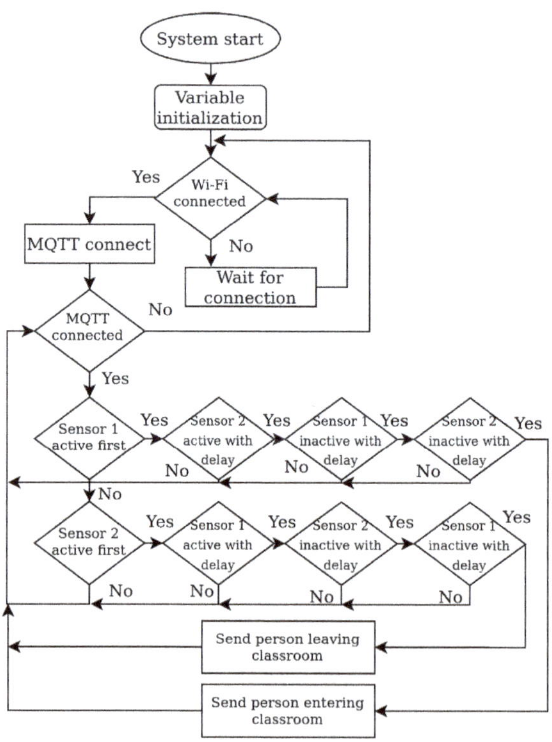

Figure 4. The Gate node algorithm implementation.

The software managing the Gate node starts with the initialization of variables and definitions. Moreover, it is necessary to set up the Wi-Fi network name (SSID), Wi-Fi password, server IP address, and MQTT credentials. When the node is connected to the Wi-Fi network as well as the MQTT broker, the Gate node starts to measure values from distance sensors. The PubSubClient library, which is available for Arduino-based boards, was used for connection and data transfer over MQTT. The MQTT topics used for this type of message are "sensors/gateEnter" and "sensors/gateExit".

The algorithm for door-crossing detection operates based on the detection of a sequence of events reported by the sensors. To explain the crossing detection algorithm, we defined two states of sensors S1 and S2. The first is the "Active sensor", which means that the measured distance from the sensor is shorter than the defined value of the door width, i.e., an active sensor means that the sensor detects the object. The second term is "Inactive sensor", which means that the measured distance from the sensor is not shorter than the defined value of the door width, i.e., there is no object detected in front of the sensor. The possible sequences of sensor states resulting in successful crossing detection by the algorithm implemented in the Gate node are shown in Table 1, where 0 represents the inactive state of the sensor and 1 stands for the active state of the sensor.

Table 1. Sequences for successful door crossing detection.

Result	**t(0)**		**t(1)**		**t(2)**		**t(3)**		**t(4)**	
	S1	**S2**	**S1**	**S2**	**S1**	**S2**	**S1**	**S2**	**S1**	**S2**
Person entered room	0	0	1	0	1	1	0	1	0	0
Person left room	0	0	0	1	1	1	1	0	0	0

In the table, steps t(0)–t(4) are defined by the time when the state of the sensor has changed. The system detects a door crossing event only when the states of the S1 and S2 sensors change according to the sequences provided in the table. In cases when other sequences are detected, the system does not detect a door crossing event, and thus does not change the number of persons in the room.

3.2. IoT Nodes for Detecting Seat Availability

Each IoT node is based on a NodeMCU board which can connect up to sixteen distance sensors. In the proposed system design, it is necessary to use just a single distance sensor per seat. Therefore, it is possible to determine seat occupancy across the classroom and automatically check whether the students in the room are keeping the desired social distance simply by evaluating data from individual distance sensors. The ultrasonic distance sensor (HC-SR04) can be placed under the PC monitor or can be attached to the bottom part of the table. In order to connect sixteen distance sensors to the NodeMCU board, it is necessary to use a 16:1 single-channel analogue multiplexer, such as CD74HC4067 in our case. An example of the sensor placement is shown in Figure 5. The implemented sensor can provide distance measurements between 2 and 400 cm with non-contact measurements and ranging accuracy up to 3 mm. Each sensor module includes an ultrasonic transmitter, a receiver, and a control circuit. During operation, it is necessary to establish whether there is a person relatively close to the sensor. The decision-making distance was set at 70 cm during the tests, however, the teacher is able to change the decision-making distance based on the conditions in the classroom using the web application interface of the server. When the measured distance is shorter than the threshold value, the seat is considered to be taken, represented by a logical one; otherwise, the seat is considered to be free (logical zero).

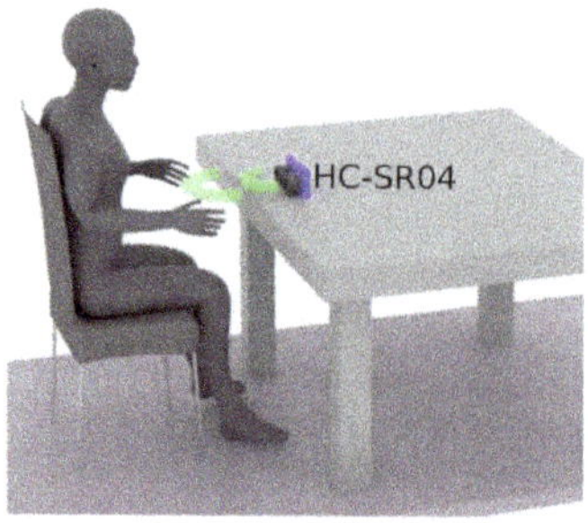

Figure 5. Sensor placement.

When the seat occupancy status changes, the updated seat occupancy information is sent to the server for evaluation and storage. A flowchart of the software implementation for the IoT node for detecting seat availability is shown in Figure 6.

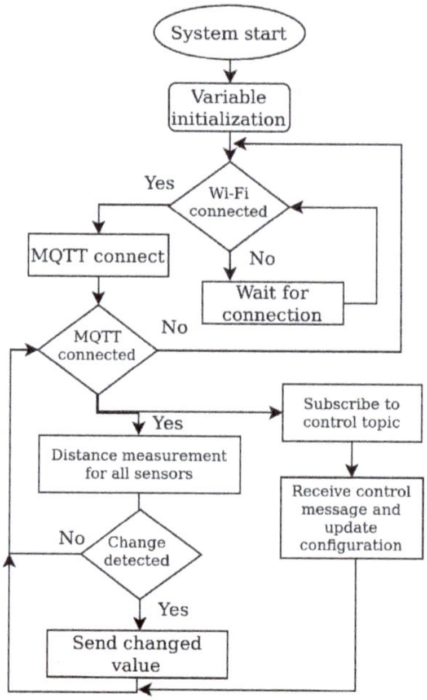

Figure 6. IoT node algorithm implementation.

The software of the IoT node starts with variable initialization and definitions. The first part of the code matches the software for the Gate node, such as Wi-Fi and MQTT connections and initialization of libraries. Afterwards, the nodes measure the distances from each sensor in the loop. When a change in the sensor value is detected, the node sends the message to the server for evaluation. The MQTT topic used for this type of message is "/sensors/distanceChanged". Useful information in messages represents JSON objects in string representation. A message with data about the occupancy sent from sensors looks like this:

{row: 1, seats: [0, 1, 0, 1, 0, 0, 0, 1]};

where the row specifies the position in the classroom and the seats represent an array of values that define the seat occupancy for individual positions in a row.

In the improved version of the IoT-based system, the option to set up the whole classroom from scratch using only the GUI was added. For this purpose, each node subscribes to the topic "nodes/setupConfig" to enable receiving of the configuration messages and is able send status data to the topic "nodes/nodeStatusUpdate". The setup message for the IoT node is as follows:

{rowPosition: 1, sensorCount: 8, thresholdDistance: 70};

where rowPosition specifies the row position in the classroom, sensorCount defines the number of seats in a particular row, and thresholdDistance is used to set up the decision-making value for seat occupancy. The status message sent by the node contains the following information:

{nodeId: "011808db24be32c5", nodeStatus: 0};

where nodeId is a string which represents the node's unique identification in the system and NodeStatus defines the status of the node in the system. Node status can have two values, zero and one. The node sends a status message with nodeStatus equal to zero when it is connected to the system. The Node-RED application evaluates the message and checks whether the node is already registered in the system. In such a case, Node-RED responds with the configuration message to this node automatically. The node responds to Node-RED with the status message with the value of nodeStatus equal to one to confirm a successful setup. On the other hand, when the node ID is not registered in the system, i.e., a new node is connected, it needs to be configured from the GUI. The process of creating the classroom and configuring the nodes is described in Section 4.

3.3. Self-Hosted Server Solution

The server, which is the central unit of the proposed system, is based on the Raspberry Pi minicomputer. This device runs all applications and services that provide connectivity to IoT nodes as well as management of the system, data evaluation, and storage. There are five primary services running on the server: Mosquitto MQTT broker, Node-RED application, React web application, Nginx server, and MongoDB database system.

MongoDB is a popular general-purpose document-based distributed database that stores all data for actual and further evaluation. Mosquitto MQTT broker is an open-source MQTT message broker which is widely used across various IoT applications. The logic of the proposed system is implemented by a Node-RED application. The schematic design of flow for handling the messages from the Gate node for processing, evaluation, and representation of results is shown in Figure 7.

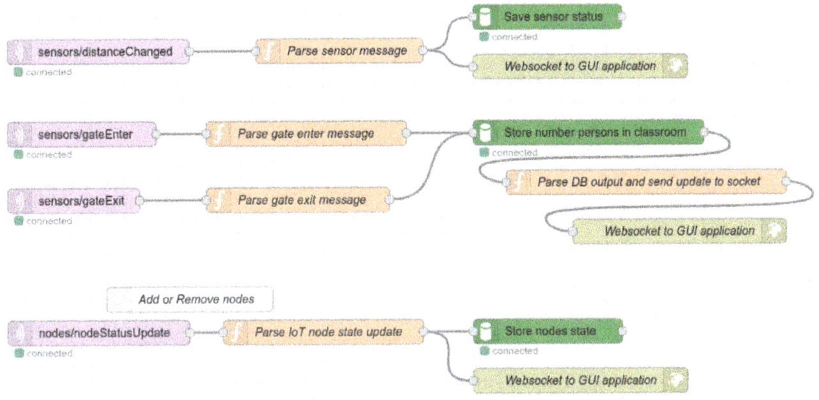

Figure 7. Schematic design of flow for handling incoming messages.

The flow starts with the MQTT broker input nodes. These input nodes listen to the topics "sensors/gateEnter" and "sensors/gateExit". The purpose of this part of the flow is to collect data from the Gate node and evaluate the number of students that are currently in the classroom. All pieces of information are stored in the Mongo DB, and thus are available to Node-RED. However, the latest values are stored in flow variables as well, and their updates are sent via WebSocket to the connected client using the React web application. The message with the number of people currently in the classroom appears as follows:

{"counter":14, "limit":20};

where the counter represents the number of people and the limit is the maximum allowed amount of people who can be inside the room due to active restrictions. This limit could be changed by a teacher via the GUI.

Another MQTT broker input node listens to the topic "/sensors/distanceChanged". This node is responsible for handling the incoming messages from the IoT nodes about changes in seat occupancy. The structure of this message was provided in the previous section.

The next part of the flow is responsible for providing all system data as the RESTful API. An example of this flow design is depicted in Figure 8. Altogether, there are eight different routes implemented in the API; three are POST routes, while the rest are GET routes. The description of the RestAPI GET routes is as follows:

- /currentPersonCount: the route is responsible for obtaining information about the actual number of persons in the classroom. The route flow acquires the data from the database and creates a response in JSON format;
- /resetConterCount: the route for resetting the current counter. For example, this can be used in cases when false detection occurs;
- /currentPersonsDistribution: the route obtains the data from the database about actual person distribution in the classroom. Data are then stored in JSON format and sent to the client. The data structure is described later in the experimental results section;
- /getDevicesList: the route for acquiring the actual list of all IoT nodes connected to the system. Data are acquired from the database as well;
- /getLessonsList: the route obtains all lessons records from database and returns them as JSON objects.

Description of RestAPI POST routes:

- /setupDevice: this route serves to set up the IoT node. It expects the JSON data in the request body with parameters specifying node position, sensor count, and threshold for decision-making distance when the seat is considered occupied;
- /startLesson: this route starts a new lesson; it expects only one parameter, e.g., lesson name;
- /finishLesson: this route finishes the current lesson, and is parameter-less.

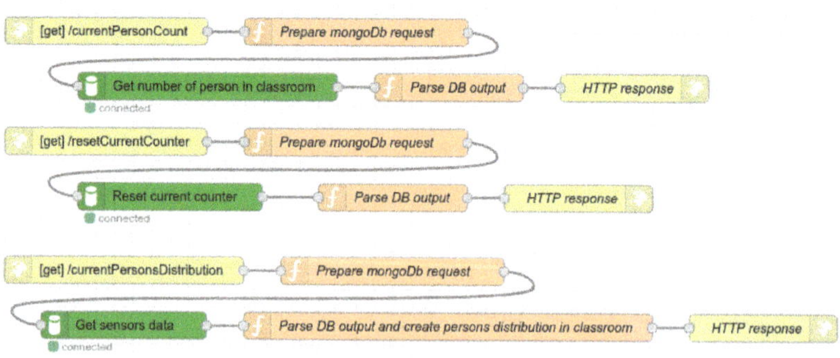

Figure 8. Schematic design of flow for providing the RESTful API.

Two of the routes shown in Figure 9 handle the IoT nodes' configuration setup from the GUI. The configuration process of the IoT nodes is described in the next section.

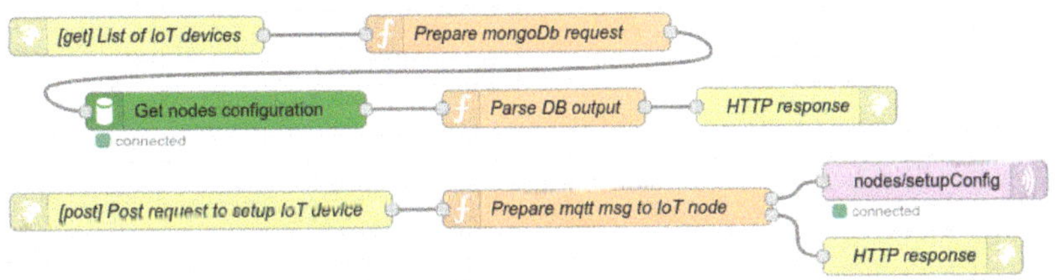

Figure 9. Schematic design of flow for providing the RESTful API for setting up the IoT nodes.

Simple algorithms were designed to evaluate whether the distribution of students across the classroom meets the requirements defined by COVID-19 restrictions. It is assumed that the students are seated in rows; however, there do not have to be the same number of seats in each row. The implemented algorithm considers student distribution to be valid when the distance between taken seats is at least 2 in both the x and y directions. Otherwise, the system shows a popup notification about the violation of the restriction.

4. Experimental Results

In this section, the GUI of the proposed IoT-based system for detecting the number of people and their distribution in the classroom is presented. The application helps the teacher to easily handle the implementation of restrictions defined due to the COVID-19 outbreak. The home screen of the developed web application is shown in Figure 10.

The teacher's daily routine will be as follows. Before the beginning of the class, the teacher enters the classroom and resets the current count of the students in the classroom. Afterwards, students can enter the classroom. The web application shows any changes in seat occupancy in real time using WebSocket communication. When students are heading to their chosen seat, they can cross other seats and temporarily change seat occupancy status. This could lead to an evaluation of person distribution that does not meet the requirements defined by COVID-19 restrictions. The system shows an alert only when the incorrect seat occupancy state holds for more than one minute. In a typical scenario, the system evaluates the data in real-time, and therefore temporarily changing seat status occupancy does not cause an alert. All other communication between the web application and the Node-RED backend is carried out via HTTP protocol. The data received from WebSocket or an HTTP GET request representing the student distribution are defined as follows:

```
{
distributionState: "Ok",
data: [
    {row: 1, seats: [0, 1, 0, 1, . . . , 0, 1]};
    {row: 2, seats: [1, 0, 0, 0, . . . , 1, 0]};
    . . .
    {row: n, seats: [0, 1, 0, 1, ..., 0, 1]};
]}
```

where "distributionState" tells whether the students' distribution across the classroom meets the requirements defined by restrictions. The entry "data" represents the real seat occupancy in the classroom. When the teacher finishes the lesson, all information gathered during the lesson is stored in the database for later analysis. The teacher can check the relevant data from the finished lessons at any time.

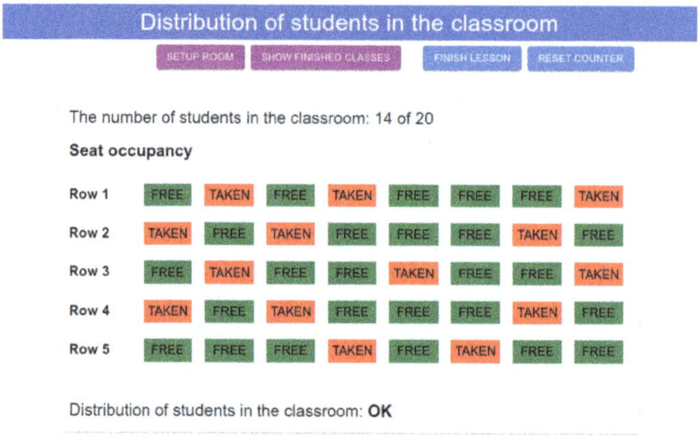

Figure 10. GUI home screen showing student distribution in the classroom.

The user window that provides the classroom setup and configuration is shown in Figure 11. The proposed system was designed to be as simple as possible from both the setup and implementation point of view. After all software is installed on the Raspberry Pi, the user is able to configure the classroom via the GUI. This configuration process is as follows. First, the user needs to set up the room limits, such as the maximum number of students in the classroom and minimum distance limits. Afterwards, the teacher turns on the first IoT node in the first row. The IoT node is registered in the system. By clicking on the button "Add IoT module", the teacher can display a window with a list of all unconfigured nodes in the system. Then, the teacher selects one of the unconfigured IoT nodes and configures it by setting the position in the classroom (e.g., the row number parameter) and the number of sensors connected to the IoT node, which is the number of seats in a particular row. Then, the IoT node is ready for use, and the teacher can add the rest of the IoT nodes. Moreover, the teacher is able to change the configuration of any IoT node at any time.

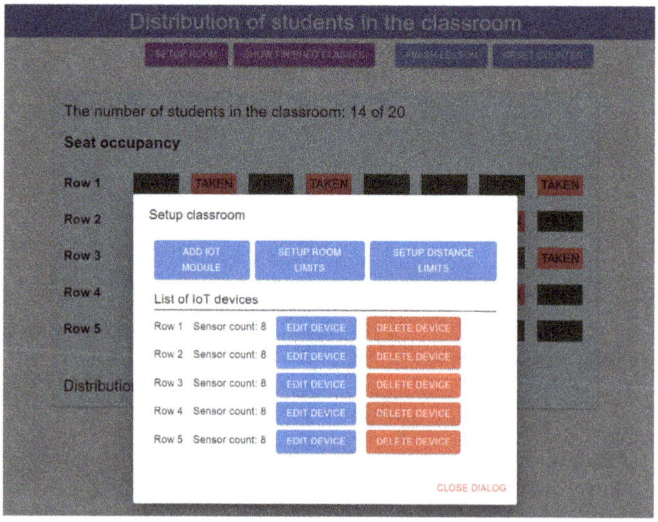

Figure 11. GUI for classroom setup.

Experiments were carried out to test the robustness and reliability of the door crossing detection algorithm at the Gate node. The tests were performed considering the following scenarios:

1. The person enters the classroom.
2. The person leaves the classroom.
3. The person enters the Gate area, stops, and then continues in the same direction.
4. The person enters the Gate area, stops, and leaves in the direction from which they entered.

Each scenario was tested 100 times, and the achieved results are presented in Table 2.

Table 2. Gate node implementation testing results.

Scenario	Number of Tests	Correct Detection	Incorrect Detection	Accuracy [%]
1.	100	100	0	100
2.	100	100	0	100
3.	100	100	0	100
4.	100	100	0	100

From the achieved results, it is obvious that the proposed system is both robust, and reliable and is able to correctly detect door crossing using the Gate node. The Gate node is able to distinguish between all four cases, i.e., a person who enters the room, leaves the room, or decides to return after entering the door from either side. The Gate can detect other crossing objects that are not human targets, such as bags, cabinets, or tables. However, it is not possible to distinguish what type of object crosses through the Gate. Moreover, the system can operate in real time and is designed to be deployed at a relatively low cost. The hardware cost of equipment required for one classroom with a capacity of 40 seats, i.e., five rows and eight seats per row, is approximately EUR 210, as can be seen in Table 3.

Table 3. Cost of system deployment.

Item	Unit Price	Count	Total Price [Eur]
Raspberry PI	40	1	40
NodeMCU	5	6	30
HC-SR04	1.4	42	58.8
74HC4067 (16-channel multiplexer)	2.3	5	11.5
Wires, pcb and secondary materials	70	1	70
Total			210.3

It is important to note here that deployment of the system does not require any preexisting infrastructure for the internet connection. The server, running on Raspberry Pi, can provide wireless connection to all IoT nodes in the room and store all the data locally.

5. Discussion

In this section, the proposed solution for counting and detecting the distribution of people around the classroom is compared with the other state-of-the-art works. As our literature review found only a single work dealing with a similar solution based on data from a sensor network, solutions based on image processing were considered in comparison with proposed system. It is important to note that in most papers a limited amount of information about the systems requirements, cost, and power consumption were presented. However, based on the provided information, several parameters for comparison could be estimated. For the comparison, implementation of the system for the classroom with a capacity of 40 seats, i.e., five rows and eight seats per row, was considered. Unfortunately, due to the lack of information provided in the literature, it is not possible to cover all comparison parameters for all solutions. The comparison of the proposed system with other solutions is shown in Table 4.

Table 4. Comparison of the proposed system with other solutions.

Solution	Proposed Solution	Zhang et al. [36]	Le et al. [4]	Hasan et al. [11]	Al-Sa'd et al. [7]
Initial investment [Eur]	210	220	80	330	minimal estimated price: 330
Power consumption [W]	8.55	14.5	4.425	300(Tesla K80+camera)	minimal estimated: 300
Privacy	respected	respected	not respected	not respected	not respected
Robustness against physical tempering	part of the system could be easily damaged	part of the system could be easily damaged	depends on the camera position	depends on the camera position	depends on the camera position
System accuracy	Very high, easy and precise algorithms	not applicable	not provided	78.3% (mAP)	F1—99.1
Provides distribution of people	yes	yes-not in GUI	no	Possible to calculate, but not provided	Possible to calculate, but not provided
Response time	Real-time	Real-time	7.5 fps	not provided	Real-time

From the comparison, it is clear that the proposed system can provide information about seat occupancy with high accuracy while maintaining the privacy of people in the monitored area. In addition to the privacy, the advantage of the proposed system over systems based on image processing is that accuracy is consistent over whole area, while in systems based on image processing the accuracy can be affected by the position of the camera. On top of that, image processing-based solutions are not designed to provide information about distribution of people in the area. Moreover, the proposed system can operate in real time with low implementation cost and lower power consumption than the IoT-based solution proposed in [36].

6. Conclusions

In the paper, an improved IoT-based system for detecting the number of people in a classroom and their distribution was proposed. The main purpose of the proposed system is to help teachers to manage their classes with respect to rules implemented due to COVID-19 restrictions. An improved system was presented in which the teacher can set up the whole classroom from a GUI. The system is more robust and much easier to extend than the previous version published in [37]. The teacher is able to configure IoT nodes from the GUI and change the configuration at any time. The system consists of a Gate node for counting people entering or leaving the classroom, IoT nodes with distance sensors placed in the room for detecting the availability of seats, and a server based on a Raspberry Pi. It is possible to connect up to sixteen HC-SR04 ultrasonic distance sensors to a single IoT node; thus, a single node is able to check the availability of sixteen seats. The Raspberry Pi server can create a Wi-Fi network, which is used to transfer the data from the IoT nodes; therefore, there is no need for an extra Wi-Fi router to provide connectivity in the classroom. Five system services running on the Raspberry Pi provide all functionalities of the proposed system, which uses an Nginx server for request routing, a React web application, the Mongo DB database, a Node-RED application for the logic part, and an eclipse-mosquitto MQTT broker. All services run on the server as Docker containers. Thanks to this setup, it is easy to deploy the proposed system in any classroom. The main idea of our system is to help teachers to handle COVID-19 restrictions. However, there may be other use cases for the system. For example, the system can check the distance between students during exams and tests in order to help prevent cheating. The main disadvantage of the proposed system is the complexity of the deployment of the IoT modules compared to camera-based systems.

When camera-based systems are deployed, it is sufficient to simply place the camera in the classroom and the system is ready immediately. However, in the proposed system it is necessary to connect cables between sensors and IoT nodes in order to provide the power supply for the individual IoT nodes. On the other hand, when privacy is considered, the proposed system has the advantage in that it cannot provide any information about the identities of people in the room.

The proposed system was deployed and tested in a classroom at the University of Zilina. The presented IoT-based system is highly reliable and robust thanks to the algorithm's simplicity and clarity. The main advantage of the system is the possibility of deployment without jeopardizing the privacy of individuals in the classroom, as there is no way to identify individuals, unlike in camera-based systems.

Author Contributions: Conceptualization, S.M., R.H., P.K. and J.M.; methodology, S.M.; software, S.M.; validation, S.M., J.M, P.K. and R.H.; formal analysis, S.M., J.M, P.K. and R.H.; investigation, S.M. and J.M.; writing—original draft preparation, S.M. and J.M.; writing—review and editing, P.K., J.M. and R.H.; visualization, S.M, J.M., P.K. and R.H.; supervision, S.M.; project administration, J.M, P.K.; funding acquisition, J.M. All authors have read and agreed to the published version of the manuscript.

Funding: This paper was supported by the project of Operational Programme Integrated Infrastructure: Independent research and development of technological kits based on wearable electronics products, as tools for raising hygienic standards in a society exposed to the virus causing the COVID-19 disease, ITMS2014+ code 313011ASK8 and by European Regional Development Fund and by Slovak Research and Development Agency under contract no. PP-COVID-20-0100: DOLORES.AI: The pandemic guard system.

Institutional Review Board Statement: Not applicable.

Informed Consent Statement: Not applicable.

Data Availability Statement: The data are available from the authors upon request.

Conflicts of Interest: The authors declare no conflict of interest.

Abbreviations

The following abbreviations are used in this manuscript:

GUI	Graphical User Interface
HTTP	Hypertext Transfer Protocol
ID	Identification
IoT	Internet of Things
MQTT	Message Queuing Telemetry Transport

References

1. Myint, E.P.; Sein, M.M. People Detecting and Counting System. In Proceedings of the 2021 IEEE 3rd Global Conference on Life Sciences and Technologies (LifeTech), Nara, Japan, 9–11 March 2021; pp. 289–290. [CrossRef]
2. Rantelobo, K.; Indraswara, M.A.; Sastra, N.P.; Wiharta, D.M.; Lami, H.F.J.; Kotta, H.Z. Monitoring Systems for Counting People using Raspberry Pi 3. In Proceedings of the 2018 International Conference on Smart Green Technology in Electrical and Information Systems (ICSGTEIS), Bali, Indonesia, 25–27 October 2018; pp. 57–60. [CrossRef]
3. Saxena, S.; Songara, D. Design of people counting system using MATLAB. In Proceedings of the 2017 Tenth International Conference on Contemporary Computing (IC3), Noida, India, 10–12 August 2017; pp. 1–3. [CrossRef]
4. Le, M.C.; Le, M.H.; Duong, M.T. Vision-based People Counting for Attendance Monitoring System. In Proceedings of the 2020 5th International Conference on Green Technology and Sustainable Development (GTSD), Ho Chi Minh City, Vietnam, 27–28 November 2020; pp. 349–352. [CrossRef]
5. Hou, Y.C.; Baharuddin, M.Z.; Yussof, S.; Dzulkifly, S. Social Distancing Detection with Deep Learning Model. In Proceedings of the 2020 8th International Conference on Information Technology and Multimedia (ICIMU), Selangor, Malaysia, 24–26 August 2020; pp. 334–338. [CrossRef]
6. Sharma, M. Open-CV Social Distancing Intelligent System. In Proceedings of the 2020 2nd International Conference on Advances in Computing, Communication Control and Networking (ICACCCN), Greater Noida, India, 18–19 December 2020; pp. 972–975. [CrossRef]

7. Al-Sa'd, M.; Kiranyaz, S.; Ahmad, I.; Sundell, C.; Vakkuri, M.; Gabbouj, M. A Social Distance Estimation and Crowd Monitoring System for Surveillance Cameras. *Sensors* **2022**, *22*, 418. [CrossRef] [PubMed]
8. Rezaei, M.; Azarmi, M. DeepSOCIAL: Social Distancing Monitoring and Infection Risk Assessment in COVID-19 Pandemic. *Appl. Sci.* **2020**, *10*, 7514. [CrossRef]
9. Cristani, M.; Bue, A.D.; Murino, V.; Setti, F.; Vinciarelli, A. The Visual Social Distancing Problem. *IEEE Access* **2020**, *8*, 126876–126886. [CrossRef]
10. Yang, D.; Yurtsever, E.; Renganathan, V.; Redmill, K.A.; Özgüner, U. A Vision-Based Social Distancing and Critical Density Detection System for COVID-19. *Sensors* **2021**, *21*, 4608. [CrossRef]
11. Hassan, N.I.; Tahir, N.M.; Zaman, F.H.K.; Hashim, H. People Detection System Using YOLOv3 Algorithm. In Proceedings of the 2020 10th IEEE International Conference on Control System, Computing and Engineering (ICCSCE), Penang, Malaysia, 21–22 August 2020. [CrossRef]
12. Gupta, P.; Sharma, V.; Varma, S. People detection and counting using YOLOv3 and SSD models. *Mater. Today Proc.* 2021; *in press*. [CrossRef]
13. Martinez, M.; Yang, K.; Constantinescu, A.; Stiefelhagen, R. Helping the Blind to Get through COVID-19: Social Distancing Assistant Using Real-Time Semantic Segmentation on RGB-D Video. *Sensors* **2020**, *20*, 5202. [CrossRef] [PubMed]
14. Rusli, M.E.; Yussof, S.; Ali, M.; Abobakr Hassan, A.A. MySD: A Smart Social Distancing Monitoring System. In Proceedings of the 2020 8th International Conference on Information Technology and Multimedia (ICIMU), Selangor, Malaysia, 24–26 August 2020; pp. 399–403. [CrossRef]
15. Choi, J.W.; Yim, D.H.; Cho, S.H. People Counting Based on an IR-UWB Radar Sensor. *IEEE Sens. J.* **2017**, *17*, 5717–5727. [CrossRef]
16. Groba, C. Demonstrations and people-counting based on Wifi probe requests. In Proceedings of the 2019 IEEE 5th World Forum on Internet of Things (WF-IoT), Limerick, Ireland, 15–18 April 2019; pp. 596–600. [CrossRef]
17. Alladi, T.; Chamola, V.; Parizi, R.M.; Choo, K.K.R. Blockchain Applications for Industry 4.0 and Industrial IoT: A Review. *IEEE Access* **2019**, *7*, 176935–176951. [CrossRef]
18. Aheleroff, S.; Xu, X.; Lu, Y.; Aristizabal, M.; Velásquez, J.P.; Joa, B.; Valencia, Y. IoT-enabled smart appliances under industry 4.0: A case study. *Adv. Eng. Inform.* **2020**, *43*, 101043. [CrossRef]
19. Paul, K.; Chatterjee, S.S.; Pai, P.; Varshney, A.; Juikar, S.; Prasad, V.; Bhadra, B.; Dasgupta, S. Viable smart sensors and their application in data driven agriculture. *Comput. Electron. Agric.* **2022**, *198*, 107096. [CrossRef]
20. Marcu, I.; Suciu, G.; Bălăceanu, C.; Vulpe, A.; Drăgulinescu, A.M. Arrowhead Technology for Digitalization and Automation Solution: Smart Cities and Smart Agriculture. *Sensors* **2020**, *20*, 1464. [CrossRef] [PubMed]
21. Cvar, N.; Trilar, J.; Kos, A.; Volk, M.; Duh, E.S. The Use of IoT Technology in Smart Cities and Smart Villages: Similarities, Differences, and Future Prospects. *Sensors* **2020**, *20*, 3897. [CrossRef] [PubMed]
22. Syed, A.S.; Sierra-Sosa, D.; Kumar, A.; Elmaghraby, A. IoT in Smart Cities: A Survey of Technologies, Practices and Challenges. *Smart Cities* **2021**, *4*, 429–475. [CrossRef]
23. Zhang, J.; Wang, Y.; Li, S.; Shi, S. An Architecture for IoT-Enabled Smart Transportation Security System: A Geospatial Approach. *IEEE Internet Things J.* **2021**, *8*, 6205–6213. [CrossRef]
24. Song, Y.; Yu, F.R.; Zhou, L.; Yang, X.; He, Z. Applications of the Internet of Things (IoT) in Smart Logistics: A Comprehensive Survey. *IEEE Internet Things J.* **2021**, *8*, 4250–4274. [CrossRef]
25. Srinivasan, P.; Anusha, B.; Reddy, K.S.K.; Reddy, N.C.S.; Maheswari, K. A Secure IoT Enabled Smart Home System. *Int. J. Early Child. Spec. Educ.* **2022**, *14*, 466–472. [CrossRef]
26. Touqeer, H.; Zaman, S.; Amin, R.; Hussain, M.; Al-Turjman, F.; Bilal, M. Smart home security: Challenges, issues and solutions at different IoT layers. *J. Supercomput.* **2021**, *77*, 14053–14089. [CrossRef]
27. Baig, A.R.; Hasanat, M.H.A.; Alsahli, A. Fog Data Analytics: Diabetic Patients Monitoring Using Wearable IoT Sensors. *Int. J. Comput. Sci. Netw. Secur.* **2022**, *22*, 603–612. [CrossRef]
28. Dian, F.J.; Vahidnia, R.; Rahmati, A. Wearables and the Internet of Things (IoT), Applications, Opportunities, and Challenges: A Survey. *IEEE Access* **2020**, *8*, 69200–69211. [CrossRef]
29. Dong, Y.; Yao, Y.D. IoT Platform for COVID-19 Prevention and Control: A Survey. *IEEE Access* **2021**, *9*, 49929–49941. [CrossRef]
30. Garg, L.; Chukwu, E.; Nasser, N.; Chakraborty, C.; Garg, G. Anonymity Preserving IoT-Based COVID-19 and Other Infectious Disease Contact Tracing Model. *IEEE Access* **2020**, *8*, 159402–159414. [CrossRef] [PubMed]
31. Tovarek, J.; Partila, P.; Jakovlev, S.; Voznak, M.; Eglynas, T.; Jusis, M. A new approach for early detection of biological self-ignition in shipping container based on IoT technology for the smart logistics domain. In Proceedings of the 2021 29th Telecommunications Forum (TELFOR), Belgrade, Serbia, 23–24 November 2021; pp. 1–4. [CrossRef]
32. Umair, M.; Cheema, M.A.; Cheema, O.; Li, H.; Lu, H. Impact of COVID-19 on IoT Adoption in Healthcare, Smart Homes, Smart Buildings, Smart Cities, Transportation and Industrial IoT. *Sensors* **2021**, *21*, 3838. [CrossRef] [PubMed]
33. Bian, S.; Zhou, B.; Lukowicz, P. Social Distance Monitor with a Wearable Magnetic Field Proximity Sensor. *Sensors* **2020**, *20*, 5101. [CrossRef]
34. Perra, C.; Kumar, A.; Losito, M.; Pirino, P.; Moradpour, M.; Gatto, G. Monitoring Indoor People Presence in Buildings Using Low-Cost Infrared Sensor Array in Doorways. *Sensors* **2021**, *21*, 4062. [CrossRef] [PubMed]

35. Lazaro, A.; Lazaro, M.; Villarino, R.; Girbau, D. Seat-Occupancy Detection System and Breathing Rate Monitoring Based on a Low-Cost mm-Wave Radar at 60 GHz. *IEEE Access* **2021**, *9*, 115403–115414. [CrossRef]
36. Zhang, C.; Long, L. Design of Intelligent Management System for Study Room. In Proceedings of the 2021 2nd International Conference on Artificial Intelligence and Information Systems, Chongqing China, 28–30 May 2021. [CrossRef]
37. Matuska, S.; Hudec, R.; Kamencay, P. IoT Based System for Detecting the Number of People and their Distribution in Classroom. In Proceedings of the 2022 ELEKTRO (ELEKTRO), Krakow, Poland, 23–26 May 2022; pp. 1–4. [CrossRef]

sensors

MDPI

Article

Towards Transfer Learning Techniques—BERT, DistilBERT, BERTimbau, and DistilBERTimbau for Automatic Text Classification from Different Languages: A Case Study

Rafael Silva Barbon and Ademar Takeo Akabane *

Postgraduate Program in Urban Infrastructure Systems and Telecommunication Networks Management, Centre for Exact Sciences, Technology and the Environment (CEATEC), Pontifical Catholic University of Campinas (PUC-Campinas), 1516 Professor Dr. Euryclides de Jesus Zerbini, Campinas 13086900, SP, Brazil
* Correspondence: ademar.akabane@puc-campinas.edu.br

Abstract: The Internet of Things is a paradigm that interconnects several smart devices through the internet to provide ubiquitous services to users. This paradigm and Web 2.0 platforms generate countless amounts of textual data. Thus, a significant challenge in this context is automatically performing text classification. State-of-the-art outcomes have recently been obtained by employing language models trained from scratch on corpora made up from news online to handle text classification better. A language model that we can highlight is BERT (Bidirectional Encoder Representations from Transformers) and also DistilBERT is a pre-trained smaller general-purpose language representation model. In this context, through a case study, we propose performing the text classification task with two previously mentioned models for two languages (English and Brazilian Portuguese) in different datasets. The results show that DistilBERT's training time for English and Brazilian Portuguese was about 45% faster than its larger counterpart, it was also 40% smaller, and preserves about 96% of language comprehension skills for balanced datasets.

Keywords: big data; pre-trained model; BERT; DistilBERT; BERTimbau; DistilBERTimbau; transformer-based machine learning

Citation: Silva Barbon, R.; Akabane, A.T. Towards Transfer Learning Techniques—BERT, DistilBERT, BERTimbau, and DistilBERTimbau for Automatic Text Classification from Different Languages: A Case Study. *Sensors* **2021**, *22*, 8184. https://doi.org/10.3390/s22218184

Academic Editors: Peter Hockicko, Róbert Hudec and Patrik Kamencay

Received: 15 September 2022
Accepted: 20 October 2022
Published: 26 October 2022

1. Introduction

It is known that the support of computational systems is in several areas of knowledge, be it in the human, exact, and biological areas. Consequently, this contributes to the accelerated increase in the generation, consumption, and transmission of data in the global network. According to the study by the Statista Research Department [1], in 2018, the total amount of data created, captured, and consumed in the world was 33 zettabytes (ZB)—equivalent to 33 trillion gigabytes. Already in 2020 it has grown to 59 ZB and is expected to reach 175 ZB by 2025.

In the Internet of Things (IoT) context, we know that these devices (e.g., virtual assistants) are connected to the Internet and generate large amounts of data. On the other hand, we also have Web 2.0 platforms, e.g., social networks, micro-blogs, and all these types of websites with massive amounts of textual information available online. It is worth mentioning that the data generated by these devices and websites are growing faster and faster. An important point worth mentioning is that the information generated from a large amount of text/data generated by users for many entrepreneurs or public agents is vital for maintaining their business. This way, one can exploit this constant and continuous feedback on a particular subject/product through these data. Due to the ever-increasing volume of online text data, the text classification task is more necessary than ever. In this context, text classification (automatically classifying textual) is an essential task.

Automatic text classification can be described as a task that automatically categorizes group documents into one or more predefined classes according to their topics. Thereby, the primary objective of text classification is to extract information from textual resources. The text classification task is the basic module for many NLP (natural language processing) applications. However, this necessitates the presence of efficient and flexible methods to access, organize, and extract useful information from different data sources. These methods can include text classification [2–4], information retrieval [5,6], summarization [7,8], text clustering [9,10], and others, collectively named text mining [2,4,6].

Many works are available in the literature on text classification tasks using various neural network models. Some typical works include convolutional neural network (CNN) models [11,12], attentional models [13,14], adversarial models [3], and recurrent neural network (RNN) models [13], which particularly outperform many statistics-based models. The previously mentioned works represent text based on words, i.e., word vectors pre-trained over a large-scale corpus are usually used as the sequence features. Such vectors are usually trained via the word2vec tool [15] or Glove [16,17] algorithm based on the presumption that similar words tend to appear in similar contexts.

In recent years, to avoid specific structures and significantly decrease the parameters to be learned from scratch, as is done in the models presented above, some researchers have contributed in another direction, highlighting the pre-training models for general language and fine-tuning them to downstream tasks. Another problem with traditional NLP approaches worth mentioning is the issue of multilingualism [18]. The Open AI group (https://openai.com/, accessed on 13 July 2022) proposes the GPT (Generative Pre-trained Transformer) using a left-to-right multi-layer Transformer architecture to learn the general language presentations from a large-scale corpus to deal with the abovementioned problems [19]. Later, Google, inspired by GPT, presented a new language representation called BERT (Bidirectional Encoder Representations from Transformers) [20]. BERT is a state-of-the-art language representation model designed to pre-train deep bidirectional representations from unlabeled text and is fine-tuned using labeled text for different NLP tasks [20]. A smaller, faster, and lighter version of BERT architecture, well-known as DistilBERT, was implemented by the HuggingFace team (https://github.com/huggingface/transformers, accessed on 13 July 2022).

This work aimed to examine an extensive dataset from different contexts, including datasets from different languages, specifically English and Brazilian Portuguese, to analyze the performance of the two models (BERT and DistilBERT). To do this, we first fine-tuned BERT and DistilBERT, then the aggregating layer was utilized as the text embedding, and then we compared the two models with several selected datasets. As a general result, we can highlight that the DistilBERT is nearly 40% smaller and around 45% faster than its larger counterpart. Yet, it preserves around 96% of language comprehension skills for both English and Brazilian Portuguese for balanced datasets.

The main contributions of the paper are as follows:

- We compare BERT and DistilBERT, demonstrating how the Light Transformer model can be very close in effectiveness compared to its larger model for different languages;
- We compared models Transformer (BERT) and Light Transformer (DistilBERT) for both English and Brazilian Portuguese.

The rest of the document is organized as follows: Section 2 presents a short summary of the necessary concepts to understand this work, while Section 3 presents the method and hyperparameter configuration for automatic text classification. The case study of this work is presented in Section 4, and the results are presented in Section 5. Thereafter, in Section 6, we discuss the performance of the two models (BERT and DistilBERT) in the different datasets used. Finally, Section 7 concludes with a discussion and recommendations for future work.

2. Theoretical Foundation

This section presents the theoretical foundation for a better understanding of the work. In Section 2.1, the Transformer architecture is described, while in Section 2.2, Bidirectional Encoder Representations from Transformers (BERT) is described. In Section 2.3, the comprehension models are presented, and finally, in Section 2.4, the BERTimbau model is introduced.

2.1. *Transformer Architecture*

It is essential to review two concepts: (i) encoder–decoder [21]; and (ii) attention [22] configurations to understand the Transformer architecture. The first concept refers to the type of training adopted to produce embeddings from input tokens.

The second is a technique to circumvent a common problem in sequential architectures applied to natural language processing problems (e.g., recurrent networks [23]). Sequential networks attempt to map the relationship between a token in the target sequence with the source sequence tokens. However, a token in the target sequence may be closer to one or more tokens in the source sequence rather than the entire source sequence. In this way, the network used to generate the representation of the tokens ends up encoding information that may not be relevant to the problem at hand. This problem occurs mainly when the input sequence is long and rich in information and selecting the essential passages is not possible.

In a few words, the idea of the attention mechanism is to make this selection explicit, consisting of a neural layer created exclusively to understand this context relationship between tokens. In this context, Vaswani et al. [19] proposed the Transformer architecture, an encoder–decoder network based on parallelization of the attention mechanism. In this network, attention mechanisms generate multiple representations of tokens, where each representation can refer to a different contextual relationship.

Transformers are based on the traditional architecture of Multilayer Perceptron, making massive use of attention mechanisms trained under the encoder–decoder configuration. Figure 1 illustrates the Transformer architecture. The Transformer receives the source and target sequences, concatenated with positional encodings that help the network understand the order between the tokens. The boxes with a light gray background on the left and right represent the encoder and decoder, respectively. Note that the encoder and decoder differ only in the presence of an additional layer of attention in the decoder. The Transformer network considers N stacked encoders–decoders, summarized in Figure 1 by the Nx notation. The embedding produced by the network is taken from its top.

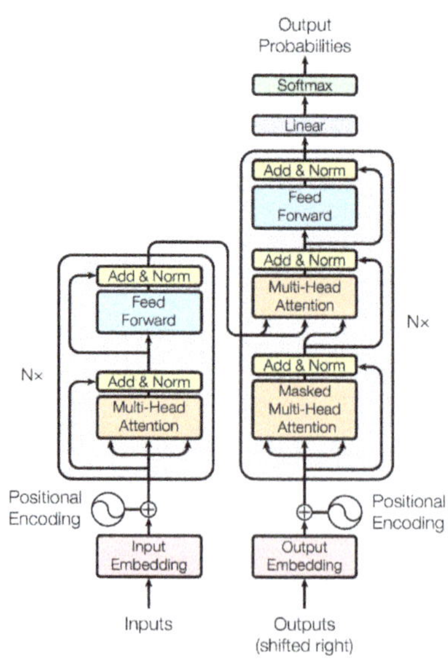

Figure 1. Transformer architecture [19].

2.2. BERT

Bidirectional Encoder Representations from Transformers is a language model based on the transformer architecture [20]. BERT is a language model designed to pre-train bidirectional deep representations from unlabeled text. Its two-way approach aims to model the context to both the right and the left of a given token. Two essential aspects of this approach are that, without substantial changes to its architecture, it can be used (i) pre-trained with or without fine-tuning; and (ii) for tasks that consider individual sentences or sentence pairs (e.g., natural language inference and semantic textual similarity [24]).

In the BERT architecture, there are two essential stages [20]: (i) pre-training; and (ii) fine-tuning. In the first stage, the model is trained on a large unlabeled corpus. While the second one, the model is initialized with the pre-trained data, and all the parameters are fine-tuned using labeled data for specific tasks.

The architecture of a BERT network can be seen in Figure 2, where the pre-trained version is shown on the left side, and fine-tuned versions adjusted for different tasks are shown on the right. The model is trained using unlabeled data from different tasks in the pre-training stage. In principle, it is possible to use pre-trained BERT models to produce contextual embeddings that can be used for (un)supervised learning tasks. The model is initialized with the pre-trained parameters in the second stage, from fine-tuning to given supervised learning tasks. Then, these parameters are readjusted using data labeled for the task to be solved. Since fine-tuning is performed by task, each task has an individual adjusted model, even if they were initialized with the same pre-trained parameters [20].

To handle various tasks, the representation of BERT input can consist of a single sentence or a pair of sentences. Both possibilities are illustrated at the bottom of the models shown in Figure 2.

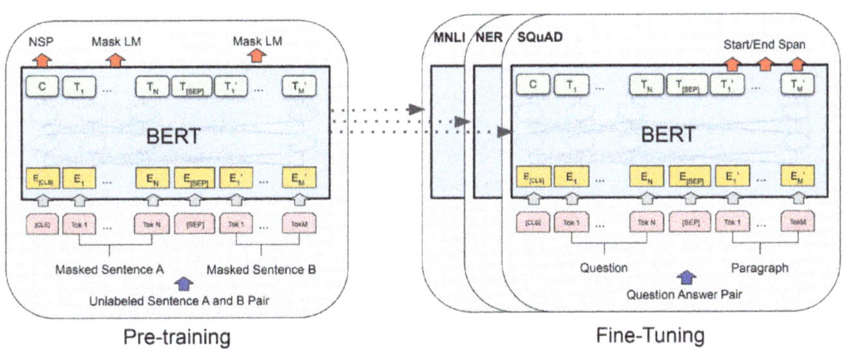

Figure 2. The BERT architecture [20] in a pre-training context (**left**) or fine-tuning for different tasks (**right**).

2.3. Compression of Deep Learning Models

Pre-trained language models (e.g., BERT) have significantly succeeded in various NLP tasks. However, high storage and computational costs prevent pre-trained language models from effectively deploying on resource-constrained devices. To overcome this issue, the compression of deep neural network techniques has been adopted to produce a model with the same robustness as the pre-training models but requires fewer computational resources. Through such a technique, it was possible to design distilled (lightweight) models known as DistilBERT [25].

The compression of the deep neural network is made using knowledge distilling. This compression technique allows a compact model to be trained to reproduce the behavior of a larger model. Dilbert (distilled BERT) is a smaller, faster, general-purpose pre-trained version of BERT that retains nearly the same language comprehension capabilities. The distillation technique [26] consists of training a model based on a larger model, called the teacher, which is used to teach the distilled model, called the student, to reproduce the behavior of the larger model. Thus, DistilBERT is a lightweight model based on the behavior of the original BERT model [25].

The main goal is to produce a smaller model able to reproduce the decisions of the robust bigger model. To do that, it is necessary to approximate the distilled model to the generated function of the bigger model. This function is used to classify a high quantity of pseudo data that show the value of each attribute on the distribution independently [27]. A faster and more compact model trained with pseudo data does not risk present overfitting and will also approximate the learned function from the bigger model [27].

The neural network produces the probability of the classes using a softmax on the output that converts the logit, z_i, calculated for each class into a probability, q_i, comparing it with the other logits.

Neural networks typically produce class probability using a $softmax$ output layer that converts the $logit$, z_i, calculated for each class into a probability, q_i, comparing it with the other $logits$, see Equation (1).

$$q_i = \frac{exp(\frac{z_i}{T})}{\sum\limits_j exp(\frac{z_j}{T})} \quad (1)$$

where the T symbol presented refers to the temperature, typically set to 1; using a more significant value for T, a more soft distributed (soft-target) over the classes is obtained.

In the simplest form of distillation, knowledge is transferred to the distilled model by training it with a transfer set. Furthermore, a soft-target distribution is used for each case of the transfer set produced by the larger model with a high value of T in its *softmax* [26]. The same T with a high value is used to train the distilled model, but temperature 1 is used after training. At low temperatures, distillation pays much less attention to matching the results of the $logit$ function, which are much more negative than the average. Thus, using temperatures more significant than 1, the distilled model extracts more relevant information from the training dataset [26].

2.4. BERTimbau: BERT Model for Brazilian Portuguese

It is known that pre-trained models such as BERT have high robustness, but this model is pre-trained with a large amount of English data. To develop a good model for another language such as Brazilian Portuguese, researchers from NeuralMind (https://neuralmind.ai/en/home-en/, accessed on 25 July 2022) developed a BERT model called BERTimbau [28].

To train the model in the Brazilian Portuguese language, the developers used an enormous Portuguese corpus called brWaC, which contains 2.68 billion tokens from 3.53 million documents on the Brazilian webpages [28,29].

Two BERTimbau versions were created: in the first one, BERTimbau Base, the weights were initialized with the checkpoint of Multilingual BERT base, a BERT version trained to 107 languages [30], and trained the model for four days on a TPU (tensor processing unit) v3-8 instance [28]. The second version is called BERTimbau Large; the weights were initialized with the checkpoint of English BERT Large. This version is more significant than the base version and took seven days to train on the same TPU [28]. The version used for evaluation in this article was BERTimbau Base. Additionally, a distilled model from BERTimbau was used and obtained on the HuggingFace Platform (https://huggingface.co/adalbertojunior/distilbert-portuguese-cased, accessed on 25 July 2022).

3. Method and Hyperparameter Configuration

This section presents the details of our proposed method for automatic text classification from different languages. The approaches we designed were mainly inspired by the works of Vaswani et al. [19] and Devlin et al. [20], in which attention mechanisms made it possible to track the relations between words across very long text sequences in both forward and reverse directions. Notwithstanding, we explore an extensive dataset from different contexts, including datasets from different languages, specifically English and

Brazilian Portuguese, to analyze the performance of the two state-of-the-art models (BERT and DistilBERT).

Our implementation follows the fine-tuning model released in the BERT project [20]. For the multi-class purpose, we use sigmoid cross entropy with logits function to replace the original softmax function, which is appropriate for one-hot classification only. To do this, we first fine-tuned the BERT and DistilBERT, used the aggregating layer as the text embedding, and compared the two models with several selected datasets.

The methodological details are organized into two subsections. The structural steps are the following: Section 3.1 presents the details of the hyperparameter configuration for fine-tuning process, while Section 3.2 presents the environment where the experiments were performed.

3.1. Hyperparameter Optimization for Fine-Tuning

In this section, we present the hyperparameter optimization for fine-tuning of our work. All the fine-tuning and evaluation steps performed on each model in this article used the Simple Transformers Library (https://simpletransformers.ai/docs/usage/ accessed on 1 August 2022). Table 1 reports the details of each hyperparameter configuration for fine-tuning process.

Table 1. Hyperparameters of the fine-tuned model.

Hyperparameter	Value
Batch size	8
Epochs	10
Learning rate	0.00004
Optimizer	AdamW
Adam_epsilon	0.00000001
Model class	Classification model
Maximum sequence length	128

The *BatchSize* is a hyperparameter that controls the number of samples from the training dataset used on each training step. On each step, the predictions are compared with the expected results, an error is calculated, and the internal parameters of the model are improved [31].

The second parameter of Table 1, *Epochs*, controls the number of times the training dataset will pass through the model during the training process. An epoch has one or more batches [31]. A high number of epochs can make the model overfit, causing it not to generalize, so when the model receives unseen data, it will not make a trustful prevision [32].

Overfitting can be detected in the evaluation step by analyzing the error of the predictions, as in Figure 3. A low number of epochs can also cause underfitting, which means that the models still need more training to learn from the training dataset.

Furthermore, the *LearningRate* is also related to underfitting or overfitting. This parameter controls how fast the model learns according to the errors obtained. Increasing the learning rate can bring the model from underfitting to overfitting [33].

The *Optimizer* determines in what measure the weight and the learning rate should be changed in order to reduce the losses of the models. The AdamW is a variant of Adam Optimizer [34]. *Adam_epsilon* is a parameter used on Adam Optimizer.

The *ModelClass* refers to the class from the Simple Transformers Library that was used to fine-tune the models. The maximum sequence length parameter refers to the maximum size of the sequence of tokens that can be inputted into the model.

Table 2 presents the hyperparameters of the pre-trained models used in this article for the performance evaluation. The distilled version of the models has six hidden layers, less than the original BERT and BERTimbau models, demonstrating how much smaller the distilled models are. Additionally, the DistilBERT model has 50 million fewer parameters than BERT. The author does not provide the number of parameters of the DistilBERTimbau model.

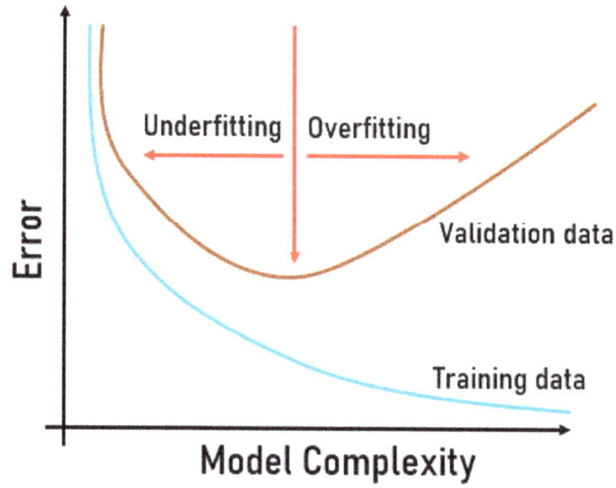

Figure 3. Overfitting example.

Table 2. Model hyperparameters.

Hyperparameter	BERT Base [20]	DistilBERT Base [25]	BERTimbau Base [1]	DistilBERTimbau Base [2]
Hidden layers	12	6	12	6
Total parameters	110 M	66 M	110 M	-

[1] https://huggingface.co/neuralmind/bert-base-portuguese-cased accessed on 1 August 2022. [2] https://huggingface.co/adalbertojunior/distilbert-portuguese-cased accessed on 1 August 2022.

3.2. Implementation

A cloud GPU environment (Google Colab Pro https://colab.research.google.com, accessed on 8 April 2022) was chosen to conduct the fine-tuning process on the models using the datasets selected; the metrics used to evaluate the models were defined. During the fine-tuning process, the (Weights and Biases https://wandb.ai/site, accessed on 8 April 2022) tool was used to monitor each training step and the models' learning process to detect some overfitting or anything that would bring about poor learning performance.

We trained our models on Google Colab Pro using the hyperparameters described in Tables 1 and 2. The results were computed and compared between each model to extract information about their performance, and graphics were built to visualize better and compare the results.

Furthermore, the K-fold cross-validation method was used, which consists of splitting the dataset into n folders so that every validation set is different from the others. The K refers to the number of approximately equal size disjoint subsets, and the fold refers to the number of subsets created. This splitting step is done by randomly sampling cases from the dataset without replacement [35].

Figure 4 represents an example from 10-fold cross-validation. Ten subsets were generated, and each subset is divided into ten parts where nine of them are used to train D_{train} and the other one to evaluate D_{val} the model.

Every evaluation part, D_{val}, differs between the subsets. The model is trained, evaluated, and then discarded for each subset or fold, so every part of the dataset will be used for training and evaluation. This allows us to see the potential of the model's generalization and prevent overfitting [35,36].

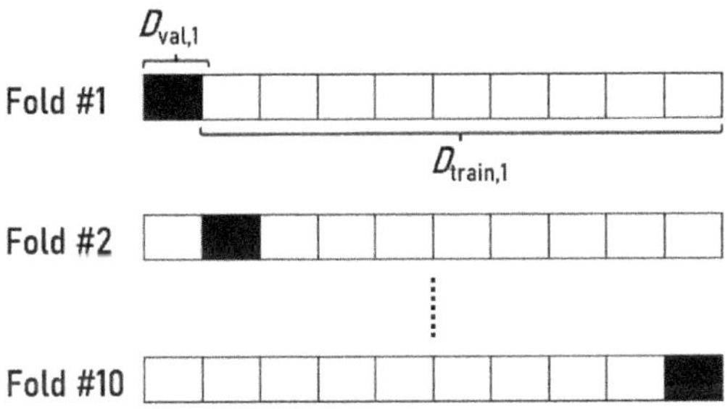

Figure 4. A 10-fold cross-validation example.

To evaluate the models, a 5-fold cross-validation was used. So five subsets were created, and each one was divided into five parts where a fourth of them (80%) are used for the fine-tuning process D_{train}, and rest (20%) to evaluate D_{val}.

4. Case Study

This section has been divided into two parts for a better presentation. The first part, Section 4.1 shows the datasets used in the experiments, while the evaluation metrics are shown in Section 4.2.

4.1. Datasets

For this case study, different datasets from the English and Portuguese languages were used. Section 4.1.1 presents the datasets used in the English language, while Section 4.1.2 presents the Brazilian Portuguese ones.

4.1.1. English Language

Three datasets were selected to evaluate the English models. The first one, called the Brexit Blog Corpus [37], contains 1682 phrases provided by a blog associated with Brexit. Those phrases are divided into nine classes, as shown in Table 3. This dataset contains a considerable number of classes and a few examples for each class. It can be seen that this dataset is unbalanced since some classes have less than 50 samples and others more than 200. The choice of an unbalanced dataset was purposeful to evaluate the performance of the chosen models.

Table 3. Brexit Blog Corpus dataset.

ID	Classes	Number of Examples
0	agreement/disagreement	50
1	certainly	84
2	contrariety	352
3	hypothetically	171
4	necessity	204
5	prediction	252
6	source of knowledge	287
7	tact/rudeness	44
8	uncertainty	196
9	volition	42
	Total	1682

The second dataset, called BBC Text, was obtained on Kaggle Platform (https://www.kaggle.com/) and built from BBC News [38], is made up of 2225 comments divided by five classes, as presented in the Table 4. Observing the number of classes and the sample numbers of each class, such a dataset is much more balanced compared to the Brexit Blog Corpus dataset.

Table 4. BBC Text dataset.

ID	Classes	Number of Examples
0	business	510
1	entertainment	386
2	politics	417
3	sport	511
4	tech	401
	Total	2225

The last English dataset selected was the Amazon Alexa Reviews Dataset, also obtained on Kaggle. This dataset contains 3150 feed-backs comments about the Amazon Virtual Assistant Alexa, containing only two classes, positive and negative, presented in Table 5.

Table 5. Amazon Alexa Reviews dataset.

ID	Classes	Number of Examples
0	positive	2893
1	negative	257
	Total	3150

This dataset contains much fewer negative samples, but contains only two classes.

4.1.2. Brazilian Portuguese Language

To evaluate the Portuguese models, two datasets were selected. The first, called PorSimples Corpus [39], is a dataset with sentences that passed through different stages of simplifications task. Table 6 contains the stages and the number of sentences produced for each stage of simplification. The Original class contains the original sentences, Natural contains the sentences produced from a Natural stage of simplification of the original sentences, and Strong has the sentences produced on a strong stage of simplification. On each stage of simplification the sentence becomes less complex. In the fine-tuning process, the model will learn the complexity of the sentences and will classify those sentences on three levels, so sentences more complex will be classified as Original, less complex sentences as Natural, and simple sentences as Strong.

Table 6. PorSimples Corpus dataset.

ID	Classes	Number of Examples
0	Original	2907
1	Natural	4066
2	Strong	4971
	Total	11,944

The second dataset selected, called Textual Complexity Corpus for School Internships in the Brazilian Educational System Dataset [40], is a dataset that contains texts divided by the stages of the Brazilian educational system. The stages of education are divided into four stages, representing the four classes presented in the Table 7.

Table 7. Textual Complexity Corpus for School Internships in the Brazilian Educational System dataset.

ID	Classes	Number of Examples
0	Elementary School—I	297
1	Elementary School—II	325
2	High School	628
3	University Education	826
	Total	2076

4.2. Metrics

Four metrics of the evaluation were used to measure the performance of the models. The first one is *accuracy* (Equation (2)), which consists of the number of correct and overall predictions. This metric is the probability that the model predicted the suitable class [41].

$$Accuracy = \frac{\sum CorrectedPredictions}{\sum AllPredictions} \quad (2)$$

The *precision* score (Equation (3)) is used to analyze the proportion of true positives that the model predicted. Precision tells how trustful the model is when predicting a particular class. The calculation is done by dividing the true positives (*TP*) by the sum of true positives (*TP*) and the false positives (*FP*) [41].

$$Precision = \frac{TP}{TP + FP} \quad (3)$$

Additionally, to measure the capability of the model to predict all the positive classes, the *recall* score (Equation (4)) is used. The recall score can be provided by dividing the true positives (*TP*) by the sum of true positives (*TP*) and the false negative (*FN*) [41].

$$Recall = \frac{TP}{TP + FN} \quad (4)$$

The last metric applied in the experiments is the *F1 score* (Equation (5)) to measure the performance of the model. This metric uses the precision score (*PS*) and recall score (*RC*) as a weighted average under the concept of harmonic mean [41].

$$F1Score = 2 \times (\frac{PS \times RS}{PS + RS}) \quad (5)$$

It is worth mentioning that all metrics presented have their best score as 1 and their worst score as 0.

5. Results

This section shows the performance assessment of the BERT, DistilBERT, BERTimbau, and DistilBERTimbau models. For a better presentation, this section was divided into two subsections. The first presented the results of the English language (Section 5.1), and the second presented the results of the Brazilian Portuguese language (Section 5.2).

It is worth mentioning that, after each K-fold iteration, an evaluation is made using the evaluation part of the dataset to measure the score of the fine-tuned model.

5.1. English Language

Brexit Blog Corpus was the first dataset evaluated. The BERT model's results are presented in Table 8 and the DistilBERT model's results are in Table 9.

Table 8. BERT—Brexit Blog Corpus dataset.

K-Fold	Accuracy	Precision	Recall	F1 Score	Evaluation Loss	Training Time	Evaluation Time
1	0.3650	0.3345	0.3290	0.3279	2.558	395	12
2	0.4214	0.4270	0.4017	0.4068	2.497	384	12
3	0.4107	0.3421	0.3495	0.3433	2.516	385	13
4	0.4286	0.3754	0.4223	0.3761	2.320	391	11
5	0.4405	0.4217	0.3909	0.3991	2.256	390	10

Table 9. DistilBERT—Brexit Blog Corpus dataset.

K-Fold	Accuracy	Precision	Recall	F1 Score	Evaluation Loss	Training Time	Evaluation Time
1	0.3561	0.3078	0.2971	0.2981	2.546	205	12
2	0.4154	0.3650	0.3557	0.3494	2.298	208	10
3	0.4137	0.3687	0.3606	0.3537	2.373	205	11
4	0.3750	0.3807	0.3442	0.3506	2.421	199	11
5	0.4048	0.3480	0.3352	0.3378	2.357	201	11

The Brexit Blog Corpus dataset obtained relatively low score results for all metrics evaluated, see Table 9. This behavior is expected since the dataset used is unbalanced. That is, many classes and few samples for each class; furthermore, some class has significantly more or fewer samples than others.

Additionally, the score results obtained by the distilled model of BERT are similar to those of its original model BERT. Still, the distilled model took around 47.7% less time on the fine-tuning process than BERT since DistilBERT is a more lightweight model than BERT.

The second English dataset evaluated was the BBC Text. The evaluation score results are presented in Table 10 for the BERT model and in Table 11 for DistilBERT.

Table 10. BERT—BBC Text dataset.

K-Fold	Accuracy	Precision	Recall	F1 Score	Evaluation Loss	Training Time	Evaluation Time
1	0.9753	0.9742	0.9747	0.9743	0.2062	435	10
2	0.9820	0.9826	0.9810	0.9816	0.1470	438	12
3	0.9820	0.9808	0.9821	0.9812	0.1116	450	13
4	0.9685	0.9685	0.9694	0.9689	0.2923	432	13
5	0.9573	0.9608	0.9540	0.9565	0.3191	438	13

Table 11. DistilBERT—BBC Text dataset.

K-Fold	Accuracy	Precision	Recall	F1 Score	Evaluation Loss	Training Time	Evaluation Time
1	0.9685	0.9680	0.9685	0.9681	0.2478	278	11
2	0.9685	0.9690	0.9705	0.9694	0.2098	269	12
3	0.9775	0.9747	0.9763	0.9754	0.2041	266	13
4	0.9753	0.9758	0.9763	0.9760	0.2039	266	13
5	0.9888	0.9877	0.9879	0.9876	0.0955	277	13

Unlike the Brexit Blog Corpus dataset, the BBC Text achieved outstanding score results. It is known that this dataset is balanced, having a good and uniform number of samples for each class. Comparing the two models, the evaluation results are very similar, but the fine-tuning time is around 37.3% lower for DistilBERT compared to BERT.

The last English dataset evaluated was Amazon Alexa Review Dataset. The BERT model's score result are presented on Tables 12 and 13 for DistilBERT model.

Table 12. BERT—Amazon Alexa Review dataset.

K-Fold	Accuracy	Precision	Recall	F1 Score	Evaluation Loss	Training Time	Evaluation Time
1	0.9651	0.8998	0.8377	0.8656	0.2050	665	12
2	0.9587	0.9191	0.7766	0.8304	0.2890	663	13
3	0.9508	0.8299	0.8109	0.8201	0.3314	663	14
4	0.9619	0.8523	0.8667	0.8593	0.2755	666	14
5	0.9508	0.9305	0.7919	0.8443	0.3640	666	14

Table 13. DistilBERT—Amazon Alexa Review dataset.

K-Fold	Accuracy	Precision	Recall	F1 Score	Evaluation Loss	Training Time	Evaluation Time
1	0.9603	0.8902	0.8065	0.8423	0.2431	324	09
2	0.9492	0.8725	0.8016	0.8323	0.3847	314	10
3	0.9413	0.8803	0.7228	0.7763	0.4456	319	11
4	0.9571	0.8369	0.8298	0.8333	0.3127	317	11
5	0.9619	0.9344	0.7966	0.8469	0.2698	318	11

The Amazon Alexa Reviews dataset reached good results. Analyzing Tables 12 and 13, it is possible to note that the precision, recall, and F1-score are a little lower than the accuracy score. Those results may occur because the dataset has fewer examples for the negative class and a very high number of samples for the positive class.

The BERT and DistilBERT score results were also very similar when compared. The DistilBERT model took around 52.1% less time to fine-tune when compared to its larger counterpart.

5.2. Brazilian Portuguese Language

In order to evaluate the Portuguese model BERTimbau and the distilled version DistilBERTimbau, the first Portuguese dataset selected was the Textual Complexity Corpus for School Internships in the Brazilian Educational System Dataset (TCIE). The BERTimbau score results are presented in Table 14 and the DistilBERTimbau results in Table 15.

Table 14. BERTimbau—TCIE dataset.

K-Fold	Accuracy	Precision	Recall	F1 Score	Evaluation Loss	Training Time	Evaluation Time
1	0.9351	0.9232	0.9280	0.9233	0.3477	418	12
2	0.9325	0.9227	0.9194	0.9206	0.4158	408	12
3	0.9301	0.9172	0.9193	0.9181	0.3752	421	13
4	0.9422	0.9375	0.9395	0.9384	0.3229	414	14
5	0.9253	0.9211	0.9168	0.9186	0.5360	241	13

Table 15. DistilBERTimbau—TCIE dataset.

K-Fold	Accuracy	Precision	Recall	F1 Score	Evaluation Loss	Training Time	Evaluation Time
1	0.9135	0.9060	0.9087	0.9072	0.5106	299	13
2	0.9277	0.9137	0.9180	0.9148	0.4370	289	15
3	0.9253	0.9025	0.9051	0.9024	0.3878	291	14
4	0.9181	0.9026	0.8990	0.9006	0.4631	307	15
5	0.9036	0.8822	0.8783	0.8797	0.6200	308	14

The TCIE dataset accomplished good results. Looking over Tables 14 and 15, it is possible to note that the distilled model had an evaluation score slightly lower than the BERTimbau model on every metric, but the fine-tuning process took around 21.5% longer on BERTimbau than the distilled version.

The second Portuguese dataset used was the PorSimples Corpus. For this dataset, the parameters used on the other datasets presented in Table 1 caused overfitting. A lower number of the learning rate hyperparameter was used to correct this issue, 0.000001 instead of 0.00004. This reduces the model's learning speed, solving the overfitting issue. The BERTimbau results are presented in Table 16 and the DistilBERTimbau evaluation score results are presented in Table 17.

The evaluation result with this dataset did not achieve very high scores in both the BERTimbau and DistilBERTimbau models. These low results may be explained because, on the PorSimples Corpus dataset, some sentences are similar to the others when passing through the simplifications process, so similar sentences are presented in each dataset class. Hence, the model has more challenges when learning the class differences. Additionally, the BERTimbau model took around 49.2% more time than the distilled model to the fine-tuning process. Furthermore, the high time results presented in Tables 16 and 17 were expected since this dataset has 11,944 samples, many more when compared to the other datasets.

Table 16. BERTimbau—PorSimples Corpus dataset.

K-Fold	Accuracy	Precision	Recall	F1 Score	Evaluation Loss	Training Time	Evaluation Time
1	0.5184	0.4948	0.5056	0.4690	0.9385	2162	25
2	0.5126	0.4824	0.4881	0.4679	0.9665	2159	25
3	0.5008	0.4737	0.4912	0.4518	0.9543	2170	26
4	0.5021	0.4813	0.4892	0.4634	0.9647	2173	24
5	0.5050	0.4681	0.4718	0.4483	0.9768	2198	27

Table 17. DistilBERTimbau—PorSimples Corpus dataset.

K-Fold	Accuracy	Precision	Recall	F1 Score	Evaluation Loss	Training Time	Evaluation Time
1	0.5038	0.4797	0.4801	0.4520	0.9635	1099	15
2	0.5067	0.4697	0.4837	0.4449	0.9559	1098	17
3	0.5226	0.4889	0.5030	0.4687	0.9311	1095	18
4	0.5251	0.4892	0.4968	0.4721	0.9428	1126	18
5	0.5008	0.4794	0.4853	0.4446	0.9746	1103	17

Table 18 contains the size of the models generated after the fine-tuning process for each dataset. Analyzing the results, it is possible to identify that the distilled models produced models around 40% smaller than their larger counterparts.

Table 18. Model Size.

Dataset	BERT	DistilBert
Amazon Alexa Review	413.3 MB	251 MB
BBC Text	413.3 MB	251 MB
Brexit Blog Corpus	413.3 MB	251 MB
TCIE	415.6 MB	253.4 MB
PorSimples Corpus	415.6 MB	253.3 MB

An important observation is that on every evaluation, the scores reached on every k-fold iteration had very similar results, which show the model's generalization capability.

The barplot presented in Figure 5 contains the arithmetic mean of each scoring metric on each k-fold iteration. In this figure, the red bars refer to BERT/BERTimbau models, and the blue ones to DistilBERT/DistilBERTimbau models.

As we can see, the score recorded by the distilled models is very similar to the ones scored by the original models. This shows the power of the compression of deep learning models technique, which produces smaller models, requires fewer computation resources, and has almost the same power as the original models.

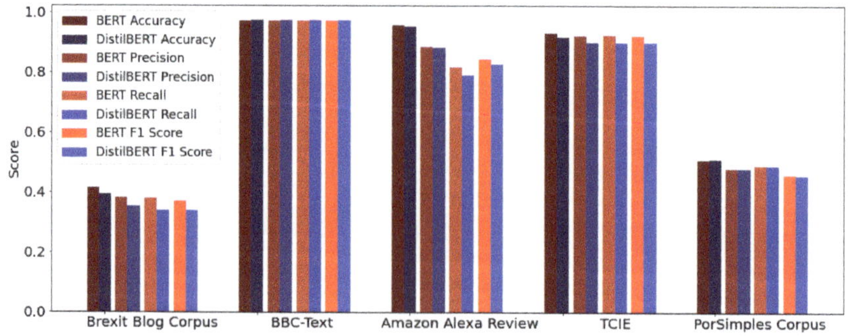

Figure 5. Bar plot comparing BERT and DistilBERT's model scores.

6. Discussion

Analyzing the results presented in Section 5 and Figure 5, the scores recorded by the distilled models are very similar to the ones scored by the original models. In our experiments, they were around 45% faster in the fine-tuning process, about 40% smaller, and also preserving about 96% of the language comprehension skills performed by BERT and BERTimbau. It is worth noting that these results are similar to the results presented on [25], where the DistilBERT models were 40% smaller, 60% faster, and retained 97% of BERT's comprehension capability.

The work presented in [42] compared BERT, DistilBERT, and other pre-trained models for emotion recognition and also achieved similar score results on BERT and DistilBERT. Furthermore, the DistilBERT model was the fastest one. These results presented in that work and also in the literature show the power of the compression of deep learning models technique, which produces smaller models, requires fewer computation resources, and has almost the same power as the original models.

Another critical point we can highlight in Figure 5 is the importance of the quality of the datasets to produce a good predicted model. In two unbalanced datasets, such as Brexit Blog Corpus and PorSimples Corpus, the accuracy was low against the other balanced datasets. The Amazon Alexa Reviews achieve good accuracy, but lower precision, recall, and F1 score since this dataset has a low number of negative samples.

Other pre-trained models have been widely developed for other languages such as BERTino [43], an Italian DistilBERT, and CamemBERT [44] for the French language based on the RoBERTa [45] model, a variation of the BERT model. The main goal of pre-trained models is to remove the necessity of building a specific model for each task and to improve the necessity of developing a pre-trained model for each language, bigger models that understand multiple languages have been developed such as BERT Multilingual [30] and also GPT-3 [46]. Still, those models are trained with more data than BERT for specific languages, especially GPT-3, and should require more computational resources.

7. Conclusions

Inspired by a state-of-the-art language representation model, this paper analyzed two state-of-the-art models, BERT and DistilBERT, for text classification tasks for both English and Brazilian Portuguese. These models have been compared with several selected datasets. The experiment results showed that the compression of neural networks responsible for the generation of the DistilBERT and DistilBERTimbau produce models around 40% smaller and take around 45% (our experiments ranged from 21.5% to 66.9%) less time for the fine-tuning process. In other words, compression models require fewer computational resources, which did not significantly impact the model's performance. Thus, the lightweight models allow being executed with low computational resources and with the performance of their larger counterparts. In addition, the distilled models preserve about 96% of language comprehension skills for balanced datasets.

Some extensions of our future work can be highlighted: *(i)* other robust models are being widely studied and developed, such as in [47] and GPT-3 [46], which can be evaluated and compared with the models mentioned in this work; and *(ii)* perform task classification for non-Western languages (e.g., Japanese, Chinese, and Korean).

In closing, the experiment results show how robust the Transformer architecture is and the possibility of using it for more languages than English, such as the Brazilian Portuguese models studied in this work.

Author Contributions: Conceptualization, R.S.B. and A.T.A.; Data curation, R.S.B.; Formal analysis, A.T.A.; Funding acquisition, A.T.A.; Methodology, R.S.B.; Project administration, A.T.A.; Resources, A.T.A.; Software, R.S.B.; Supervision, A.T.A.; Validation, R.S.B.; Visualization, R.S.B.; Writing—original draft, R.S.B. and A.T.A.; Writing—review and editing, A.T.A. All authors have read and agreed to the published version of the manuscript.

Funding: This work was funded by National Council for Scientific and Technological Development (CNPq 405498/2021-7) and was financed in part by the Coordenação de Aperfeiçoamento de Pessoal de Nível Superior—Brasil (CAPES)—Finance Code 001.

Institutional Review Board Statement: Not applicable.

Informed Consent Statement: Not applicable.

Data Availability Statement: All the datasets used in this work can be found on the links below: Brexit Blog Corpus dataset-https://snd.gu.se/en/catalogue/study/snd1037#dataset; BBC Text dataset-https://www.kaggle.com/code/yufengdev/bbc-text-categorization/data; Amazon Alexa Reviews dataset-https://www.kaggle.com/datasets/sid321axn/amazon-alexa-reviews; PorSimples Corpus dataset-https://github.com/sidleal/porsimplessent; Textual Complexity Corpus for School Internships in the Brazilian Educational System dataset-https://github.com/gazzola/corpus_readability_nlp_portuguese.

Conflicts of Interest: The authors declare no conflict of interest.

References

1. Statista Research Department. Amount of Data Created, Consumed, and Stored 2010–2025. 2018. Available online: https://www.statista.com/statistics/871513/worldwide-data-created/ (accessed on 8 July 2022).
2. Kowsari, K.; Jafari Meimandi, K.; Heidarysafa, M.; Mendu, S.; Barnes, L.; Brown, D. Text classification algorithms: A survey. *Information* **2019**, *10*, 150. [CrossRef]
3. Miyato, T.; Dai, A.M.; Goodfellow, I. Adversarial training methods for semi-supervised text classification. *arXiv* **2016**, arXiv:1605.07725.
4. Chen, J.; Yang, Z.; Yang, D. Mixtext: Linguistically-informed interpolation of hidden space for semi-supervised text classification. *arXiv* **2020**, arXiv:2004.12239.
5. Jiang, H.; Nie, L.; Sun, Z.; Ren, Z.; Kong, W.; Zhang, T.; Luo, X. Rosf: Leveraging information retrieval and supervised learning for recommending code snippets. *IEEE Trans. Serv. Comput.* **2016**, *12*, 34–46. [CrossRef]
6. Guo, J.; Fan, Y.; Pang, L.; Yang, L.; Ai, Q.; Zamani, H.; Wu, C.; Croft, W.B.; Cheng, X. A deep look into neural ranking models for information retrieval. *Inf. Process. Manag.* **2020**, *57*, 102067. [CrossRef]
7. Li, H.; Zhu, J.; Zhang, J.; Zong, C. Ensure the correctness of the summary: Incorporate entailment knowledge into abstractive sentence summarization. In Proceedings of the 27th International Conference on Computational Linguistics, Santa Fe, NM, USA, 20–26 August 2018; pp. 1430–1441.
8. Rothe, S.; Maynez, J.; Narayan, S. A thorough evaluation of task-specific pretraining for summarization. In Proceedings of the 2021 Conference on Empirical Methods in Natural Language Processing, Punta Cana, Dominican Republic, 7–11 November 2021; pp. 140–145.
9. Aggarwal, C.C.; Zhai, C. A survey of text clustering algorithms. In *Mining Text Data*; Springer: Berlin/Heidelberg, Germany, 2012; pp. 77–128.
10. Jánez-Martino, F.; Fidalgo, E.; González-Martínez, S.; Velasco-Mata, J. Classification of spam emails through hierarchical clustering and supervised learning. *arXiv* **2020**, arXiv:2005.08773.
11. Conneau, A.; Schwenk, H.; Barrault, L.; Lecun, Y. Very deep convolutional networks for text classification. *arXiv* **2016**, arXiv:1606.01781.
12. Johnson, R.; Zhang, T. Deep pyramid convolutional neural networks for text categorization. In Proceedings of the 55th Annual Meeting of the Association for Computational Linguistics, Vancouver, BC, Canada, 30 July–4 August 2017; pp. 562–570.
13. Kokkinos, F.; Potamianos, A. Structural attention neural networks for improved sentiment analysis. *arXiv* **2017**, arXiv:1701.01811.
14. Yang, Z.; Yang, D.; Dyer, C.; He, X.; Smola, A.; Hovy, E. Hierarchical attention networks for document classification. In Proceedings of the 2016 Conference of the North American Chapter of the Association for Computational Linguistics: Human Language Technologies, San Diego, CA, USA, 12–17 June 2016; pp. 1480–1489.
15. Mikolov, T.; Sutskever, I.; Chen, K.; Corrado, G.S.; Dean, J. Distributed representations of words and phrases and their compositionality. In *Advances in Neural Information Processing Systems 26*; Curran Associates Inc.: Red Hook, NY, USA, 2013.
16. Pennington, J.; Socher, R.; Manning, C.D. Glove: Global vectors for word representation. In Proceedings of the 2014 Conference on Empirical Methods in Natural Language Processing (EMNLP), Doha, Qatar, 25–29 October 2014; pp. 1532–1543.
17. Brochier, R.; Guille, A.; Velcin, J. Global vectors for node representations. In Proceedings of the World Wide Web Conference, San Francisco, CA, USA, 13–17 May 2019; pp. 2587–2593.
18. Van Nguyen, M.; Lai, V.D.; Veyseh, A.P.B.; Nguyen, T.H. Trankit: A light-weight transformer-based toolkit for multilingual natural language processing. *arXiv* **2021**, arXiv:2101.03289.
19. Vaswani, A.; Shazeer, N.; Parmar, N.; Uszkoreit, J.; Jones, L.; Gomez, A.N.; Kaiser, Ł.; Polosukhin, I. Attention is all you need. In *Advances in Neural Information Processing Systems 30*; Curran Associates Inc.: Red Hook, NY, USA, 2017.
20. Devlin, J.; Chang, M.W.; Lee, K.; Toutanova, K. Bert: Pre-training of deep bidirectional transformers for language understanding. *arXiv* **2018**, arXiv:1810.04805.
21. Sutskever, I.; Vinyals, O.; Le, Q.V. Sequence to sequence learning with neural networks. In *Advances in Neural Information Processing Systems 27*; Curran Associates Inc.: Red Hook, NY, USA, 2014.

22. Bahdanau, D.; Cho, K.; Bengio, Y. Neural machine translation by jointly learning to align and translate. *arXiv* **2014**, arXiv:1409.0473.
23. Rumelhart, D.E.; Hinton, G.E.; Williams, R.J. *Learning Internal Representations by Error Propagation;* Technical Report; California University, San Diego La Jolla Institute for Cognitive Science: La Jolla, CA, USA, 1985.
24. Jurafsky, D. *Speech & Language Processing;* Pearson Education: Noida, India, 2000.
25. Sanh, V.; Debut, L.; Chaumond, J.; Wolf, T. DistilBERT, a distilled version of BERT: Smaller, faster, cheaper and lighter. *arXiv* **2019**, arXiv:1910.01108.
26. Hinton, G.; Vinyals, O.; Dean, J. Distilling the knowledge in a neural network. *arXiv* **2015**, arXiv:1503.02531.
27. Buciluă, C.; Caruana, R.; Niculescu-Mizil, A. Model compression. In Proceedings of the 12th ACM SIGKDD International Conference on Knowledge Discovery and Data Mining, Philadelphia, PA, USA, 20–23 August 2006; pp. 535–541.
28. Souza, F.; Nogueira, R.; Lotufo, R. BERTimbau: Pretrained BERT models for Brazilian Portuguese. In Proceedings of the Brazilian Conference on Intelligent Systems, Rio Grande, Brazil, 20–23 October 2020; Springer: New York, NY, USA, 2020; pp. 403–417.
29. Wagner Filho, J.A.; Wilkens, R.; Idiart, M.; Villavicencio, A. The brWaC corpus: A new open resource for Brazilian Portuguese. In Proceedings of the Eleventh International Conference on Language Resources and Evaluation (LREC 2018), Miyazaki, Japan, 7–12 May 2018.
30. Pires, T.; Schlinger, E.; Garrette, D. How multilingual is multilingual BERT? *arXiv* **2019**, arXiv:1906.01502.
31. Brownlee, J. What Is the Difference between a Batch and an Epoch in a Neural Network. Machine Learning Mastery. 2018. Available online: https://deeplearning.lipingyang.org/wp-content/uploads/2018/07/What-is-the-Difference-Between-a-Batch-and-an-Epoch-in-a-Neural-Network_.pdf (accessed on 1 August 2022).
32. Perin, G.; Buhan, I.; Picek, S. Learning When to Stop: A Mutual Information Approach to Fight Overfitting in Profiled Side-Channel Analysis. Cryptology ePrint Archive, Paper 2020/058. 2020. Available online: https://eprint.iacr.org/2020/058 (accessed on 1 August 2022).
33. Smith, L.N. A disciplined approach to neural network hyper-parameters: Part 1–learning rate, batch size, momentum, and weight decay. *arXiv* **2018**, arXiv:1803.09820.
34. Llugsi, R.; Yacoubi, S.E.; Fontaine, A.; Lupera, P. Comparison between Adam, AdaMax and Adam W optimizers to implement a Weather Forecast based on Neural Networks for the Andean city of Quito. In Proceedings of the 2021 IEEE Fifth Ecuador Technical Chapters Meeting (ETCM), Quito, Ecuador, 11–14 October 2021; pp. 1–6. [CrossRef]
35. Berrar, D. Cross-Validation. In *Encyclopedia of Bioinformatics and Computational Biology;* Elsevier: Amsterdam, The Netherlands, 2019.
36. Refaeilzadeh, P.; Tang, L.; Liu, H. Cross-validation. *Encycl. Database Syst.* **2009**, *5*, 532–538.
37. Simaki, V.; Paradis, C.; Skeppstedt, M.; Sahlgren, M.; Kucher, K.; Kerren, A. Annotating speaker stance in discourse: The Brexit Blog Corpus. *Corpus Linguist. Linguist. Theory* **2020**, *16*, 215–248. [CrossRef]
38. Greene, D.; Cunningham, P. Practical Solutions to the Problem of Diagonal Dominance in Kernel Document Clustering. In Proceedings of the 23rd International Conference on Machine Learning (ICML'06), Pittsburgh, PA, USA, 25–29 June 2006; ACM Press: New York, NY, USA, 2006; pp. 377–384.
39. Caseli, H.M.; Pereira, T.F.; Specia, L.; Pardo, T.A.; Gasperin, C.; Aluísio, S.M. Building a Brazilian Portuguese parallel corpus of original and simplified texts. *Adv. Comput. Linguist. Res. Comput. Sci.* **2009**, *41*, 59–70.
40. Gazzola, M.; Leal, S.E.; Aluisio, S.M. Predição da Complexidade Textual de Recursos Educacionais Abertos em Português. In Proceedings of the Brazilian Symposium in Information and Human Language Technology, Salvador, Brazil, 15–18 October 2019.
41. Grandini, M.; Bagli, E.; Visani, G. Metrics for multi-class classification: An overview. *arXiv* **2020**, arXiv:2008.05756.
42. Acheampong, F.A.; Nunoo-Mensah, H.; Wenyu, C. Comparative Analyses of BERT, RoBERTa, DistilBERT, and XLNet for Text-based Emotion Recognition. In Proceedings of the 2020 17th International Computer Conference on Wavelet Active Media Technology and Information Processing (ICCWAMTIP), Chengdu, China, 18–20 December 2020.
43. Muffo, M.; Bertino, E. Bertino: An italian distilbert model. In Proceedings of the Seventh Italian Conference on Computational Linguistics CLiC-It 2020, Bologna, Italy, 1–3 March 2021.
44. Martin, L.; Muller, B.; Suárez, P.J.O.; Dupont, Y.; Romary, L.; de La Clergerie, É.V.; Seddah, D.; Sagot, B. CamemBERT: A tasty French language model. *arXiv* **2019**, arXiv:1911.03894.
45. Liu, Y.; Ott, M.; Goyal, N.; Du, J.; Joshi, M.; Chen, D.; Levy, O.; Lewis, M.; Zettlemoyer, L.; Stoyanov, V. Roberta: A robustly optimized bert pretraining approach. *arXiv* **2019**, arXiv:1907.11692.
46. Floridi, L.; Chiriatti, M. GPT-3: Its nature, scope, limits, and consequences. *Minds Mach.* **2020**, *30*, 681–694. [CrossRef]
47. Radford, A.; Wu, J.; Child, R.; Luan, D.; Amodei, D.; Sutskever, I. Language models are unsupervised multitask learners. *OpenAI Blog* **2019**, *1*, 9.

sensors

MDPI

Article

Multi-Delay Identification of Rare Earth Extraction Process Based on Improved Time-Correlation Analysis

Rongxiu Lu [1,2,*], Hongliang Liu [1,2], Hui Yang [1,2], Jianyong Zhu [1,2] and Wenhao Dai [1,2]

[1] School of Electrical and Automation Engineering, East China Jiaotong University, Nanchang 330013, China
[2] Key Laboratory of Advanced Control and Optimization of Jiangxi Province, Nanchang 330013, China
* Correspondence: rxlu@ecjtu.edu.cn

Abstract: The rare earth extraction process has significant time delay characteristics, making it challenging to identify the time delay and establish an accurate mathematical model. This paper proposes a multi-delay identification method based on improved time-correlation analysis. Firstly, the data are preprocessed by grey relational analysis, and the time delay sequence and time-correlation data matrix are constructed. The time-correlation analysis matrix is defined, and the H_∞ norm quantifies the correlation degree of the data sequence. Thus the multi-delay identification problem is transformed into an integer optimization problem. Secondly, an improved discrete state transition algorithm is used for optimization to obtain multi-delay. Finally, based on an Neodymium (Nd) component content model constructed by a wavelet neural network, the performance of the proposed method is compared with the unimproved time delay identification method and the model without an identification method. The results show that the proposed algorithm improves optimization accuracy, convergence speed, and stability. The performance of the component content model after time delay identification is significantly improved using the proposed method, which verifies its effectiveness in the time delay identification of the rare earth extraction process.

Keywords: rare earth extraction; time delay identification; grey correlation analysis; time-correlation; discrete state transition algorithm; wavelet neural network

Citation: Lu, R.; Liu, H.; Yang, H.; Zhu, J.; Dai, W. Multi-Delay Identification of Rare Earth Extraction Process Based on Improved Time-Correlation Analysis. *Sensors* **2023**, *23*, 1102. https://doi.org/10.3390/s23031102

Academic Editors: Peter Hockicko, Róbert Hudec and Patrik Kamencay

Received: 13 December 2022
Revised: 12 January 2023
Accepted: 12 January 2023
Published: 18 January 2023

1. Introduction

The rare earth extraction process includes dozens or even hundreds of extraction tanks. The mixing speed and time of each group of agitators are different, which affects the reaction and transmission time of materials, leading to multi-delay, so a large amount of data cannot be effectively utilized. Current modeling studies of the rare earth extraction process do not consider time delay or take it as a constant [1–3], resulting in a particular gap between the model and the extraction site. Therefore, it is significant to study how to identify multi-delay.

A series of solutions have been proposed for the problem of time delay identification. The step response method was used in the time delay identification of systems [4,5], but this method is susceptible to noise and requires a filter to remove high-frequency noise. Rad et al. [6] estimated the time delay of input and output signals based on the cross-correlation function, which cannot reflect their specific relationship, so the results are not ideal. Some scholars used a recursive least squares algorithm to identify the system with time delay [7–9], which is assumed to be known and may not be established in engineering practice. Neural networks were used to identify time delay, which have the problem of long training times and easily falling into local optima [10–12]. Liu et al. [13] developed a compressed sensing recovery algorithm for the multiple input single output finite impulse response systems with unknown time delay, but it was not easy to choose the optimal threshold. Chen et al. [14] proposed an effective identification model based on the Bayesian theorem for systems with unknown time delay. Wang et al. [15] proposed a parameter identification method of a fractional-order time delay system based on the Legendre wavelet, which reduces the effect of noise on the accuracy of parameter identification. Hofmann et al. [16] proposed an offline time-delay identification strategy based on falling film evaporator pilot plant experiments and obtained good results in both validation experiments, with and without evaporation. Prasad et al. [17] used fractional-order modeling technique to identify the parameters of Hammerstein structured nonlinear systems with discontinuous asymmetric (two segment piecewise-linear with a dead-zone) nonlinearity and input time delay. Li et al. [18] designed a discrete-time robust adaptive estimator to identify the time delay and sandwich system parameters. Meanwhile, they reconstructed the observation and augmented data to obtain the explicit expression of the delay parameter. To achieve an effective system control strategy and accurate response prediction, Liu et al. [19] proposed a new method to identify the parameters of linear time-delay differential systems by analyzing the frequency domain response of complex systems. To solve the influence of time delay on HVAC systems, Li et al. [20] introduced transfer entropy and proposed a model-free identification method based on the information theory framework. Ni et al. [21] studied the parameter estimation problem for a class of linear time-delay systems. Based on the frequency responses and harmonic balance methods and by means of the gradient search, a two-stage stochastic gradient and gradient-based iterative algorithm was developed by using the collected data under the sinusoidal excitation. The maximum likelihood method [22], variable structure observer [23], particle swarm optimization [24], and other methods have also been applied.

Industrial processes have become complex with the rapid development of science and technology, resulting in multi-delay, and the abovementioned methods have been unable to meet practical requirements. Xie et al. [25] improved the genetic algorithm, applied it to the identification of multi-delay, and obtained their optimal estimate, but the method requires an accurate system model. Huang et al. [26] proposed an improved cross-correlation function method and realized multi-delay identification of the alumina carbon separation process. Wang et al. [27] proposed a trend similarity analysis method and realized the multi-delay identification of the hydrocracking process. All the abovementioned methods have the problems of high computational redundancy and time consumption.

This paper draws on the successful application of the time-correlation analysis method [28] in the alumina carbon separation and evaporation processes. We propose an improved time-correlation analysis method for the rare earth extraction process. Based on field data and the improved method, the multi-delay identification problem is transformed into an unconstrained integer optimization problem without changing the time delay relationship. Since the discrete state transition algorithm [29] can effectively solve the unconstrained integer optimization problem [30], we adopt the improved algorithm to solve it. Experimental comparison and analysis show that the improved time-correlation analysis method has high speed, high accuracy, and good stability, and is suitable for the multi-delay identification of the rare earth extraction process.

2. Improvement of Time-Correlation Analysis Method

Time-correlation analysis is a time delay identification method based on the relationship between data sequences, which has the advantage of high efficiency. This paper improves its shortcomings in data preprocessing and selection.

2.1. Grey Relational Analysis

Grey relational analysis (GRA) is derived from the grey system theory in system science [31]. Its basic idea is to judge the tightness of sequence connections according to the similarity between the geometric shapes of sequence curves. The closer the curves, the more significant the correlation between sequences.

Compared with traditional multi-factor analysis methods (such as canonical correlation analysis and multiple linear regression), this method has lower data requirements and less computational burden. It is suitable for quantitative analysis of the dynamic development process and hence can be used to analyze the correlation between a large amount of process data obtained from a production site and the content of rare earth elements.

Let the time base of a critical process variable in N work units be $d = [d_1, d_2, \ldots, d_i, \ldots, d_N]$, where $i = 1,2,\ldots,N$, $d_i = \tau_i/T$. T is the sampling period, the delay sequence is $\Gamma = [\tau_1, \tau_2, \ldots, \tau_i, \ldots, \tau_N]$, and τ_i is the time delay of the ith work unit. The content of n rare earth elements and m process variables is obtained by k times sampling. The element component content $U'_j(t) = [u'_1(t), u'_2(t), \ldots, u'_j(t), \ldots, u'_n(t)]$ are used as the reference sequence in the correlation analysis, where $1 \le j \le n$, $1 \le t \le k$, and $u'_j(t)$ represents the jth rare earth element component content. The process variable data $E'_l(t) = [e'_1(t), e'_2(t), \ldots, e'_l(t), \ldots, e'_m(t)]$ are used as the comparison sequence, where $1 \le l \le m, 1 \le t \le k$, and $e'_l(t)$ represents the lth process variable data.

The original data are normalized to eliminate the influence of different dimensions on the results. Standard processing methods include initialization and averaging. This paper adopts averaging,

$$\begin{cases} U_j(t) = U'_j(t)/\frac{1}{k}\sum_{t=1}^{k} U'_j(t) \\ E_l(t) = E'_l(t)/\frac{1}{k}\sum_{t=1}^{k} E'_l(t) \end{cases} \tag{1}$$

where $U_j(t)$ is the processed reference sequence data, and $E_l(t)$ is the processed comparison sequence data.

The correlation coefficient is used to express the degree of closeness between the index values of the comparison and reference sequence in grey relational analysis. The higher the value, the greater the degree of proximity,

$$\xi_{lj}(t) = \frac{\min_l \min_j |U_j(t) - E_l(t)| + \rho \max_l \max_j |U_j(t) - E_l(t)|}{|U_j(t) - E_l(t)| + \rho \max_l \max_j |U_j(t) - E_l(t)|} \tag{2}$$

where $\xi_{lj}(t)$ is the correlation coefficient of the lth characteristic variable corresponding to the content of the jth rare earth element component, and $\rho \in [0,1]$ is the resolution coefficient, which we take as 0.5.

According to the correlation coefficient, the correlation degree between each process variable and the content of rare earth elements can be obtained as

$$r_{lj}(t) = \frac{1}{k}\sum_{t=1}^{k} \xi_{lj}(t) \tag{3}$$

The correlation degree is sorted from large to small. If $r_{11} < r_{21}$, then the correlation degree between the comparison sequence $e_2(t)$ and the content of the first rare earth element component is greater than that of comparison sequence $e_1(t)$.

2.2. Time Delay Identification Method Based on Time-Correlation Analysis

The process variable with the highest gray correlation is taken as the key process variable, and its standardized data $e_l(t)$ are used to form the time-correlation data matrix,

$$E = \begin{vmatrix} e_{0,t} & e_{1,t+\tau_1} & \cdots & e_{i,t+\tau_1+\cdots+\tau_i} & \cdots & e_{N,t+\tau_1+\cdots+\tau_N} \\ e_{0,t+T} & e_{1,t+\tau_1+T} & \cdots & e_{i,t+\tau_1+\cdots+\tau_i+T} & \cdots & e_{N,t+\tau_1+\cdots+\tau_N+T} \\ \vdots & \vdots & \ddots & \vdots & \ddots & \vdots \\ e_{0,t+(F-1)T} & e_{1,t+\tau_1+(F-1)T} & \cdots & e_{i,t+\tau_1+\cdots+\tau_i+(F-1)T} & \cdots & e_{N,t+\tau_1+\cdots+\tau_N+(F-1)T} \end{vmatrix} \tag{4}$$

where $e_{0,*}$ and $e_{i,*}$ are the time series of the inlet process variables and the ith unit outlet process variables, respectively. $F \ge \sum_1^N d_i$, so that the data in the time-correlation matrix contain information about the entire process cycle, from inlet to outlet.

The matrix E describes the degree of correlation between the data corresponding to the time sequences. Multiple time sequences are described by the time-correlation analysis matrix,

$$R = \frac{cov(E)}{\prod_{i=1}^{N} \sigma_i} \tag{5}$$

where $cov(E)$ is the covariance matrix of E, whose standard deviation of column i is σ_i.

R is a time-correlation analysis matrix, which reflects the correlation degree of multiple groups of sequences under multi-delay. The H_∞ norm can quantify it and is expressed as

$$\beta = max(||R||_\infty) \tag{6}$$

A maximum value of the H_∞ norm indicates the maximum correlation between each time sequence in the data matrix. At this time, the corresponding delay sequence Γ consists of the multi-delay to be solved.

According to the extraction production site experience, each unit's time delay during operation will fluctuate within a fixed range. Based on this, the time-base value range of key process variables can be determined as $d_i \in [d_{imin}, d_{imax}]$. Thus, the time delay identification problem of Equation (6) can be transformed into an unconstrained integer optimization problem,

$$\begin{cases} \beta = \max(\|R\|_\infty) \\ s.t \quad d_i \in [d_{imin}, d_{imax}] \end{cases} \tag{7}$$

In summary, the steps of the improved method of combining grey correlation analysis with time-correlation analysis are as follows.

Step 1: Based on the original data of the rare earth extraction process, the multi-delay sequence Γ is constructed;

Step 2: Grey correlation analysis is used to identify key process variables and construct a time-correlation data matrix according to Equation (4);

Step 3: According to Equation (5), the time-correlation analysis matrix is defined, and the correlation degree of the data sequence is quantified by the H_∞ norm;

Step 4: The time delay identification results are obtained using Equation (7).

There are many methods to solve the time delay identification. The discrete state transition algorithm (DSTA) has been successfully applied to typical discrete optimization problems such as Boolean integer programming [32] and staff assignment [30]. We use this method to solve Equation (7).

3. Adaptive Chaotic Discrete State Transition Algorithm

Discrete state transition algorithm is an individual-based optimization algorithm. Its basic idea is to regard the solution of an optimization problem as a state, and the process of updating the solution is called state transition. The standard form of the discrete state transition algorithm can be described as

$$\begin{cases} x_{s+1} = A_s x_s \oplus B_s u_s \\ y_{s+1} = f(x_{s+1}) \end{cases} \tag{8}$$

where $x_s \in \mathbb{Z}$ is a current state; A_s, B_s are transformation operators; $u_s \in \mathbb{Z}$ is a control variable; $\oplus$ is an operation; $f(\cdot)$ is the evaluation function, which is used to measure the quality of x_s.

The four special transformation operators [32] are as follows.

(1) Swap transformation:

$$x_{s+1} = A_s^{swap}(m_a)x_s \tag{9}$$

where $A_s^{swap} \in R^{n \times n}$ is a random 0–1 matrix with swap action, called a swap transformation matrix, and m_a is a swap factor that can control the number of swap elements in the solution. Swap transformation is called local exploration and global exploration when $m_a = 2$ and $m_a \geq 3$, respectively.

(2) Shift transformation:

$$x_{s+1} = A_s^{shift}(m_b)x_s \tag{10}$$

where $A_s^{shift} \in R^{n \times n}$ is a random 0–1 matrix with shift action, called a shift transformation matrix, and a shift factor, m_b, can control the continuous shift of elements in the solution. If $m_b = 1$, the shift transformation is regarded as local exploitation, and if $m_b \geq 2$, the shift transformation is regarded as global exploration.

(3) Symmetry transformation:

$$x_{s+1} = A_s^{sym}(m_c)x_s \tag{11}$$

where $A_s^{sym} \in R^{n \times n}$ is a random 0–1 matrix with symmetry action, called a symmetry transformation matrix, and a symmetry factor, m_c, can control the continuous symmetry of elements in the solution. Symmetry transformation is intrinsically called global exploration.

(4) Substitute transformation:

$$x_{s+1} = A_s^{sub}(m_d)x_s + B_s^{sub}(m_d)u_s \tag{12}$$

where A_s^{sub}, $B_s^{sub} \in R^{n \times n}$ is a substitute transformation matrix, and m_d is a constant integer, called a substitute factor, to control the maximum number of positions to be substituted. If $m_d = 1$, the substitute transformation is regarded as local exploitation, and if $m_d \geq 2$, the substitute transformation is regarded as global exploration.

The initial solution is given randomly in the discrete state transition algorithm, and the solution significantly affects the convergence performance of the algorithm. DSTA easily falls into local optima in the iterative process. Therefore, an adaptive chaotic discrete state transition algorithm (ACDSTA) is proposed by introducing the opposition-based learning strategy, chaotic perturbation strategy, and adaptive recovery strategy.

3.1. Initialization Method Based on Opposition-Based Learning Strategy

The initialization method of the discrete state transition algorithm has the problem of uneven distribution, which somewhat affects its optimization efficiency. Therefore, an opposition-based learning strategy (OBL) is introduced to initialize the discrete state transition algorithm. OBL [33] is a machine-learning method whose idea is to generate a reverse solution based on the forward solution, compare their fitness values, and select the optimal solution as the initial solution, thereby improving the optimization speed of the algorithm.

Let $X = [x_1, x_2, \ldots, x_c, \ldots, x_D]$ be an entity in D-dimensional space. The reverse solution based on OBL is $X' = [x'_1, x'_2, \ldots, x'_c, \ldots, x'_D]$, where $x_c, x'_c \in [L_a, L_b], c = 1, 2, \ldots, D$ calculated as

$$X' = L_a + L_b - X \tag{13}$$

where L_a and L_b are the upper and lower bounds, respectively, of the value range of the target vector.

Based on the idea of OBL, the best initial solution is selected as

$$y_o = \max(f(X), f(X')) \tag{14}$$

where $f(\cdot)$ is the fitness function, y_o is the fitness value corresponding to X_o, and X_o is the initial solution into the subsequent iterative process.

3.2. Chaos Perturbation Strategy

Chaos comes from nonlinear dynamic systems. Because of its unique randomness, ergodicity, and complexity, it can effectively prevent the algorithm from falling into local optima, and it is widely used in optimization problems [34]. We propose the chaotic perturbation strategy to improve the problem that the discrete state transition algorithm can easily fall into local optima and is described as follows. During the algorithm iteration, when a value recurs a certain number of times, it indicates that the algorithm has fallen into a local optimum. Chaotic perturbation is applied to obtain a chaotic variable sequence, which is inversely mapped to the original solution space to obtain the perturbation solution, which is substituted in the next iteration so that the calculation exits the local extremum. When the termination condition is satisfied, the algorithm ends the iteration, and the final solution is the global optimum.

There are many rules to generate chaos, among which logistic mapping is common, but it has the problem of uneven frequency distribution. Zhou et al. [35] combined logistic mapping and tent mapping based on the uniform ergodicity of tent mapping to realize logistic-tent mapping,

$$X_{n+1} = \begin{cases} (\alpha X_n(1-X_n) + (4-\alpha)X_n/2)mod1 & X_n < 0.5 \\ (\alpha X_n(1-X_n) + (4-\alpha)(1-X_n)/2)mod1 & X_n \geq 0.5 \end{cases} \tag{15}$$

where $X_n \in [0,1]$ is a chaotic variable, and the $mod1$ operation ensures that its output data are in the range of $[0,1]$, and $\alpha \in (0,4]$ is a chaotic factor, which we take as 3.99.

The chaotic variables generated by Equation (15) cannot be directly used for iterative calculation of the algorithm. So, chaotic variables are mapped to the solution space of the objective function,

$$X_{new} = round(L_b + (L_a - L_b)X_n) \tag{16}$$

where X_{new} is a new solution generated after chaotic perturbation.

3.3. Adaptive Recovery Strategy

In the iterative process of the discrete state transition algorithm, the chaos perturbation strategy is introduced to generate new solutions, which can effectively improve its ability to jump out of local optima. However, the new solution directly enters the next iteration, which will decrease the algorithm's convergence performance. To only use the greedy criterion can no longer meet the convergence requirements. We propose adaptive recovery, adopting the greedy criterion to ensure the general convergence of the algorithm. The current value is restored to the preserved historical best value with adaptive probability to further improve convergence performance.

The discrete state transition algorithm is in the stage of rapid optimization in the early stages of iteration, which greatly decreases fitness. The recovery probability should be small at this time, so as not to affect the searchability of the algorithm in the early stage of iteration. In the middle and later stages, the searchability decreases, and the optimal historical value should be restored with a large probability. We use a nonlinear adaptive adjustment method [36],

$$P = (p_a - p_b) \times (1 - \sin(\frac{\pi}{2} \cdot (\frac{iter}{iter_{max}})^{\mu})) \tag{17}$$

where $P \in [0,1]$, P_a is the maximum value of the recovery probability P, P_b is the minimum value of P, $iter$ is the current number of iterations, $iter_{max}$ is the maximum number of iterations, and μ is the adaptive factor, which we take as 2.

The steps of the proposed adaptive chaotic discrete state transition algorithm are as follows.

Step 1: Set relevant parameters such as swap factor m_a, shift factor m_b, symmetry factor m_c, substitution factor m_d, chaos factor α, and adaptive factor μ;

Step 2: Generate the reverse solution using the initialization method according to Equation (13). The fitness of the forward and reverse solutions is compared by Equation (14) to select the best initial solution;

Step 3: Update the solution by swap, shift, symmetry, and substitution transformations (Equations (9)–(12)) in turn;

Step 4: According to Equation (17), judge whether the adaptive recovery strategy is satisfied, and if so, assign the optimal historical value to the current solution;

Step 5: Determine whether the condition of the chaotic perturbation strategy is satisfied. If so, generate the chaotic variable according to Equation (15), and the perturbation solution X_{new} according to Equation (16) for the next iteration. Otherwise, go to step 6.

Step 6: Determine whether the iteration termination condition is satisfied. If so, terminate the search and output the final optimization result. Otherwise, return to step 3.

3.4. Validation of ACDSTA

Many applications in economics, chemistry, manufacturing, and other fields can be transformed into unconstrained integer optimization problems. To verify the feasibility, superiority, and applicability of ACDSTA, three functions of unconstrained integer optimization problems [37] are selected for experiments, denoted by $EXP1$, $EXP2$, and $EXP3$, and expressed as follows, with respective optimal values of −620, −70,429, and −1,439,658.

$$[EXP1]\begin{cases} \min f(x) = \frac{1}{2}x^T Q x + c^T x \\ s.t \quad x \in \{0,1,2,\ldots,10\}^8 \end{cases} \tag{18}$$

where

$$Q = \begin{pmatrix} 4 & -2 & -3 & 0 & 1 & 4 & 5 & -2 \\ -2 & -4 & 0 & 0 & 2 & 2 & 0 & 0 \\ -3 & 0 & 8 & -2 & 0 & 3 & 4 & 0 \\ 0 & 0 & -2 & -4 & 4 & 4 & 0 & 1 \\ 1 & 2 & 0 & 4 & 100 & 2 & 0 & -2 \\ 4 & 2 & 3 & 4 & 2 & 100 & 1 & 0 \\ 5 & 0 & 4 & 0 & 0 & 1 & 200 & 4 \\ -3 & 0 & 0 & 1 & -2 & 0 & 4 & 10 \end{pmatrix},$$

$$c^T = (-4 \quad 1 \quad -8 \quad 3 \quad -100 \quad -10 \quad -20 \quad 0).$$

$$[EXP2]\begin{cases} \min f(x) = x^T Q x \\ s.t \quad x \in \{0,1,2,\ldots,49\}^{10} \end{cases} \tag{19}$$

where

$$Q = \begin{pmatrix} -1 & -2 & 2 & 8 & -5 & 1 & -4 & 0 & 0 & 8 \\ -2 & 2 & 0 & -5 & 4 & -4 & -4 & -5 & 0 & -5 \\ 2 & 0 & 2 & -3 & 7 & 0 & -3 & 7 & 5 & 0 \\ 8 & -5 & -3 & -1 & -3 & -1 & 7 & 1 & 7 & 2 \\ -5 & 4 & 7 & -3 & 1 & 0 & -4 & 2 & 4 & -2 \\ 1 & -4 & 0 & -1 & 0 & 1 & 9 & 5 & 2 & 0 \\ -4 & -4 & -3 & 7 & -4 & 9 & 3 & 1 & 2 & 0 \\ 0 & -5 & 7 & 1 & 2 & 5 & 1 & 0 & -3 & -2 \\ 0 & 0 & 5 & 7 & 4 & 2 & 2 & -3 & 2 & 3 \\ 8 & -5 & 0 & 2 & -2 & 0 & 0 & -2 & 3 & 3 \end{pmatrix}.$$

$$[EXP3]\begin{cases} \min f(x) = x^T Q x + c^T x \\ s.t \quad x \in \{0,1,2,\ldots,99\}^{20} \end{cases} \tag{20}$$

where

$$Q = \begin{pmatrix}
-3 & 7 & 0 & -5 & 1 & 1 & 0 & 2 & -1 & -1 & -9 & 3 & 5 & 0 & 0 & 1 & 7 & -7 & -4 & -6 \\
7 & 0 & -5 & 1 & 1 & 0 & 2 & -1 & -1 & -9 & 3 & 5 & 0 & 0 & 1 & 7 & -7 & -4 & -6 & -3 \\
0 & -5 & 1 & 1 & 0 & 2 & -1 & -1 & -9 & 3 & 5 & 0 & 0 & 1 & 7 & -7 & -4 & -6 & -3 & 7 \\
-5 & 1 & 1 & 0 & 2 & -1 & -1 & -9 & 3 & 5 & 0 & 0 & 1 & 7 & -7 & -4 & -6 & -3 & 7 & 0 \\
1 & 1 & 0 & 2 & -1 & -1 & -9 & 3 & 5 & 0 & 0 & 1 & 7 & -7 & -4 & -6 & -3 & 7 & 0 & -5 \\
1 & 0 & 2 & -1 & -1 & -9 & 3 & 5 & 0 & 0 & 1 & 7 & -7 & -4 & -6 & -3 & 7 & 0 & -5 & 1 \\
0 & 2 & -1 & -1 & -9 & 3 & 5 & 0 & 0 & 1 & 7 & -7 & -4 & -6 & -3 & 7 & 0 & -5 & 1 & 1 \\
2 & -1 & -1 & -9 & 3 & 5 & 0 & 0 & 1 & 7 & -7 & -4 & -6 & -3 & 7 & 0 & -5 & 1 & 1 & 0 \\
-1 & -1 & -9 & 3 & 5 & 0 & 0 & 1 & 7 & -7 & -4 & -6 & -3 & 7 & 0 & -5 & 1 & 1 & 0 & 2 \\
-1 & -9 & 3 & 5 & 0 & 0 & 1 & 7 & -7 & -4 & -6 & -3 & 7 & 0 & -5 & 1 & 1 & 0 & 2 & 1 \\
-9 & 3 & 5 & 0 & 0 & 1 & 7 & -7 & -4 & -6 & -3 & 7 & 0 & -5 & 1 & 1 & 0 & 2 & 1 & 2 \\
3 & 5 & 0 & 0 & 1 & 7 & -7 & -4 & -6 & -3 & 7 & 0 & -5 & 1 & 1 & 0 & 2 & 1 & 2 & 3 \\
5 & 0 & 0 & 1 & 7 & -7 & -4 & -6 & -3 & 7 & 0 & -5 & 1 & 1 & 0 & 2 & 1 & 2 & 3 & 9 \\
0 & 0 & 1 & 7 & -7 & -4 & -6 & -3 & 7 & 0 & -5 & 1 & 1 & 0 & 2 & 1 & 2 & 3 & 9 & 4 \\
0 & 1 & 7 & -7 & -4 & -6 & -3 & 7 & 0 & -5 & 1 & 1 & 0 & 2 & 1 & 2 & 3 & 9 & 4 & -1 \\
1 & 7 & -7 & -4 & -6 & -3 & 7 & 0 & -5 & 1 & 1 & 0 & 2 & 1 & 2 & 3 & 9 & 4 & -1 & -3 \\
7 & -7 & -4 & -6 & -3 & 7 & 0 & -5 & 1 & 1 & 0 & 2 & 1 & 2 & 3 & 9 & 4 & -1 & -3 & 9 \\
-7 & -4 & -6 & -3 & 7 & 0 & -5 & 1 & 1 & 0 & 2 & 1 & 2 & 3 & 9 & 4 & -1 & -3 & 9 & 7 \\
-4 & -6 & -3 & 7 & 0 & -5 & 1 & 1 & 0 & 2 & 1 & 2 & 3 & 9 & 4 & -1 & -3 & 9 & 7 & -9 \\
-6 & -3 & 7 & 0 & -5 & 1 & 1 & 0 & 2 & 1 & 2 & 3 & 9 & 4 & -1 & -3 & 9 & 7 & -9 & 8
\end{pmatrix},$$

$$c^T = (-5\ 2\ -1\ -3\ 5\ 4\ -1\ 0\ 9\ 4\ 7\ -4\ 3\ 5\ 8\ -1\ 1\ 5\ -6\ 9).$$

The algorithm was simulated using an Intel Core i5-11300H CPU at 3.10 GHz, with 16.00 GB RAM, Windows 10, and MATLAB R2018a, and compared with particle swarm optimization (PSO), DSTA, and DSTAI [32], based on the results of each group of experiments running 20 times.

The maximum numbers of iterations of $EXP1$–$EXP3$ were set to 100, 500, and 1000, respectively. The remaining parameters were set as follows: PSO learning probability $c_1 = c_2 = 1.5$, initial population 120, inertia weight 0.8. The ACDSTA parameters were set to $SE = 30$, $m_a = 2$, $m_b = 1$, $m_c = 0$, $m_d = 1$, and $m_{max} = 20$. The parameter settings of DSTA and DSTAI were the same as in Zhou et al.'s study [32].

The average error, average value, and accuracy of optimization $(S/20)$ were selected as performance evaluation indices of each algorithm. Table 1 compares the results of quadratic integer programming problems $EXP1$, $EXP2$, and $EXP3$, where the optimal values of each function index are in boldface.

Table 1. Results of unconstrained integer optimization.

Algorithm	Index	EXP1	EXP2	EXP3
PSO	average error average value S/20	**0** **−620** **20/20**	7.6% −64,907 11/20	18.7% −1,170,653 2/20
DSTA	average error average value S/20	82.9% −106 0/20	62.6% −26,264 0/20	58.1% −603,975 0/20
DSTAI	average error average value S/20	**0** **−620** **20/20**	1.6% −69,148 18/20	8.4% −1,319,270 7/20
ACDSTA	average error average value S/20	**0** **−620** **20/20**	**0** **−70,429** **20/20**	**2.2%** **−1,408,107** **15/20**

Figure 1 compares the convergence curves of $EXP1$, $EXP2$, and $EXP3$ under four optimization algorithms.

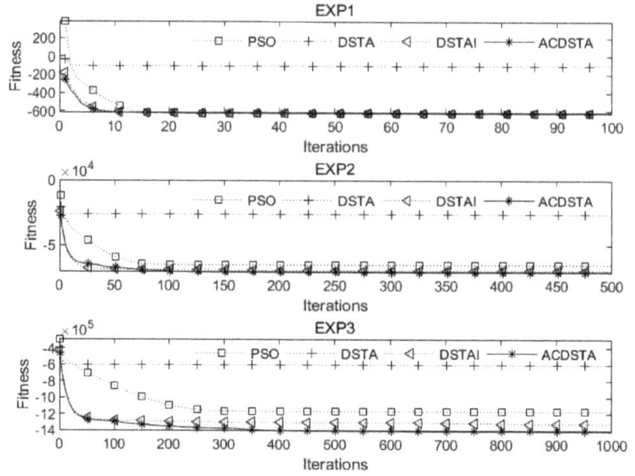

Figure 1. Convergencecurves of EXP1, EXP2, and EXP3 on four algorithms.

From Table 1 and Figure 1, we can see the following: (1) For the unconstrained integer optimization problem $EXP1$ with a low dimension, PSO, DSTAI, and ACDSTA can all find the optimal solution, and all the indices are the best; (2) For the unconstrained integer optimization problems $EXP2$ and $EXP3$ with higher dimensions, although PSO and DSTAI can also find the optimal solution, each index of ACDSTA is better. Among them, the optimization accuracies of $EXP3$ were improved by 65%, 75%, and 40% compared with PSO, DSTA, and DSTAI, respectively, and the respective average values were 20.3%, 133.1%, and 6.7% higher. Moreover, the average error was only 2.19%, which is significantly lower than for the other comparison algorithms. This indicates that the higher the dimensionality of the integer optimization problem and the larger the search space of the solution, the more prominent the advantage of ACDTSA; (3) Whether low-dimensional $EXP1$ or high-dimensional $EXP3$, ACDSTA was superior to the other three algorithms in terms of initial average fitness and convergence speed and could find the optimal value faster. This shows that ACDSTA can approach the optimal value faster and improve convergence performance through the opposition-based learning strategy and adaptive recovery strategy; (4) For $EXP3$, the optimization accuracy of ACDSTA was better than that of the other three algorithms. Although there was a tendency to fall into local optima in late iterations, it could effectively jump out of local optima and obtain better optimization accuracy because of the chaos perturbation strategy.

In summary, ACDSTA was superior to the other algorithms in the test of unconstrained integer optimization problems, especially when the dimension and optimization range were extensive, which can better reflect the advantages of ACDSTA. This demonstrates the feasibility, superiority, and applicability of ACDSTA when solving practical engineering problems.

4. Application of ACDSTA in Rare Earth Extraction Process

4.1. Rare Earth Extraction Process Analysis

Rare earth extraction is the obtaining of a single rare earth product from rare earth liquid, extractant, and detergent through specific equipment. We use the praseodymium/ neodymium (Pr/Nd) extraction and separation process as an example, as shown in Figure 2.

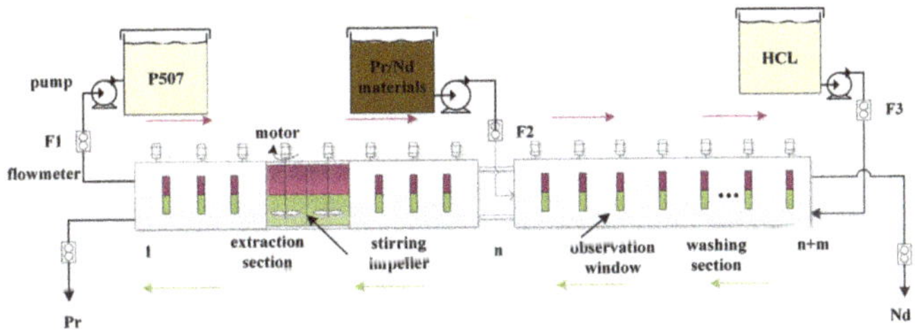

Figure 2. Schematic Diagram of Rare Earth Extraction Process.

The extractant is added from the first stage, and it flows from left to right through the stirrer. The detergent is added from the $n + m$ stage and flows from right to left. The organic phase is added to the feed liquid from the nth stage. The solution in the tank is divided into two layers by stirring and clarifying. The upper layer is the organic phase, and the lower layer is the water phase. The difficult extraction product Pr is obtained in the lower layer of the first stage and the easy extraction product Nd from the upper layer of the $n + m$ stage.

There are many extraction stages, and it is necessary to control the content distribution of Pr/Nd components in the tank to ensure the stability of product quality. It takes several hours or more to cause changes in the content of the export grade Pr/Nd component due to changes in the flow rate of the feed solution, extractant, and detergent. Therefore, it is necessary to obtain the residence time of materials in each unit group for timely control.

4.2. Time Delay Identification of Rare Earth Extraction Process

To verify the ability of the proposed method to solve the engineering delay problem, the delay identification of the 30-stage Pr/Nd production process in a rare earth extraction plant was carried out. During the Pr/Nd extraction process, the content of Pr/Nd components in different tanks changes with time, which leads to a change in the color of the solution. Therefore, the characteristic color variable of the solution image can be selected as an auxiliary variable to identify the time delay.

The sampling period was 5 min, and 220 groups of data of Nd component content and color characteristic variables (H, S, I, R, G, B) were selected in the continuous stable production process. The grey correlation method was used to analyze the correlation between Nd component content and color characteristic variables. The results are shown in Table 2, where the B component has the highest correlation degree, and the H component has the lowest correlation degree. Therefore, the B component is regarded as the critical process variable. In the actual extraction site, every five stages of the extraction tank share a set of agitators, i.e., every five stages constitute a unit group. Therefore, the 30-stage extraction tank was constructed as six groups of units for identification. According to the flow direction of the extractant, the first-stage inlet sampling data and each group of outlet sampling data were recorded as $e_0, \ldots, e_6$, and the original data matrix E was obtained.

Table 2. Results of GRA.

Color Feature	R	G	B	H	S	I
correlation degree	0.6179	0.5832	**0.6734**	0.543	0.5706	0.6123

According to the operation experience of the extraction separation site, the time delay range of stirring and clarification of each stage extraction tank is $[3, 8]$ min. Given the abovementioned construction method, the time delay range of each unit group is $[15, 40]$ min. Therefore, the value range of the time-base sequence is $[3, 8]$. According to Equation (4), the time-correlation data matrix E is constructed, and the time delay sequence solution is transformed into the optimization problem by Equations (5) and (6), as shown in Equation (7).

ACDSTA is used to solve the abovementioned optimization problem. A certain number of time-based sequences is generated according to the range of time-based values, and the fitness function is constructed according to Equation (7). A new sequence is generated after the operation of the time-based sequence by the swap, shift, symmetry, and substitution operators. At the same time, the fitness value is calculated, and the current optimal value and the optimal time-based sequence are retained. The abovementioned operation is repeated until the iteration termination condition is satisfied, and the global optimal fitness value and global optimal time-based sequence are obtained. The solution process of ACDSTA is shown in Figure 3. Here, the maximum number of iterations is set to 100, $SE = 5$, and the remaining parameters are set according to Section 3.4.

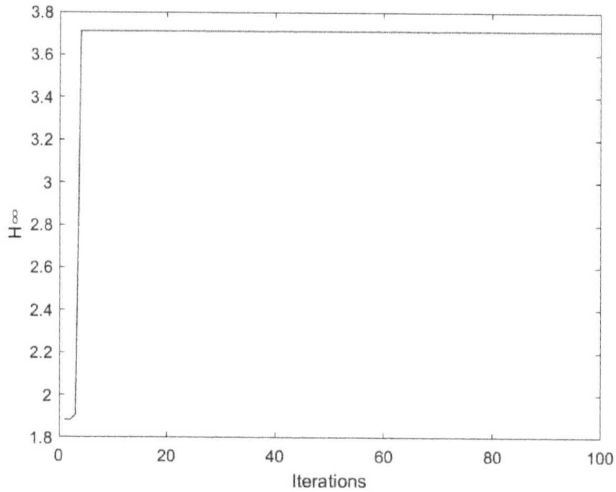

Figure 3. Iterative curve of improved discrete state transition algorithm.

As seen from Figure 3, ACDSTA can obtain the optimal value $H_\infty = 3.711$ in the early stage of iteration, and the corresponding time-base sequence is [3 6 4 5 4 4]. Since the sampling period is 5 min, the time delay of the identified unit group is [15 30 20 25 20 20], i.e., the 30-stage Pr/Nd extraction production process and the time delay identification result of each stage of the extraction tank are [3 3 3 3 3 6 6 6 6 6 4 4 4 4 4 5 5 5 5 5 4 4 4 4 4 4 4 4 4 4].

4.3. Time Delay Identification Results and Method Verification

To verify the accuracy of the improved time delay identification method, we conducted the following experiments. Firstly, the characteristic components H, S, and I of the solution image of the rare earth extraction tank are considered auxiliary variables. A prediction model of Nd component content that meets the requirements of the extraction site is established by the wavelet neural network and used as a verification model. Then the maximum relative error ($MAXRE$), mean relative error ($MEANRE$), and mean absolute error (MAE) are determined to measure the performance of the model,

$$MAXRE = MAX\left(\frac{|z - z'|}{z'} \times 100\%\right) \tag{21}$$

$$MEANRE = \frac{1}{n}\sum_{i=1}^{n} \frac{|z - z'|}{z'} \times 100\% \tag{22}$$

$$MAE = \frac{1}{n}\sum_{i=1}^{n} |z - z'| \tag{23}$$

where z is the predicted component content value of the wavelet neural network, and z' is the actual component content value. Finally, the following comparative experiments are designed. Based on the data processed by the improved time delay identification method (Improved method), time-correlation analysis method (Original method), and Unused method, the performance of the component content model based on wavelet neural network (WNN) is compared and analyzed.

WNN is a neural network based on wavelet analysis theory that combines the excellent time-frequency localization property of wavelet function and the powerful self-learning function of the neural network. WNN uses the wavelet basis function as the activation function, which has a strong prediction ability than the back propagation neural network. Lu et al. [38] showed a nonlinear relationship between the color characteristic components H, S, and I of the rare earth extraction solution image and Nd component content. Therefore, we use the WNN to model the component content of the rare earth extraction process, using the Morlet function as the wavelet basis function.

$$y = cos(1.75x)^{\left(-0.5x^2\right)} \tag{24}$$

The parameter settings are as follows. The learning probability is 0.01, the momentum factor is 0.001, and the maximum number of iterations is 1000. The data samples consist of 220 groups, and 190 groups are randomly selected for model training. The remaining are used to verify the model, as shown in Figure 4.

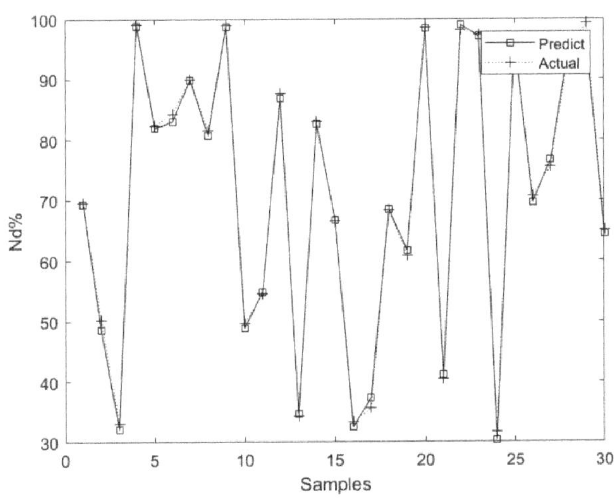

Figure 4. Prediction model result of wavelet neural network.

The maximum relative error of the model prediction is 4.76%, which meets the accuracy requirements of the rare earth extraction site for the prediction model, so it can be used as a verification model for time delay identification.

Table 3 shows the model evaluation indices based on the Unused, Original, and Improved methods, where the bold data are the optimal values, and the corresponding relative error curve is shown in Figure 5.

Table 3. Model evaluation indices of different methods.

Method	MAXRE%	MEANRE%	MAE
Unused	7.96	1.63	0.9183
Original	5.08	1.33	0.8738
Improved	**1.69**	**0.48**	**0.362**

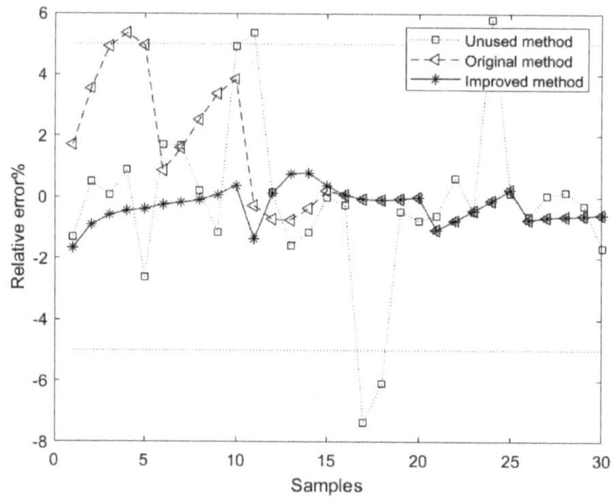

Figure 5. Forecast relative error of different methods.

From Table 3 and Figure 5, we can see the following: (1) The performance of the component content model based on the Original method is better than that based on the Unused method. However, its maximum relative error is greater than 5%, which does not meet actual requirements. This shows that although the Original method can improve the model's performance to a certain extent, due to the randomness of its data selection and the lack of data preprocessing, the method fails to accurately identify the real-time delay. (2) The mean relative error of the component content model based on the Improved method is better than that based on the Unused and Original methods, which is reduced by 70.1% and 63.9%, respectively. This shows that the model based on the improved method is more stable. (3) Compared with the Unused and Original methods, the mean absolute error of the model based on the improved method is reduced by 60.6% and 58.6%, respectively, indicating that the prediction accuracy of the model is higher.

In summary, the improved method significantly improved the model's performance. This shows that the improved method based on grey correlation analysis can effectively select the data closest to the real-time delay and improve the accuracy of the delay identification results. At the same time, the maximum relative error of the model based on the improved method is less than 5%, which meets the actual requirements. This shows that the improved time delay identification method proposed in this paper is suitable for the time delay identification of rare earth extraction process.

5. Conclusions

Rare earth extraction is a typical nonlinear, large-time-delay industrial process. The existence of time delay precludes the effective use of much field data and leads to a gap between the model describing the rare earth extraction process and the actual situation. We did the following work in response to this problem: Based on the standard discrete state transition algorithm, an improved algorithm (ACDSTA) was proposed, using an opposition-based learning strategy to initialize and accelerate the convergence of the algorithm, and an adaptive recovery strategy to improve its convergence performance. A chaotic perturbation strategy can improve the ability of the algorithm to jump out of local optima. An experimental comparison with three unconstrained integer optimization problems showed that ACDSTA can effectively solve such problems, and verified its effectiveness, superiority, and applicability; An improved time delay identification method was proposed to solve the problem that the data are not preprocessed and are randomly selected in the time-correlation analysis method. The method was applied to the time delay identification of a rare earth extraction process. The superiority and effectiveness of the proposed improved time delay identification method were verified by comparing the time-correlation analysis method and the data without the identification method under the same Nd component content. To sum up, the proposed time delay can provide a reference for modeling the rare earth extraction process and has guiding significance for the improvement of the online detection accuracy of component content.

Author Contributions: H.L., R.L. and H.Y. conceived and designed the study. H.L. and R.L. conducted the theoretical analysis and wrote the manuscript. H.L. performed the numerical simulation and the experiment. J.Z. and W.D. helped to revise the manuscript. All authors have read and agreed to the published version of the manuscript.

Funding: This research was funded by the National Key R&D Program of China (No. 2020YFB1713700), the National Natural Science Foundation of China (No. 61863014, No. 61733005, and No. 61963015), the Open Fund of State Key Laboratory of Synthetical Automation for Process Industries, Northeastern University (No. 2021-KF-21-01).

Institutional Review Board Statement: Not applicable.

Informed Consent Statement: Not applicable.

Data Availability Statement: Not applicable.

Conflicts of Interest: The authors declare no conflict of interest.

References

1. Jia, W.; Chai, T. Bilinear model of rare earth cascade extraction process and its parameter identification. *Control Theory Appl.* **2006**, *23*, 717–723.
2. Chan, Y.Y.; Lee, C.; Lee, G.G.; Jo, S.; Su, W.S. Modeling and simulation of multicomponent solvent extraction processes to purify rare earth metals. *Hydrometallurgy* **2016**, *159*, 40–45.
3. Yang, H.; Dai, W.; Lu, R.; Zhu, J. Simulation of rare earth extraction process based on separation coefficient correction. *CIESC J.* **2020**, *71*, 3180–3190.
4. Wang, Q.; Zhang, Y. Robust identification of continuous systems with dead-time from step responses. *Automatica* **2001**, *37*, 377–390. [CrossRef]
5. Salim, A. Step response-based identification of fractional order time delay models. *Circuits Syst. Signal Process.* **2020**, *39*, 1–17.
6. Rad, A.B.; Lo, W.L.; Tsang, K.M. Simultaneous online identification of rational dynamics and time delay: A correlation-based approach. *IEEE Trans. Control Syst. Technol.* **2003**, *11*, 957–959. [CrossRef]
7. Li, J.; Zheng, Y.; Lin, Z. Recursive identification of time-varying systems: Self-tuning and matrix RLS algorithms. *Syst. Control Lett.* **2014**, *66*, 104–110. [CrossRef]
8. Wang, X.; Ding, F. Recursive parameter and state estimation for an input nonlinear state space system using the hierarchical identification principle. *Signal Process.* **2015**, *117*, 208–218. [CrossRef]
9. You, K. Recursive algorithms for parameter estimation with adaptive quantizer. *Automatica* **2015**, *52*, 192–201. [CrossRef]
10. Lu, Y.; Du, J.; Li, C. Neural networks compensate control for unknown systems with time delay. *J. Tsinghua Univ. (Sci. Technol.)* **1998**, *38*, 67–69.
11. Saki, S.; Fatehi, A. Neural network identification in nonlinear model predictive control for frequent and infrequent operating points using nonlinearity measure. *ISA Trans.* **2020**, *97*, 216–229. [CrossRef]
12. Zhao, X.; Shen, S.; Su, L.; Yin, X. Elman neural network–based identification of rate-dependent hysteresis in piezoelectric actuators. *J. Intell. Mater. Syst. Struct.* **2020**, *31*, 980–989. [CrossRef]
13. Liu, Y.; Tao, T. A CS recovery algorithm for model and time delay identification of MISO-FIR systems. *Algorithms* **2015**, *8*, 743–753. [CrossRef]
14. Chen, J.; Ma, J.; Liu, Y.; Ding, F. Identification methods for time-delay systems based on the redundant rules. *Signal Process.* **2017**, *137*, 192–198. [CrossRef]
15. Wang, Z.; Wang, C.; Ding, L.; Wang, Z.; Liang, S. Parameter identification of fractional-order time delay system based on Legendre wavelet. *ISA Trans.* **2022**, *163*, 108141. [CrossRef]
16. Hofmann, J.; Ponomarev, A.; Hagenmeyer, V.; Gröll, L. Time-delay identification and validation of a liquid film transport model based on pilot plant experiments. *IFAC-PapersOnLine* **2021**, *54*, 401–408. [CrossRef]
17. Prasad, V.; Mehta, U. Modeling and parametric identification of Hammerstein systems with time delay and asymmetric dead-zones using fractional differential equations. *Mech. Syst. Signal Process.* **2022**, *167*, 108568. [CrossRef]
18. Li, L.; Zhang, H.; Ren, X. Robust adaptive identification for sandwich systems with unknown time-delay. *ISA Trans.* **2020**, *100*, 289–298. [CrossRef]
19. Liu, G.; Yu, M.; Wang, L.; Yin, Z.; Liu, J.; Lu, Z. Rapid parameter identification of linear time-delay system from noisy frequency domain data. *Appl. Math. Model.* **2020**, *83*, 736–753. [CrossRef]

20. Li, Z.; Wang, P.; Zhang, J.; Guan, H. A model-free method for identifying time-delay characteristics of HVAC system based on multivariate transfer entropy. *Build. Environ.* **2022**, *217*, 109072. [CrossRef]
21. Ni, J.; Zhang, Y.; Deng, F.; Zhan, X.; Hayat, T. Parameter estimation algorithms of linear systems with time-delays based on the frequency responses and harmonic balances under the multi-frequency sinusoidal signal excitation. *Signal Process.* **2021**, *181*, 107904. [CrossRef]
22. Ishida, A.; Miyazaki, S. Maximum likelihood identification of a posture control system. *IEEE Trans. Biomed. Eng.* **1987**, *34*, 1–5. [CrossRef] [PubMed]
23. Drakunov, S.V.; Perruquetti, W.; Richard, J.P.; Belkoura, L. Delay identification in time-delay systems using variable structure observers. *Annu. Rev. Control* **2006**, *30*, 143–158. [CrossRef]
24. Tang, Y.; Guan, X. Parameter estimation for time-delay chaotic system by particle swarm optimization. *Chaos Solitons Fractals* **2017**, *40*, 1391–1398. [CrossRef]
25. Xie, Y.; Zhang, J.; Huang, C.; Yang, C. An improved genetic algorithm for multi-delay times identification. In Proceedings of the 2011 Chinese Control and Decision Conference, Mianyang, China, 23 May 2011; pp. 3290–3293.
26. Huang, C.; Gui, W.; Xie, Y.; Yang, C. Multi-delays identification for alumina carbonation decomposition process based on improved cross-correlation function. *Chin. J. Nonferrous Met.* **2011**, *21*, 1186–1191.
27. Wang, Y.; Xia, H.; Yuan, X.; Gui, W. Multi-delay identification by trend-similarity analysis and its application to hydrocracking process. *CIESC J.* **2018**, *69*, 1149–1157.
28. Wang, F.; Wang, X.; Xie, Y.; Xie, S.; Yang, C. Multi-delays identification for alumina evaporation process based on time-correlation analysis. *CIESC J.* **2017**, *68*, 992–997.
29. Yang, C.; Tang, X.; Zhou, X.; Gui, W. A discrete state transition algorithm for traveling salesman problem. *Control Theory Appl.* **2013**, *30*, 1040–1046.
30. Zhou, X.; Yang, C.; Gui, W. The principle of state transition algorithm and its applications. *Acta Autom. Sin.* **2020**, *46*, 2260–2274.
31. Zhang, S.; Guo, J.; Luo, N.; Zhang, D.; Wang, W.; Wang, L. A calibration-free method based on grey relational analysis for heterogeneous smartphones in fingerprint-based indoor positioning. *Sensors* **2019**, *19*, 3885. [CrossRef]
32. Zhou, X.; Gao, D.Y.; Yang, C.; Gui, W. Discrete state transition algorithm for unconstrained integer optimization problems. *Neurocomputing* **2016**, *173*, 864–874. [CrossRef]
33. Wang, M.; Chen, L.; Chen, H. Multi-strategy learning boosted colony predation algorithm for photovoltaic model parameter identification. *Sensors* **2022**, *22*, 8281. [CrossRef]
34. Zheng, F.; Liu, G. An adaptive sinusoidal-disturbance-strategy sparrow search algorithm and its application. *Sensors* **2022**, *22*, 8787. [CrossRef]
35. Zhou, Y.; Bao, L.; Chen, C.P. A new 1D chaotic system for image encryption. *Signal Process.* **2014**, *97*, 172–182. [CrossRef]
36. Dong, Y.; Zhang, H.; Wang, C. State transition algorithm with strategy adaptation. *Control Decis.* **2022**, *37*, 574–582.
37. Wu, Z.Y.; Li, G.Q.; Quan, J. Global optimality conditions and optimization methods for quadratic integer programming problems. *J. Glob. Optim.* **2011**, *51*, 549–568.
38. Lu, R.; Lai, L.; Yang, H.; Zhu, J. Prediction of CePr/Nd component content based on virtual sample generation. *Transducer Microsyst. Technol.* **2022**, *41*, 152–156.

sensors MDPI

Article

Enhancing the Generalization for Text Classification through Fusion of Backward Features

Dewen Seng *,† and Xin Wu †

School of Computer Science and Technology, Hangzhou Dianzi University, Hangzhou 310005, China
* Correspondence: sengdw@hdu.edu.cn
† These authors contributed equally to this work.

Abstract: Generalization has always been a keyword in deep learning. Pretrained models and domain adaptation technology have received widespread attention in solving the problem of generalization. They are all focused on finding features in data to improve the generalization ability and to prevent overfitting. Although they have achieved good results in various tasks, those models are unstable when classifying a sentence whose label is positive but still contains negative phrases. In this article, we analyzed the attention heat map of the benchmarks and found that previous models pay more attention to the phrase rather than to the semantic information of the whole sentence. Moreover, we proposed a method to scatter the attention away from opposite sentiment words to avoid a one-sided judgment. We designed a two-stream network and stacked the gradient reversal layer and feature projection layer within the auxiliary network. The gradient reversal layer can reverse the gradient of features in the training stage so that the parameters are optimized following the reversed gradient in the backpropagation stage. We utilized an auxiliary network to extract the backward features and then fed them into the main network to merge them with normal features extracted by the main network. We applied this method to the three baselines of TextCNN, BERT, and RoBERTa using sentiment analysis and sarcasm detection datasets. The results show that our method can improve the sentiment analysis datasets by 0.5% and the sarcasm detection datasets by 2.1%.

Keywords: deep learning; text classification; two-stream networks; feature fusion; sentiment classification; sarcasm detection

Citation: Seng, D.; Wu, X. Enhancing the Generalization for Text Classification through Fusion of Backward Features. *Sensors* **2023**, *23*, 1287. https://doi.org/10.3390/s23031287

Academic Editors: Peter Hockicko, Róbert Hudec and Patrik Kamencay

Received: 3 December 2022
Revised: 16 January 2023
Accepted: 18 January 2023
Published: 23 January 2023

1. Introduction

Text classification is an essential and vital branch of the natural language process (NLP). It has received widespread attention from many scholars who utilize neural networks to extract high-quality semantic features from inputs such as sentences and documents. The most classic model among the neural network models is the convolutional neural network [1]. This model can extract highly representative semantic features for classification. Although the CNN-based model can effectively capture local- and fixed-position features, its accuracy still needs to be improved. The most recent model is BERT, proposed by Kenton et al. [2]. BERT and RoBERTa [3] use pretrained technology on large datasets to capture the universal information to improve the generalization ability.

However, even if these algorithms could achieve a state-of-the-art performance, there would still be room for improvement. From the perspective of features, these algorithms make mistakes in specific sentences. For example, when the trained RoBERTa judges the sentence, a "charming and funny (but ultimately silly) movie", RoBERTa classifies it as a negative comment. However, it is a positive comment. We analyzed the attention heat map and can assume that RoBERTa pays more attention to the phrase "but ultimately silly" rather than "charming and funny", which is the reason for the mistake. The heat map of RoBERTa's analysis of this sentence is shown in Figure 6.

To address this point, our study follows the research ideas and processes presented in this paper to allow models to be able to scatter the attention of opposite sentiment words given to core sentiment words. In this paper, we utilize a two-stream network structure to further study the relative differences between the backward features of the auxiliary network and normal features of the main network to improve the representation ability of feature vectors. For example, we first extract backward features through an auxiliary network and then use a feature projection layer to obtain the extra vector, which has the same direction as the normal feature. After that, to leverage the balance between normal feature vectors and extra vectors, we study a method to aggregate normal features and extra vectors that are projected after the projection layer. We assume that such a design has the potential to analyze more information about the context through end-to-end training.

Inspired by the feature purification network (FP-Net) [4], we propose a method called the feature augmentation network (FA-Net). Separately, F-Net refers to the main network, and A-Net refers to the auxiliary network. The A-Net uses the gradient reversal layer [5] to extract the backward features that contain backward contextual information. Meanwhile, the F-Net is a normal network such as a CNN or BERT. This means the major work carried out by the F-Net is meant to extract normal feature vectors. Before feeding the normal vector into the classifier, we calculate an extra vector through feature projection that is in the same direction as the normal vectors. After that, we concatenate normal vectors and extra vectors together, creating a new feature vector. Finally, the model feeds a new vector into the classifier. This study makes three main contributions:

- The parameters of our model are acceptable. Even in the BERT-based model, our model only has one or two additional encoder layers.
- Our algorithm is efficiently utilized at different benchmarks, such as with the CNN and BERT, and it is not conflicted with other operations that improve generalization capabilities.
- We analyze the influence of the auxiliary network on the attention score of the main network, expressing the efficiency of the auxiliary network through the attention heat map.

To better explain the proposed methods, we introduce relevant research on text classification and feature fusion and briefly describe their practices in Section 2. In Section 3, we describe the six open-source datasets and three open-source models that are used as the experiment materials. We also focus on introducing our model's structure in Section 3. Later, we list our experimental data in Section 4. To show the effectiveness of our method, we implement it on sentiment analysis and sarcasm detection datasets. We list the average results under five seeds and illustrate the stability of the model through deviation. We also analyze our experimental results and discuss the projection type, the number of subnetworks, and the type of subnetworks. We also prove that our idea is consistent with the hypothesis from the perspective of the attention heat map. In Section 5, we explain our conclusions and future prospects.

2. Related Works

The well-known RNN model used for text classification is long short-term memory (LSTM) [6,7]. LSTM uses the forget gate to choose whether to retain the previous information or not. Thus, LSTM is good at processing long-term-dependent input. However, compared to CNNs, LSTM does not run fast, causing some scholars to turn their attention to the CNN models, which can operate fast and parallel to the training stage.

The TextCNN [1] sets fixed filter sizes that work on embedded vectors to capture context information. Then, the maxpooling focus is used on salience features. To obtain more information on these features, Wang et al. [8] proposed a method that uses a concentration mechanism to pick out the key features for short text classification. However, the problem with TextCNN is that it is hard to obtain the long-term information because of the n-gram mechanism of convolution filters, which can only operate on several consecutive words simultaneously. Therefore, Lai et al. [9] combined the advantages of the TextCNN and RNN and designed the TextRCNN algorithm. They utilized the Bi-RNN to build the left and right contexts, then concatenated those vectors with embedding vectors to feed them into one neural layer and generate the latent semantic vector and finally extracted the feature through max pooling, as is performed via the TextCNN. After the TextCNN, many advent approaches related to neural networks were proposed, such as, e.g., the DCNN [10], HAN [11]. The DCNN uses dynamic k-max pooling to extract long-distance features that were separated dynamically. The HAN uses a multilayer recurrent neural network and an attention mechanism to extract long-sentence semantic features.

Although some people are still studying the RNN and CNN model structures for text classification, the naive attention [12] mechanism was proposed. After this, many variants of attention mechanisms appeared, such as local attention, global attention, and soft attention. The most effective attention mechanism is self-attention. As proposed by Vaswani et al. [13], this model effectively reduces the calculation cost by parallel computing the attention score of each word in the text or document. Furthermore, based on transformer and attention mechanisms, some scholars utilized other fields' technologies to improve the model's performance. BERT combines pretrained technology, transformer encoders, and the training of a vast corpus to extract more comprehensive feature vectors and achieve an SOTA performance for a wide range of tasks. Based on BERT, some scholars proposed more effective models, such as RoBERTa. This method changed static masking to the dynamic masking of sentences and removed next-sentence prediction (NSP); thus, its main difference with BERT is in the pretraining stage. Indeed, this way improves the performance of BERT.

Due to the highly expressive ability of BERT's encoder, many scholars take the features extracted by the encoder as their research focus. Qin et al. [4] proposed a feature projection layer to eliminate the redundant information of features and improve the quality of features. G Niu et al. [14] proposed a new Encoder1–Encoder2 structure, where Encoder1 is a global information extractor and Encoder2 is a local information extractor. The global information vectors are merged with the local information vectors for a higher performance. Ying et al. [15] proposed an unsupervised saliency detection approach, which utilizes an elastic-net-based hypergraph model to discover the group structure relationships of salient regional points. They also use a saliency map to obtain high-level semantic features. Then, they fused the low-level deep and high-level semantic features into a similarity matrix. Wang et al. [16] proposed a novel structure comprising three modules. One of the modules is responsible for multiscale feature alignment fusion. The other modules are focused on different scale channels and the adaptive weighted fusion of spatial locations, as well as the multiscale fusion of global and local features. Long et al. [17] proposed a method for mining the relationships between labeled and unlabeled data. They used the co-occurrence of words in all documents to build a neighbor table and use multidimensional scaling (MDS) to extract the feature representation of the adjacency table. Then, they integrated the new graph-based representation and the document–term representation as the new hybrid augmented feature representation. Huang et al. [18] added other cheap modules, called Ghost Modules, to capture more semantic information and then fused them with normal features extracted by the base model. They studied the 2D convolutional operation and 1D convolutional operation of the Ghost Modules. Additionally, they also analyzed which position was better when inserting their Ghost Modules within the transformer encoder.

3. Materials and Methods

3.1. Methods

In this paper, we mainly studied the deep learning approach to improve the quality of feature vectors. Our method has a built-in feature projection layer and gradient reversal layer in the auxiliary network. For the projection layer, in our implementation, we decomposed the backward features in two directions and chose the one that has the same direction as the normal feature, keeping in line with our assumption. Another direction is not suitable for our method because the effectiveness is not apparent. The gradient reversal layer is vital to help the model scatter the attention score of a phrase whose semantic meaning is opposite to that of the sentence. Because the gradient reversal layer can reverse the gradient of backward features, a normal feature fused with the extra feature generated by the projection layer can contain the reversed gradient information to force the model to focus on core words other than the contradictory words. We conducted our experiments under the different types and sizes of the subnetwork to show the different results. For example, we cut down the number of filter sizes in the CNN model to find an appropriate and time-effective extractor for the auxiliary network. Additionally, we did not use the convolution layer to extract extra features except in the CNN models. In contrast, we utilized the transformer encoder to extract additional features, as it can generate more explainable features than the convolution layer. The overview of the structure of our FA-Net is described in Figure 1.

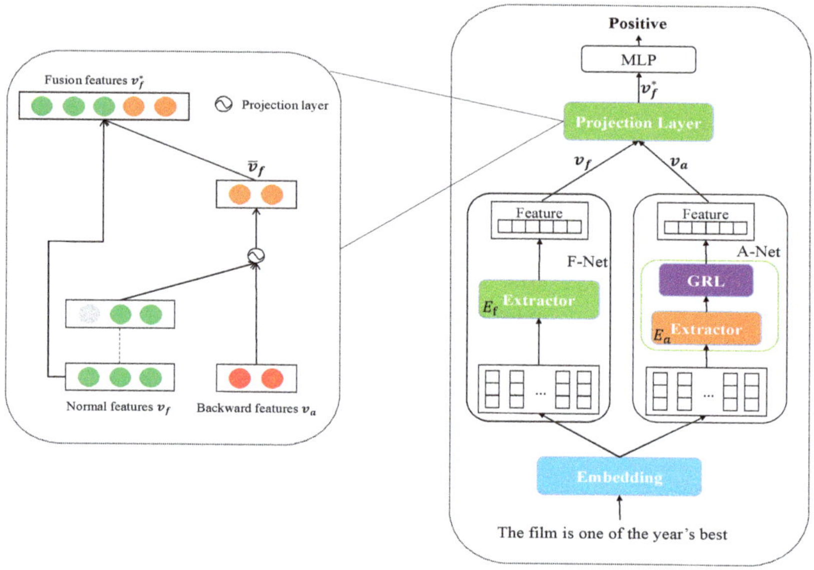

Figure 1. The architecture of FA-Net. The right structure shows the detail of the projection layer. The left one displays the whole network as part of the right boxed figure. Each component is the same except for the extractor of F-Net and A-Net. Note that the "Embedding" concludes with the blue box, but it is not the embedding layer. Rather, it is an abstract layer. The outputs of "Embedding" may be produced by the embedding layer or by one of the transformer encoder layers.

As shown in Figure 1, our network consists of two networks. The F-Net attentively extracts the normal features v_f by utilizing the extractor E_f with perturbation from the A-Net. On the other side, the A-Net is focused on extracting backward features v_a through the gradient reversal layer. In our implementations, the type of E_a is the same as that of E_f, which means the layer type of the A-Net follows that of the F-Net. If the F-Net utilizes the convolutional layer to extract the features, the A-Net does the same. The other exciting settings are the size and inputs of E_a. The size of extractor E_a is smaller than the size of E_f. The input of E_a is the cloned output from one of the encoders of E_f. After feeding the cloned feature vector into the A-Net, the output of the A-Net is entered into the projection layer with the output of the F-Net. The final feature vector is obtained through the projection layer and fed into the classification layer to generate the output of the whole network.

In the backpropagation stage, we initialized two optimizers. The optimizer of the F-Net is responsible for updating the F-Net's parameters and the parameter of the embedding layer, and the optimizer of the A-Net is accountable for updating the A-Net's parameters. To explain the algorithms directly, we list the algorithm procedure in Algorithm 1.

Algorithm 1: Feature Augmentation Network

input : Dataset $D = \{(x_i, y_i)_{i=1}^{N}\}$, and $\mathbf{X}_i$ is the embedding outputs of x_i
Initialize the paramaters θ of the model;
for *each iteration* $b = 1, 2, \ldots, M$ **do**
  Sample one batch of data from D;
  Use BERT tokenizer to tokenize the batched data;
  F-Net part:
  Generate all hidden states $\boldsymbol{V_f}$ via E_f;
  if *model is BERT-based,* **then** clone one of the hidden states $\boldsymbol{v}$ from $\boldsymbol{V_f}$;
  else clone the embedding outputs $\mathbf{X}_i$ as $\boldsymbol{v}$;
  C-Net part:
  Feed $\boldsymbol{v}$ to E_a to genrate the hidden state $\boldsymbol{v_a}$ by using Equation (6);
  $\boldsymbol{v_a}$ go through GRL by using Equations (7)–(9);
  Projection part:
  Generate $\overline{\boldsymbol{v}}_f$ by using Equation (10);
  Concatenate $\boldsymbol{v_f}$ and $\overline{\boldsymbol{v}}_f$ and then feed into Equation (13);
  Update parameters:
  Backpropagation of the gradient according to loss $\mathcal{L}$;
  Update the parameters θ_f of the F-Net;
  Update the parameters θ_a of the A-Net;
end

Although our network consists of two networks, both networks share one loss function $\mathcal{L}$. The advantage of this is that the amount of time consumed by the model is reduced. The loss function of the whole network is as follows:

$$\mathcal{L} = \mathcal{L}_f = \mathcal{L}_a \tag{1}$$

We introduced the proposed method by following the structure of the FA-TextCNN as an example. The FA-TextCNN is a model that applies our method to the TextCNN. Each part is as follows:

F-Net Module: For the FA-TextCNN, a dataset $D = \{(x_i, y_i) \mid i \in 1, \ldots, N\}$ is given, where x_i is a sentence or document with the corpus length L (after padding or cutting), N is the size of the training data, and y_i is the label of x_i. Here, x_i feeds into the embedding layer with a fixed embedding size e to generate the embedded output $\mathbf{X}_i \in \mathbb{R}^{L \times e}$. Whereafter, $\mathbf{X}_i$ feeds into the feature extractor E_f with convolutional filters and n-gram to generate features v_f as follows:

$$c_i^j = f(\mathbf{W} \cdot \mathbf{X}_i[j : j + n - 1, :] + b) \tag{2}$$

$$c_i = [c_i^0, c_i^1, \ldots, c_i^{L-n}] \tag{3}$$

where $j \in 0, \ldots, L - n$ and $\mathbf{W} \in \mathbb{R}^{n \times e}$ is the weight of the convolution filter, and n is the n-gram size of each convolutional filter. Moreover, f is the active function, similar to *ReLU*. The outcome of feature f_c under the n-gram and a filter is as follows:

$$f_c = [c_0, c_1, c_2, \ldots, c_{L-n}] \tag{4}$$

After this, we used the maxpooling operation over the feature map and took a maximum value $m_f = max\{f_c\}$ as the most characteristic feature under the one filter. In our experiments, we initialized q filters, and each filter initialized m kinds of parameters. Therefore, a filter can generate one of the most characteristic values of a parameter. Finally, we concatenated those characteristic values. Obviously, the $v_f \in \mathbb{R}^{q \cdot m}$ is extracted by E_f as follows:

$$v_f = CNN_f(\boldsymbol{X_i}) \tag{5}$$

A-Net module: This module is our designed network. The output of the embedding layer $\boldsymbol{X_i}$ is cloned, and then $\boldsymbol{X_i}$ is fed into E_a, which is similar to E_f. Because the E_a is also a convolution filter that can set the n-gram and kinds of parameters, it can generate the feature v_a of $\boldsymbol{X_i}$ under a specific convolution filter:

$$v_a = CNN_a(\boldsymbol{X_i}) \tag{6}$$

Upon E_a, we innovatively stacked the GRL on the extractors of the A-Net and used the projection layer to eliminate the harmful semantic information of backward features v_a. The procedure of the gradient reversal layer is as follows:

$$GRL_\lambda(\boldsymbol{x}) = \widetilde{\boldsymbol{x}} \tag{7}$$

$$\frac{\partial GRL}{\partial \boldsymbol{x}} = -\lambda I \tag{8}$$

where λ is a hyperparameter of the gradient reversal layer, and $\widetilde{x}$ is a new feature vector passed through the gradient reversal layer. We noted that the classification of our whole model is mainly completed through the F-Net. Therefore, to reduce the influence of the A-Net at the beginning of training, we gradually increased the λ as follows:

$$\lambda = \frac{2}{1 + exp(-\gamma \cdot p)} - 1 \tag{9}$$

where γ was set to 10 in all experiments, and p represents the iteration ratio of training from 0 to 1.

When ready, v_f and v_a were both fed into the projection layer to generate $\overline{v}_f$ and $\overline{v}_a$:

$$\overline{v}_f = Proj(v_a, v_f) \tag{10}$$

$$\overline{v}_a = Proj(v_f, v_a) \tag{11}$$

where $\overline{v}_f$ is the projected feature suited for us, and $\overline{v}_a$ is the tested feature that we concatenated with the normal feature, as is explained in the Discussion section of this paper. Additionally, *Proj* is a projection function that projects a vector to another:

$$Proj(v_x, v_y) = \frac{v_x \cdot v_y}{|v_y|^2} \cdot v_y \tag{12}$$

After the projection layer, we concatenated two features, v_f and $\overline{v}_f$, as a new vector v_f^*, then fed v_f^* into the classifier. Finally, we utilized the *Softmax* function to achieve the classification and used the *CrossEntropy* function as our *Loss* function:

$$v_f^* = concat(v_f, \overline{v}_f) \tag{13}$$

$$Y_f = Softmax(v_f^* \cdot W_f + b_f) \tag{14}$$

$$Loss_f = CrossEntropy(Y_{truth}, Y_f) \tag{15}$$

where $W_f \in \mathbb{R}^{q \cdot m \oplus q_a \cdot m_a \times C}$, q_a is the number of filters of E_a, m_a is the kind of initialized filter for each filter, and $b_f \in \mathbb{R}^C$, C is the number of labels. Y_{truth} is the marked label, and Y_f

is the label predicted by the F-Net. The entire model that is used in the experiments on the FA-TextCNN is shown in Figure 2. Furthermore, we used a two-dimensional vector projection process to more intuitively express what the feature projection layer does in Figure 3.

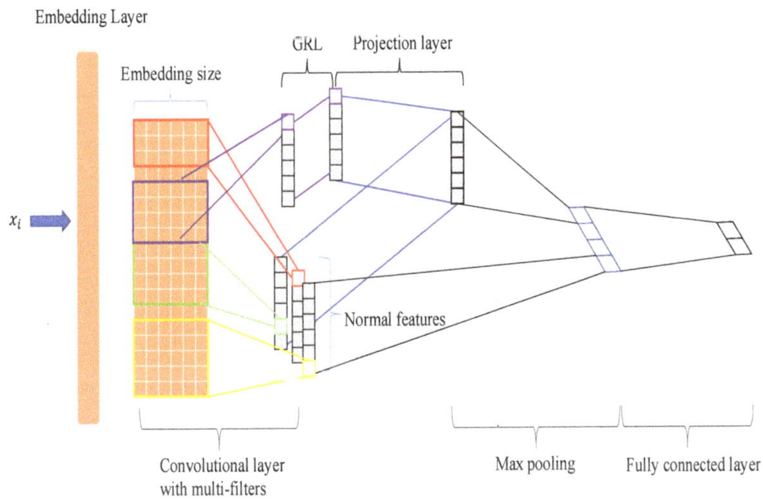

Figure 2. The total structure of the *FA-TextCNN*. The purple line is the extractor of the auxiliary network that utilizes a gradient reversal layer. As shown in the figure, we concatenated the features after maxpooling and fed the features together to form a fully connected layer.

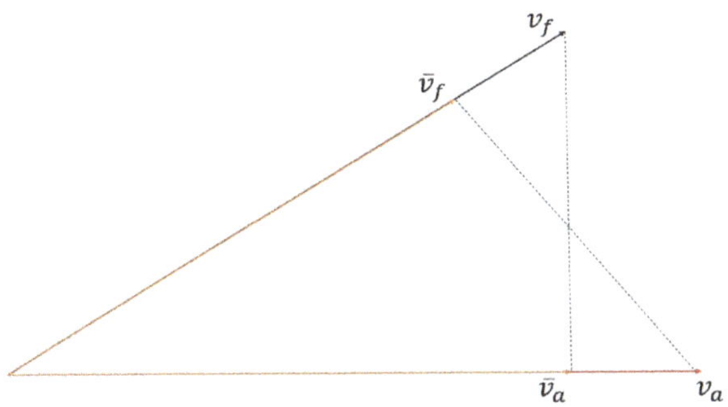

Figure 3. Two-dimensional progress of projection function. v_a is the feature vector extracted by A-Net, and v_f is the feature vector extracted by F-Net. $\overline{v}_f$ is the new feature vector that v_f projected to v_a, and $\overline{v}_a$ is the new feature vector that v_a projected to v_f.

Keep in mind that we implemented our method not only on CNN-based, but also on BERT-based models. To express our idea more clearly, we also drew a figure of the FA-Net being used on the BERT-based model, as shown in Figure 4. In the BERT-based model, some details do not align with the TextCNN. We weighed the time consumption and accuracy of the FA-Net by referred to the analysis of Sun et al. [19]. Consequently, we identified the high-level encoder from the base BERT as the encoder of E_a because its layers' output had a good classification ability. We noticed that the gradient of the A-Net is climbed up and then went back to the F-Net, influencing the parameters of the F-Net further. We decided to feed different kinds of the output of extractor E_f to extractor E_a to study the different types of projection operations. Therefore, we chose different n-gram sizes in the FA-TextCNN model and different encoder layers in the FA-BERT and FA-RoBERTa models. One reason we tried different inputs of E_a is that there are two projection types. For each type, we tried the same input to verify which is better. Another reason is that the input of E_a is cloned from the F-Net. Naturally, the A-Net gradient can go back to the F-Net. The higher the number of encoder outputs picked up by E_a, the more the encoders of the F-Net are influenced by the gradient of the A-Net.

On the one hand, we want to feed high-quality feature vectors to E_a. On the other hand, we want the gradient of the A-Net to affect the F-Net as little as possible. Thus, experiments on different inputs of E_a are necessary.

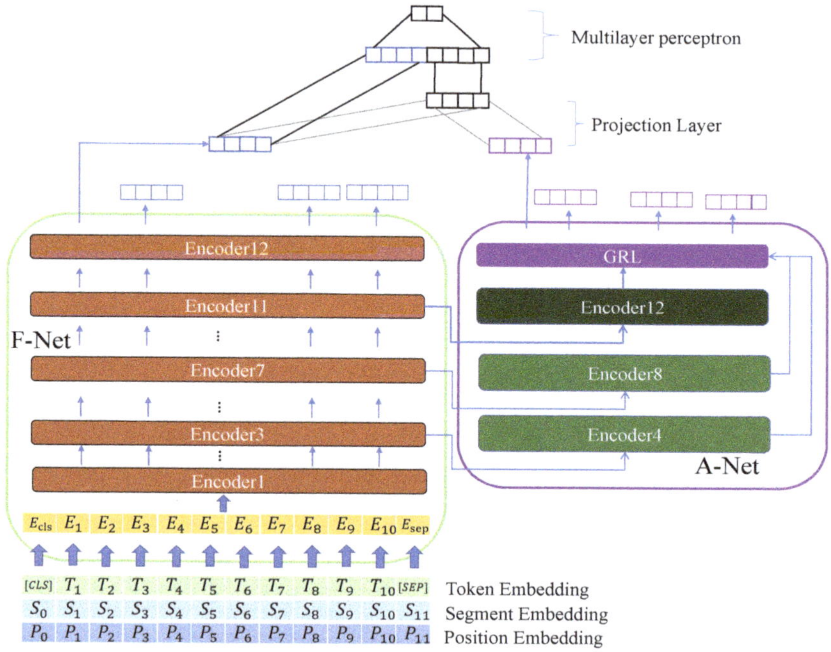

Figure 4. The figure shows the entire structure of the *FA-Net + BERT*. We implemented it based on our assumptions; thus, there are inputs for E_a in the auxiliary network. In this figure, A-Net can extract different features according to different inputs of A-Net. For example, when the A-Net obtains the output of Encoder7, the backward features are extracted by Encoder8 rather than Encoder12. Additionally, the projection layer only works on the CLS token of two features. The encoder in dark green is the best encoder of E_a that we tested in *FA-Net + RoBERTa* models.

3.2. Materials

To verify the effectiveness of our algorithm, we experimented with it by using six corpora, including a multilabel corpus and a binary-label corpora. The summarization of each corpus is shown in Table 1.

MR (https://www.cs.cornell.edu/people/pabo/movie-review-data/(accessed on 17 January 2023)): This corpus contains data on a document level, sentence level, sentiment scale, and subjectivity level. In our algorithm, we chose a sentence-level dataset to conduct the experiment on. It contains 4796 positive samples and 4796 negative samples.

SST2 (https://nlp.stanford.edu/sentiment/(accessed on 17 January 2023)): This corpus contains 67350 positive and negative samples in the training dataset and 1821 samples in the testing dataset [20]. To conduct the experiment faster, we determined the difference between the benchmarks and our algorithm. We cut it down to 6920 training samples and 1821 testing samples.

SemEval-2018 task 3 (https://github.com/Cyvhee/SemEval2018-Task3(accessed on 17 January 2023)): This task is named "Irony detection in English tweets." [21] The task is part of the 12th workshop on semantic evaluation. It contains two labels: non-irony and irony. There are 1916 non-irony and 1901 irony samples in the downloaded training dataset and 472 non-irony and 310 irony samples in the downloaded test dataset.

Sem-2017 task 4 (https://github.com/cbaziotis/datastories-semeval2017-task4 (accessed on 17 January 2023)): This task is the subtask of SemEval-2017 [22]. Task 4 contains five tasks. This research only needs three tasks: A, C, and E. Task 4A contains 1188 negative, 2724 neutral, and 4088 positive samples. The label of task 4CE is the five-point scale for sets of tweets and topics. It contains 4159 *1point*, 2237 *0point*, 901 *-1point*, 585 *2point*, and 118 *-2point* sets.

Waimai-10k (https://gitee.com/sprite0153/ChineseNlpCorpus/tree/master/datasets (accessed on 17 January 2023)): This is a Chinese corpus. The data come from user reviews of a particular food delivery platform. It is a binary-label database containing 3612 positive examples and 7176 negative examples.

Table 1. The summarization of each corpus.

Corpus	Language	Labels	Samples of Training Data	Samples of Test Data
MR	English	2	7460	2132
SST2	English	2	6920	1820
SemEval-2018 task 3	English	2	3817	782
Sem-2017 task 4A	English	3	6000	2000
Sem-2017 task 4CE	English	5	6000	2000
Waimai-10k	Chinese	2	8390	2398

Note that some of those corpora have not separated data into training datasets. Therefore, we split the data into 80% training and 20% testing datasets if the original data was not split. We kept the split ratio if the original data had been split.

3.3. Experimental Benchmarks

To expressively verify the effectiveness of our model, we conducted experiments with three benchmarks to obtain the discrepant results. As BERT achieved state-of-the-art results for productive tasks, we utilized the vocabulary table of BERT to tokenize the original sentence at each benchmark to decrease the preprocessing time of the corpus.

TextCNN: As the extractor, we utilized the most representative model, CNN-rand, which is a TextCNN [1]. It uses a set of filters to capture the semantic feature maps and pool them. Then, it concatenates the features of different filter sizes to make the classification.

BERT: We utilized a pretrained BERT-based model, which includes 12 layers and 756 hidden sizes, to fine-tune the parameters of our datasets.

RoBERTa: The RoBERTa model also has huge pretrained models. Like BERT, we fine-tuned our corpus in the pretrained RoBERTa-based model.

Although we implemented three benchmarks, there is no difference in the settings between our algorithm and benchmarks, such that the batch size and convolution filters or other settings were not changed between the benchmarks and our FA-Nets model. Moreover, as we know, the initial seed greatly influences the model. We further considered the influence of seeds on the model, and thus, we calculated the average result under five seeds.

3.4. Experimental Settings

We fixed the embedding size to be 128-dimensional for each experiment, except for when using BERT and RoBERTa because the pretrained models cannot be changed. The setting details at each benchmark of our experiments are as follows:

FA-Net of TextCNN: The filter sizes are 3, and the n-grams are set to be (3, 4, and 5). The parameter of *L2-norm* is 0.001. The parameter of *Dropout* [23] is 0.5. In the A-Net, we empirically fixed the length of the filter sizes to 1 and set n-gram to be 4.

FA-Net of RoBERTa: Because our experiments contain four corpora, including English and Chinese datasets, we not only used the English pretrained model, which is called roberta-base (https://huggingface.co/roberta-base(accessed on 17 January 2023)), but we also chose the Chinese pretrained model, which is called hfl/chinese-roberta-wwm-ext (https://huggingface.co/hfl/chinese-roberta-wwm-ext(accessed on 17 January 2023)). However, we utilized the cardiffnlp/twitter-roberta-base-sentiment (https://huggingface.co/cardiffnlp/twitter-roberta-base-sentiment(accessed on 17 January 2023)) pretrained model for the sarcasm detection datasets.

FA-Net of BERT: Keeping it the same as RoBERTa, we utilized bert-base-uncased (https://huggingface.co/bert-base-uncased(accessed on 17 January 2023)) and bert-base-chinese (https://huggingface.co/bert-base-chinese(accessed on 17 January 2023)) pre-trained models to conduct our experiments.

In the training stage, we fixed some hyperparameters corresponding to datasets at all benchmarks, such as batch size and the length of samples. The length of samples in all experiments was fixed to be 32 in waimai_10k, 64 in SST2, 64 in MR, and 256 in R8. In addition, the batch size was set to 32. For the other parameters in the BERT-based models, such as attention dropout, dropout, and weight decay, we kept the default setting that is applied by *Hugging Face*.

Because we utilized the different optimizers for the backpropagation stage and set different learning rates, except for the CNN-based models, the optimizer of the F-Net is Adam [24] with $\beta_1 = 0.9, \beta_2 = 0.999$. However, in the A-Net, the optimizer is the SGD optimizer where $moment = 0.9$. The difference between the CNN-based and BERT-based models and our FA-Nets have two optimizers and networks. Thus, the setting of the learning rate differs between models. Therefore, we set the learning rate to 0.001 in both optimizers in the FA-TextCNN models. However, in the FA-BERT-based models, we tried three kinds of learning rates (1e-5, 2e-5, and 3e-5) for the F-Net's optimizer and set the learning rates to 0.001 for the optimizer of the A-Net.

4. Results and Discussion

4.1. Results

The evaluation indicator of the multicategory dataset is *F1-score*, and the evaluation indicator of the binary classification dataset is *accuracy* because all the datasets are classification corpora. The total parameters of the models are shown in Table 2 and the experiment results are shown in Table 3.

Table 2. Comparison of total parameters between base models and our models: +1En means the auxiliary network has one transformer encoder, and +2En means two transformer encoders exist in the auxiliary network.

	TextCNN	**FA-Net + TextCNN**	
#param	40M	41M	
	RoBERTa	**FA-Net + RoBERTa**	
#param	125M	134M(+1En)	141M(+2En)
	BERT	**FA-Net + BERT**	
#param	109M	119M(+1En)	126M(+2En)

The models that start with "FA-" mean we added our auxiliary network to the original models. Additionally, we carried out different kinds of experiments to analyze the influences of different extractor sizes of the A-Net. The columns represent different datasets, and indices represent different algorithms.

Table 3. The contrast between benchmarks and our models is shown in this table. The results of BERT-based models are conducted under the five seeds, including benchmarks and FA-Nets. Task 3 means SemEval-2018 task 3, task 4A means Sem-2017 task 4A, and task 4CE means Sem-2017 task 4CE. In the TextCNN model, n_gram=n means the filter is set to n when initializing the A-Net's convolution layer. In the BERT and RoBERTa models, s means the sth F-Net's encoder is copied as the first encoder at A-Net, and e means the eth F-Net's encoder is copied as the last encoder at A-Net. The input of A-Net is the output of the $(s-1)$th encoder of F-Net. Furthermore, the "OGRL" means that we removed the GRL in the A-Net. The *Avg* is the average result from the sentiment analysis and sarcasm detection datasets. The best results in our implementation are marked with bold font.

Model	**SST2**	**MR**	**waimai_10k**	**Avg**
TextCNN [1]	82.17 (±0.48)	76.68 (±0.27)	90.10 (±0.20)	82.98
BERT [2]	91.59 (±0.23)	86.38 (±0.22)	91.27 (±0.27)	89.74
RoBERTa [3]	94.43 (±0.31)	88.66 (±0.28)	88.73 (±0.34)	90.61
FA-TextCNN (n_gram = 3)	81.86 (±0.88)	76.68 (±0.56)	90.14 (±0.37)	82.89
FA-TextCNN (n_gram = 4)	**82.67** (±0.37)	**77.28** (±0.51)	**90.48** (±0.26)	**83.47**
FA-TextCNN (n_gram = 5)	82.56 (±0.62)	76.99 (±0.42)	90.32 (±0.21)	83.29
FA-BERT (s = 12, e = 12)	91.93 (±0.29)	86.91 (±0.45)	91.63 (±0.30)	90.16
FA-BERT (s = 11, e = 12)	91.79 (±0.35)	86.91 (±0.24)	91.39 (±0.23)	90.03
FA-BERT (s = 8, e = 8)	91.85 (±0.30)	87.14 (±0.22)	**91.69** (±0.20)	90.23
FA-BERT (s = 7, e = 8)	91.81 (±0.41)	86.88 (±0.21)	91.39 (±0.20)	90.03
FA-BERT (s = 5, e = 5)	**91.97** (±0.22)	**87.21** (±0.24)	91.60 (±0.16)	**90.26**
FA-BERT (s = 4, e = 5)	91.59 (±0.34)	86.82 (±0.33)	91.18 (±0.50)	89.98
FA-RoBERTa (s = 12, e = 12)	95.05 (±0.39)	**89.11** (±0.46)	**89.36** (±0.24)	**91.17**
FA-RoBERTa (s = 11, e = 12)	94.61 (±0.40)	88.91 (±0.37)	89.20 (±0.24)	90.91
FA-RoBERTa (s = 8, e = 8)	**95.11** (±0.32)	88.96 (±0.24)	89.25 (±0.17)	91.11
FA-RoBERTa (s = 7, e = 8)	94.62 (±0.15)	88.97 (±0.50)	89.23 (±0.24)	90.94
FA-RoBERTa (s = 5, e = 5)	95.03 (±0.34)	88.93 (±0.27)	89.35 (±0.29)	91.10
FA-RoBERTa (s = 4, e = 5)	94.82 (±0.14)	88.97 (±0.62)	89.34 (±0.09)	91.04
Model	**task 3**	**task 4A**	**task 4CE**	**Avg**
TextCNN	70.56 (±0.73)	45.04 (±1.07)	27.09 (±0.44)	47.56
BERT	71.17 (±1.58)	60.02 (±0.73)	39.99 (±0.90)	57.06
RoBERTa	72.70 (±0.71)	64.53 (±0.46)	43.50 (±1.62)	60.97
FA-TextCNN (n_gram = 4)	**71.62** (±0.87)	**46.39** (±0.73)	**27.85** (±0.89)	**48.62**
FA-TextCNN (n_gram = 4, OGRL)	71.32 (±1.19)	44.83 (±1.48)	26.07 (±1.45)	47.40
FA-BERT (s = 5, e = 5)	**72.42** (±1.06)	**60.95** (±0.52)	**41.62** (±0.65)	**58.33**
FA-BERT (s = 5, e = 5, OGRL)	71.48 (±0.79)	60.69 (±0.36)	41.03 (±0.56)	57.73
FA-RoBERTa (s = 12, e = 12)	**73.45** (±1.08)	**65.52** (±0.27)	**45.67** (±1.52)	**61.55**
FA-RoBERTa (s = 12, e = 12, OGRL)	72.16 (±0.93)	65.10 (±0.55)	44.44 (±1.24)	60.57

Following the different results in Table 3 for each line, we made some observations as follows:

In the sentiment analysis datasets, the *accuracy* of the TextCNN in three of the corpora is 82.17%, 76.68%, and 90.10%. However, the *accuracy* can become higher, reaching 82.67%, 77.28%, and 90.48%, and the average *accuracy* under those corpora received a 0.5% boost when the model added our auxiliary network. This shows that our auxiliary network can boost the original models by adding acceptable parameters in the TextCNN. The same results occurred in BERT and RoBERTa. The average *accuracy* of BERT and RoBERTa can also receive a 0.5% boost due to our auxiliary network. The *accuracy* can receive a 0.8% boost for MR and SST2 when our auxiliary network is added to BERT and RoBERTa.

We utilized the best hyperparameters from the experiments on the sentiment analysis datasets to implement our methods on the sarcasm detection datasets. The effectiveness of our proposed method is higher than that of the sentiment analysis. Compared with the TextCNN, the FA-TextCNN models have an average improvement of 1% for three datasets.

Compared with BERT, FA-BERT has an average improvement of 1.3% in the SemEval-2018 task 3, 0.9% in the Sem-2017 task 4A, and 1.6% in the Sem-2017 task 4CE. Compared with RoBERTa, FA-RoBERTa has an average improvement of 0.8% in the SemEval-2018 task 3, 1% in the Sem-2017 task 4A, and 2.1% in the Sem-2017 task 4CE. Even if we removed the GRL, our proposed algorithms still cause an improvement, but the magnitude of the improvement is lower than in the models that are stacked with the GRL. The difference that boosts the effectiveness between the sentiment analysis and sarcasm detection datasets is caused by the proposed method, since its core idea is to scatter the semantic information of contradictory words to other words in a sentence. Additionally, the sarcasm detection datasets have many of the sentences that contain contradictory words. Therefore, the boosted effectiveness of our proposed method on the sarcasm detection datasets is higher than on the sentiment analysis datasets. As seen in Table 3, the results show that even in different settings, our algorithm still can improve the benchmarks' performance. In the CNN-based model, our FA-TextCNN is more stable than the TextCNN when the A-Net chooses a filter that fixed the n-gram = 4 as an extractor. However, compared to the other settings, the algorithms are not stable. Our algorithms are worse than CNN benchmarks when the n-gram = 3. We assumed that unstable conditions are due to the *dropout* because the corresponding deviation of the CNN benchmark is close to 0.5% in the SST2 and MR datasets. This means that the original CNN model is unstable, as it is simply uninterpretable. Therefore, when we fused the projected features that were extracted from normal features, the property further influences the stability of our FA-TextCNN models.

The results echo our previous assumption that semantic information can still be exploited in the backward features at high-level layers in BERT-based and RoBERTa-based models. After conducting the different experiments, we also studied the influence of the FA-Net on the attention heat map, which is explained in the Discussion.

4.2. Discussion

We studied the influence of the number of encoders and the projection types on the FA-Net. As the CNN is poor at extracting high-quality features, fusing the backward features directly through Equation (10) can hurt the performance of the CNN model. Thus, this discussion omits the CNN models. Both further experiments are focused on the BERT-based and RoBERTa-based models. Equation (10) projects v_a to v_f. In other words, the result of Equation (10) acquires the subfeature of v_f. From this point, we can discuss another projection type, which projects v_f to v_a, to further study which projection type is better for our idea.

Projection type: Figure 5 shows the experiments that change the projection type to Equation (11). According to Table 3, the best results appear in different types of the A-Net. In the *FA-Net + BERT* model, the most effective focuses on the encoder in the middle, but in *FA-Net + RoBERTa*, the most effective is the encoder at the top. Thus, we implemented more experiments by using Equation (11) for SST2 with different encoders.

As seen in Figure 5, there is improvement when the first encoder of the A-Net is between four and six, but the improvements are minor for Equation (10). Additionally, the accuracy becomes lower when the A-Net's first encoder is copied from the top of the F-Net. This means that Equation (11) is too unstable to conflict with our idea. We assumed that the gradient of v_a can influence several encoders of the F-Net when x is larger than nine and then further influence the model's performance even if we fuse a part of v_a.

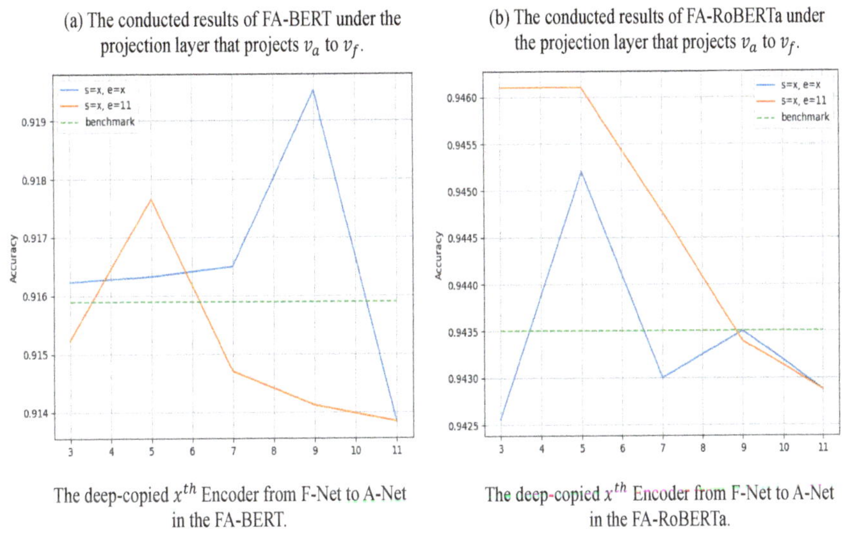

Figure 5. In this figure, we concatenate v_f with $\overline{v}_a$ to implement our FA-Nets. Both models are implemented in the SST2 dataset. Chart *a* is the BERT-based model, and chart *b* is the RoBERTa-based model. All results shown in this figure are averaged from the five seeds. The **blue** line is labeled by $s = x$, $e = x$, which means the first encoder and the last encoder in A-Net are copied from the xth encoder in F-Net. Furthermore, the **orange** line means the last encoder of A-Net is fixed to *11*, but the first encoder is changed with x. Moreover, the **dotted green** line results from corresponding benchmark models.

However, if we feed the outputs of the three to nine layers to the A-Net, our network's accuracy is also competitive. We discovered that the influence is lower when feeding low-level outputs to the A-Net because the gradient of v_a only affects a few layers of the low-level encoders. The influence becomes higher as the A-Net obtains the high-level outputs. Because of the gradient of v_a, the feedback to the F-Net is early. According to the projection type analysis and discussion, we empirically fused a part of v_f to concatenate more semantic information rather than fuse part of v_a in order to achieve our ideas. We assumed that the normal vector v_f is augmented by concatenating with $\overline{v}_f$, as it contains the essential abstract information relative to high-gradient features. Concatenating with $\overline{v}_f$ also can avoid the A-Net feedback that reverses the gradient of v_a to the F-Net.

The number of encoders of the A-Net: The BERT and RoBERTa models are perplexing; thus, the BERT-based model needs to consider how many encoders should be copied to the A-Net. According to Table 3, the best results occurred when the A-Net had one encoder. Furthermore, we can obtain average results for MR, SST2, and waimai_10k, as shown in Table 3. The average results of the multi-encoder that exist in the A-Net are lower than using a single encoder in the A-Net. This shows that the gradient reversal layer with a multi-encoder can learn a higher gradient feature than a single encoder. Therefore, a multi-encoder is more unsuitable than the single encoder in these corpora.

The attention heat map of an example sentence in FA-RoBERTa: We tested the sample we mentioned previously and drew an attention heat map. Compared to Figure 6, in Figure 7, although attention is paid to the phrase "ultimately silly", it does not focus on just this phrase anymore. FA-RoBERTa focuses on the core phrase, "charming and funny". Then, FA-RoBERTa makes the right classification. This proves that our idea could scatter the attention score of contradictory words to other words to reduce the destructive influence of contradictory words. As long as the model spreads the attention score to other words, the semantic information of other words can fuse with the normal features after concatenating.

Thus, the classifier can capture more information about other words rather than focusing on the contradictory words.

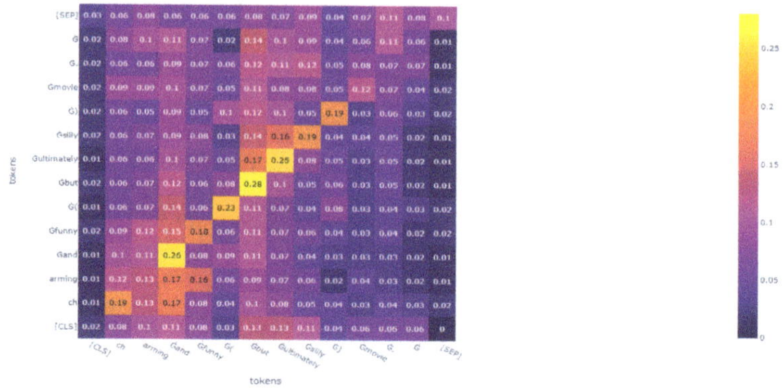

Figure 6. The attention heat map of phrase, "charming and funny (but ultimately silly) movie.", in the RoBERTa model.

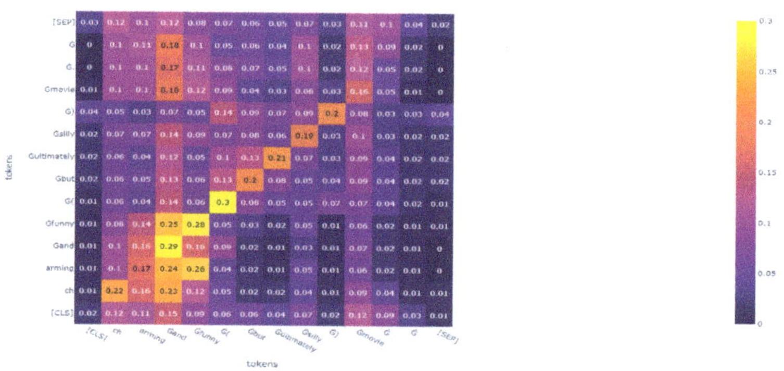

Figure 7. The attention heat map of phrase, "charming and funny (but ultimately silly) movie.", in our FA-RoBERTa model.

5. Conclusions

In this paper, we proposed a concise two-stream network that combines the extracted feature vectors to absorb more semantic information, allowing a sentence that contains special phrases which are opposite to the semantic information of the whole sentence to be classified correctly. In addition, we utilized the feature projection layer, gradient reversal layer, and vector concatenate to achieve this goal. Based on the original model, we considered the size of the parameters and the accuracy of the model, and we developed a proper model to avoid huge parameters and low accuracy. To prove the effectiveness of our algorithm, we conducted contrast experiments on sentiment analysis and sarcasm detection datasets. Ultimately, the results show that our algorithms are effective for those benchmarks. After implementing extra experiments, we further studied the influence of projection type and the number of encoders of the A-Net. We determined the general hyperparameters for three benchmarks in several datasets.

The current algorithm focuses on text classification. In the future, we will examine other NLP tasks using the short- and long-length sentence corpus or document corpus, and we will try to study the impact of data granularity in meta-learning.

Author Contributions: Conceptualization, D.S. and X.W.; methodology, D.S. and X.W.; software, D.S. and X.W.; validation, X.W.; formal analysis, D.S. and X.W.; investigation, X.W.; resources, D.S. and X.W.; data curation, X.W.; writing—original draft preparation, X.W.; writing—review and editing, X.W.; visualization, X.W.; supervision, D.S.; project administration, D.S. All authors have read and agreed to the published version of the manuscript.

Funding: This work has received no external funding.

Institutional Review Board Statement: Not applicable.

Informed Consent Statement: Not applicable.

Data Availability Statement: Not applicable.

Conflicts of Interest: We declare that we have no financial and personal relationships with other people or organizations that can inappropriately influence our work, and there is no professional or other personal interest of any nature or kind in any product, service and/or company that could be construed as influencing the position presented in, or the review of, the manuscript presented here.

Abbreviations

The following abbreviations are used in this manuscript:

GRL	gradient reversal layer
FA-Net	feature augmentation network
CNN	convolutional neural network
RNN	recurrent neural network
NSP	next-sentence prediction
SOTA	state-of-the-art

References

1. Kim, Y. Convolutional neural networks for sentence classification. In Proceedings of the EMNLP, Doha, Qatar, 25–29 October 2014.
2. Kenton, J.D.M.-W.C.; Toutanova, L.K. BERT: Pretraining of deep bidirectional transformers for language understanding. In Proceedings of the NAACLHLT, Minneapolis, MN, USA, 2–7 June 2019; pp. 4171–4186.
3. Liu, Y.; Ott, M.; Goyal, N.; Du, J.; Joshi, M.; Chen, D.; Levy, O.; Lewis, M.; Zettlemoyer, L.; Stoyanov, V. Roberta: A robustly optimized bert pretraining approach. *arXiv* **2019**, arXiv:1907.11692.
4. Qin, Q.; Hu, W.; Liu, B. Feature projection for improved text classification. In Proceedings of the 58th Annual Meeting of the Association for Computational Linguistics, Virtual, 5–10 July 2020; pp. 8161–8171.
5. Ganin, Y.; Lempitsky, V. Unsupervised domain adaptation by backpropagation. In Proceedings of the International Conference on Machine Learning, PMLR, Lille, France, 7–9 July 2015; pp. 1180–1189.
6. Xiao, L.; Wang, G.; Zuo, Y.. Research on patent text classification based on word2vec and LSTM. In Proceedings of the 2018 11th International Symposium on Computational Intelligence and Design (ISCID), Hangzhou, China, 8–9 December 2018; Volume 1; pp. 71–74.
7. Nowak, J.; Taspinar, A.; Scherer, R. LSTM recurrent neural networks for short text and sentiment classificatio. In Proceedings of the International Conference on Artificial Intelligence and Soft Computing, Zakopane, Poland, 11–15 June 2017; Springer: Cham, Switzerland, 2017; pp. 553–562.
8. Wang, H.; He, J.; Zhang, X.; Liu, S. A short text classification method based on N-gram and CNN. *Chin. J. Electron.* **2020**, *29*, 248–254. [CrossRef]
9. Lai, S.; Xu, L.; Liu, K.; Zhao, J. Recurrent convolutional neural networks for text classification. In Proceedings of the 29th AAAI Conference on Artificial Intelligence, Austin, TX, USA, 25–30 January 2015.
10. Conneau, A.; Schwenk, H.; Barrault, L.; Lecun, Y. Very deep convolutional networks for text classification. In Proceedings of the 15th Conference of the European Chapter of the Association for Computational Linguistics: Volume 1, Long Papers, Association for Computational Linguistics, Valencia, Spain, 3–7 April 2017; pp. 1107–1116.
11. Yang, Z.; Yang, D.; Dyer, C.; He, X.; Smola, A.; Hovy, E. Hierarchical attention networks for document classification. In Proceedings of the 2016 Conference of the North American Chapter of the Association for Computational Linguistics: Human Language Technologies, San Diego, CA, USA, 12–17 June 2016; pp. 1480–1489.
12. Bahdanau, D.; Cho, K.H.; Bengio, Y. Neural machine translation by jointly learning to align and translate. In Proceedings of the 3rd International Conference on Learning Representations, ICLR, San Diego, CA, USA, 25–29 April 2015.
13. Vaswani, A.; Shazeer, N.; Parmar, N.; Uszkoreit, J.; Jones, L.; Gomez, A.N.; Kaiser, Ł.; Polosukhin, I. Attention is all you need. *Adv. Neural Inf. Process. Syst.* **2017**, *30*, 5998–6008.

14. Niu, G.; Xu, H.; He, B.; Xiao, X.; Wu, H.; Gao, S. Enhancing local feature extraction with global representation for neural text classification. In Proceedings of the 2019 Conference on Empirical Methods in Natural Language Processing and the 9th International Joint Conference on Natural Language Processing (EMNLP-IJCNLP), Hong Kong, China, 3–7 November 2019; pp. 496–506.
15. Ying, Y.; Hjw, B.; Xdl, B.; Ping, Z.A. Capturing the grouping and compactness of high-level semantic feature for saliency detection. *Neural Netw.* **2021**, *142*, 351–362.
16. H. Wang, L. Jiang, Q. Zhao, H. Li, K. Yan, Y. Yang, Li, S.; Zhang, Y.; Qiao, L.; Fu, C.; Yin, H.; Hu, Y.; Yu, H. Progressive structure network-based multiscale feature fusion for object detection in real-time application. *Eng. Appl. Artif. Intell.* **2021**, *106*, 104486. [CrossRef]
17. Long, G.; Jiang, J. Graph based feature augmentation for short and sparse text classification. In Proceedings of the International Conference on Advanced Data Mining and Applications, Hangzhou, China, 14–16 December 2013; pp. 456–467.
18. Huang, Z.; Hou, L.; Shang, L.; Jiang, X.; Chen, X.; Liu, Q. GhostBERT: Generate more features with cheap operations for BERT. In Proceedings of the 59th Annual Meeting of the Association for Computational Linguistics and the 11th International Joint Conference on Natural Language Processing (Volume 1: Long Papers), Virtual, 1–6 August 2021; pp. 6512–6523.
19. Sun, C.; Qiu, X.; Xu, Y.; Huang, X. How to finetune BERT for text classification? In Proceedings of the China National Conference on Chinese Computational Linguistics, Kunming, China, 18–20 October 2019; pp. 194–206.
20. Socher, R.; Perelygin, A.; Wu, J.; Chuang, J.; Manning, C.D.; Ng, A.Y.; Potts, C. Recursive deep models for semantic compositionality over a sentiment treebank. In Proceedings of the 2013 Conference on Empirical Methods in Natural Language Processing, Seattle, WA, USA, 18–21 October 2013; pp. 1631–1642.
21. Van Hee, C.; Lefever, E.; Hoste, V. Semeval-2018 task 3: Irony detection in english tweets. In Proceedings of the 12th International Workshop on Semantic Evaluation, New Orleans, LO, USA, 5–6 June 2018; pp. 39–50.
22. Rosenthal, S.; Farra, N.; Nakov, P. SemEval-2017 task 4: Sentiment analysis in Twitter. *arXiv* **2019**, arXiv:1912.00741.
23. Srivastava, N.; Hinton, G.; Krizhevsky, A.; Sutskever, I.; Salakhutdinov, R. Dropout: A simple way to prevent neural networks from overfitting, *J. Mach. Learn. Res.* **2014**, *15*, 1929–1958.
24. Duchi, J.; Hazan, E.; Singer, Y. Adaptive subgradient methods for online learning and stochastic optimization. *J. Mach. Learn. Res.* **2011**, *12*, 2121–2159.

sensors MDPI

Article

Impact of Packet Loss Rate on Quality of Compressed High Resolution Videos

Juraj Bienik, Miroslav Uhrina, Lukas Sevcik * and Anna Holesova

University of Zilina, Univerzitna 1, 010 26 Zilina, Slovakia
* Correspondence: lukas.sevcik@uniza.sk

Abstract: Video delivered over IP networks in real-time applications, which utilize RTP protocol over unreliable UDP such as videotelephony or live-streaming, is often prone to degradation caused by multiple sources. The most significant is the combined effect of video compression and its transmission over the communication channel. This paper analyzes the adverse impact of packet loss on video quality encoded with various combinations of compression parameters and resolutions. For the purposes of the research, a dataset containing 11,200 full HD and ultra HD video sequences encoded to H.264 and H.265 formats at five bit rates was compiled with a simulated packet loss rate (PLR) ranging from 0 to 1%. Objective assessment was conducted by using peak signal to noise ratio (PSNR) and Structural Similarity Index (SSIM) metrics, whereas the well-known absolute category rating (ACR) was used for subjective evaluation. Analysis of the results confirmed the presumption that video quality decreases along with the rise of packet loss rate, regardless of compression parameters. The experiments further led to a finding that the quality of sequences affected by PLR declines with increasing bit rate. Additionally, the paper includes recommendations of compression parameters for use under various network conditions.

Keywords: H.264/AVC; H.265/HEVC; QoE; QoS; packet loss rate; video quality

Citation: Bienik, J.; Uhrina, M.; Sevcik, L.; Holesova, A. Impact of Packet Loss Rate on Quality of Compressed High Resolution Videos. *Sensors* **2023**, *23*, 2744. https://doi.org/10.3390/s23052744

Academic Editors: Róbert Hudec, Patrik Kamencay and Peter Hockicko

Received: 10 January 2023
Revised: 19 February 2023
Accepted: 20 February 2023
Published: 2 March 2023

1. Introduction

A high standard of living, education, and more opportunities, along with people's demands for such better conditions, has been constantly increasing over the last decades. Naturally, this applies to expectations in all areas of life, including entertainment and technology. Much like the rest of the world, Internet service providers, as well as broadcasting companies, want to bring their customers the best possible experiences. This individual user impression, also referred to as quality of experience (QoE) is highly dependent on quality of service (QoS) parameters. In the field of video transmission, these further depend on factors such as latency, jitter, or packet loss rate (PLR). However, although such factors are easily measurable, QoE cannot be quantified so effortlessly. Nowadays, one of the most popular network services is live video streaming, which is growing on a huge scale [1,2]. However, the video delivered over an IP network in real-time applications, is often exposed to various types of quality degradation. One of the reasons is that this type of application operates on RTP over unreliable UDP. The major sources of impairments are video compression and transmission over the noisy communication channel. Of the latter, the factor which affects video quality the most is undoubtedly PLR. It is due to consider that even a small percentage of lost packets can cause a massive decline in perceived video quality. This is proven in the following sections of the paper.

This paper is focused primarily on evaluating the influence of PLR on the quality of H.264/AVC and H.265/HEVC-encoded full HD and ultra HD video sequences. The different combinations of the resolution and compression parameters were used. Objective and subjective evaluation methods, namely peak signal-to-noise ratio (PSNR), Structural Similarity Index (SSIM), and absolute category rating (ACR) were employed for video-quality quantification. Additionally, the principle of the subjective ACR method enabled further research into the combined effect of PLR and compression parameters on perceived video quality. As a result, this paper also suggests the most suitable combination of resolution, bit rate, and codec for use under different conditions.

The paper is organized as follows. Section 3 illustrates the dataset compilation procedure, and briefly describes the source sequences and the process of video encoding and packet loss simulation. Section 4 deals with the acquisition and analysis of the results of objective video quality assessment. Analogously, Section 5 covers subjective video-quality evaluation. In addition, the analysis of the subjective results is extended by the calculation of the correlation with the objective assessment results. Moreover, further examination of the combined impact of PLR and compression on subjective video quality is described. Section 6 concludes the paper.

2. State of the Art

There are several research papers that deal with QoS and QoE evaluation, especially the performance assessment of various compression standards or the effect which has the transmission over noisy communication channel on the video quality. Coding structures, syntax, various tools, and settings relevant to coding effectivity were described in that paper. Human perception of compression, spatial and temporal information was examined in [3]. The authors have compiled a large-scale database of video sequences whose quality was subjectively evaluated. The coding efficiency of HEVC standard was compared with earlier codecs in [4]. The authors in [5] examine the effects of network impairment on HEVC video streaming. The impact of packet loss on the video quality and the main factors observed to be most perceived by the end-users were evaluated. In [6], a detailed, quantitative analysis of the video-quality degradation in a homogeneous HEVC video transcoder together with the analysis of the origin of these impairments and the influence of quantization step alignment on transcoding were presented. Differences between video transcoding and direct compression were described as well. The authors have also found the dependence between the quality degradation caused by transcoding and bit rate changes of the transcoded data stream.

A parametric planning model combining characteristics of the channel and the video is introduced in [7]. The video distortion due to the packet loss can be estimated by this model. Authors in [8] presented the first empirical study of the impact of loss-related errors on TV viewing engagement. They compared different platforms and delivery technologies. Video sequences and information about quality of delivery obtained from the service provider were used in the experiments. The length of the sequences, content, and connection type were compared. In [9], 16 types of metrics for the quality evaluation were compared. Packet loss was simulated in the video coding, and losses were subsequently concealed by using various error-concealment techniques. The purpose was to show that the subjective video quality cannot be predicted only from the visual quality of a frame, when some concealed error occurs. A new objective measure XLR (piXel Loss Rate) indicator was proposed in [10]. It evaluates the packet loss rate in video streaming. This method achieved comparable results with full-reference metrics and a very high correlation with MOS. Authors in [11] provided an overview of packet losses in wi-fi networks, mainly multimedia and real-time applications. A method for estimation of packet delivery ratio, which depends on the length of the packet, where longer packets are more likely to be damaged, is described in [12]. The loss of a compressed video packet results in an estimate of the expected mean square error distortion. In [13], the authors examined whether the packet loss model and primarily length of series have an impact on the accuracy of the error estimation.

The authors of [14] claimed that the highly textured video content is difficult to compress, as it is essential to achieve the compromise between bit rate and perceived video quality. Based on this, they introduced a synthetic video texture dataset that was generated by using a computer-generated imagery environment. It was named BVI-SynTex video dataset and was conformed from 196 video sequences clustered in three different texture types. It includes five-second-long FHD video scenes with a frame rate of 60 fps, and 8 bits depth.

In [15] a method for the optimal classification of the packets was proposed. They got different priority assigned when the transmission conditions became poor. The network transmits the segments with the greatest contribution toward the quality of perception if limited network conditions occur. The results showed that the proposed method can achieve a higher MOS compared to nonselective packet drop.

The purpose of this paper is not to point out the difference between QoS and QoE. These concepts, their correlation, and the derived mapping function were described in our previous paper [16]. The differences between the concepts of QoE and QoS are described in the Table 1. PLR is one of the basic QoS parameters, and the purpose of this paper is to show how this purely technical parameter can affect the perception of QoE. Thus, the main output is the mapping function of one of the QoS parameters to the metrics which represent the QoE. The selected relevant works were compared with our paper. HD video sequences based on motion in pictures were analyzed in the paper [17]. All sequences were encoded into H.264 and H.265 compression standards. The video quality was objectively measured by PSNR and SSIM metrics and subjectively assessed by using DMOS value. Compressed video files were streamed by an emulated 5G network. This network has a specific type of PLR based on different characteristics of radio channels compared with a wired metallic network. Two types of dynamic sequences (low motion and high motion) were compared. The authors did not use various sequences from the aspect of contrast. Paper [18] is focused on streamed videos based on TCP segments via a 5G network. Our paper analyzes video transmission over the metallic network with services based on RTP protocol and UDP segments. Authors in the paper [18] used more network parameters in their experiments, like bandwidth and delay. These parameters are especially critical for wireless networks, except for ISDN and low-speed xDSL networks. In paper [19], a QoE smart algorithm based on using a machine learning model is proposed. The model is based on the wireless network. It takes into account the GOP size, and technical parameters of endpoint devices (CPU, RAM, and screen size). In this manuscript, QoS parameters were fixed—280Mbps of throughput, 3 ms of two-ways packet latency, and 0.001% for packet loss and jitter. The authors describe a comparison between the traditional and adaptation approaches. Results using the proposed algorithm improved the quality. The transmission of videos via the network with various parameters of PLR was compared in our paper. The difference from our paper is that the parameters of the network are constant in the paper [19]. The mentioned parameters are changing dynamically and randomly in a real 5G radio network.

Table 1. The differences between the concepts of QoE and QoS.

Distinguishing Factors	QoS	QoE
Scope	Typically, telecom services	Broader domain, (not necessarily network-based)
Focus	Performance aspects of physical systems	Users' assessment of system performance
Method	Technology oriented, empirical, or simulation measurements	Multi-disciplinary and multi-methodological approach

3. Dataset Compilation

Selection of Source Sequences

The basic requirements for reference sequences intended for further processing are uncompressed format, high bit depth, resolution and frame rate, suitable aspect ratio, and sufficient diversity in terms of scene content. Meeting all these standards, eight high-quality video sequences were selected from the Shanghai Jiao Tang University dataset [20]. Common parameters of these reference sequences are presented in Table 2. Original sequences vary in temporal (TI) and spatial perceptual information (SI) as shown in Figure 1, colors, contrast, and other features.

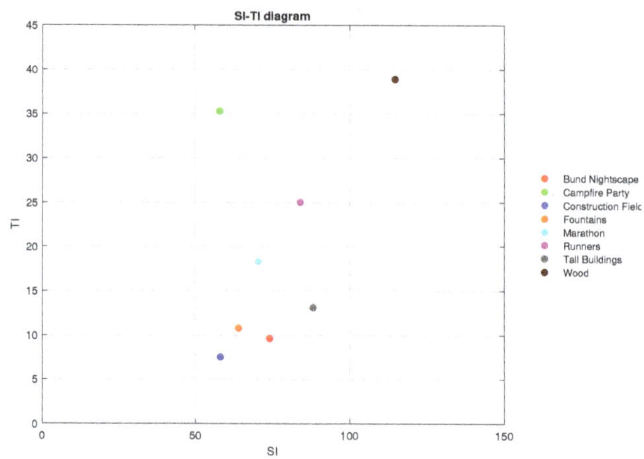

Figure 1. SI-TI diagram of reference sequences.

Table 2. Common parameters of reference sequences.

Resolution	Bit Depth	Aspect Ratio	Chroma Subsampling	Framerate [fps]	Length [s]
3840 × 2160 Ultra HD	10 bits per channel	16:9	4:4:4	30	10

Spatial perceptual information (SI) indicates the amount of detail in the video frames. The equation for calculating SI, based on the Sobel filter, is defined as

$$SI = max_{time}[std_{space}[Sobel(F_n)]], \tag{1}$$

where $Sobel(F_n)$ stands for the video frame at time n filtered by using the Sobel filter, std_{space} is the standard deviation over the pixels of the previously filtered frame, and max_{time} represents the maximum spatial information value of the entire video sequence. Temporal perceptual information (TI) expresses the degree of temporal activity, i.e., the change of the corresponding pixels values in consecutive frames of the video sequence. This change is called the motion difference feature $M_n(i,j)$ and is defined as a function of time,

$$M_n(i,j) = F_n(i,j) - F_{n-1}(i,j), \tag{2}$$

where i,j are the pixel coordinates in the adjacent frames n and $n-1$. The TI is then computed as

$$TI = max_{time}[std_{space}[M_n(i,j)]], \tag{3}$$

where max_{time} is the maximum standard deviation std_{space} of the $M_n(i,j)$ over time.

Both SI and TI are calculated for the luminance component of the frames only. Higher TI indicates more motion in the scene, whereas higher SI is generated by frames with more edges and disparity in luminance [21]. Previews of the employed sequences are displayed in Figure 2, followed by a brief description of the scene content and the scenario of each video.

Figure 2. Previews of used sequences: Bund Nightscape, Campfire Party, Construction Field, Fountains (top row from the left), Marathon, Runners, Tall Buildings and Wood (bottom row from left).

The Bund Nightscape is a time-lapse video sequence capturing cars driving through a night city near a river. The video was shot from a high angle with a static camera. The only sources of movement are passing cars, people in the streets, and flags hoisted on the roofs of buildings. The Campfire Party is a nighttime video sequence, picturing a group of people posing for a photograph by a campfire. The camera is almost still, except for the slight zoom in the end of the sequence. Most of the motion is generated by the blazing flames in the foreground. The Construction Field is a relatively static video sequence depicting an excavator digging a foundation pit on a construction site. Apart from this, the only other movement is caused by several people walking in the background. Fountains is a video sequence showing a large fountain with several jets situated in a small square between trees and buildings in the background. The video was recorded with a still camera; hence only the water gushing from the fountain adds dynamism to the scene. Marathon is a video sequence shot from a bird's eye view by an almost motionless camera. The scene depicts many people dressed in colorful raincoats competing in a race on a wet asphalt road during heavy rain. Runners is a video sequence capturing dozens of people in numbered jerseys racing on a road lined with trees. Even though the scene is filmed with a static camera, its considerable dynamism is added to by the fast movement of the runners and the wind in the leaves of the trees. Tall Buildings is a video sequence, recorded from a bird's eye perspective, that captures the tallest Shanghai skyscrapers. The camera pans steadily, revealing a view of the driving cars in the distance, which are the only moving objects in the video. Wood is a video sequence recorded in bright daylight, picturing a sunlit forest. The camera moves with a fairly swift, panning motion. It is the most complex sequence, considering the amount of detail and motion in the video.

Video Encoding and Packet Loss Simulation

First, all eight reference video sequences were subjected to chroma subsampling from the original YUV 4:4:4 to YUV 4:2:0 format. It is followed by a change in a bit of depth from 10 to 8 bits per channel. Afterward, these modified sequences were encoded by using two conventional standards, H.264 and H.265 at full HD and ultra HD resolutions. The bit rate ranged from 1 to 15 Mbps, employing the well-known FFmpeg software [22]. A description of codecs settings is possible to find in the Table 3. The used settings of both compression standards were the same. The target bit rate of 1 Mbps for the videos coded by H.264 could cause some artifacts which may not meet users' expectations, but we used these extreme values due to codec quality comparison—it is well known that the highest codec efficiency is in low bit rates, especially in newer codecs, as, for instance, H.265/HEVC. The frame rate remained set at 30 fps, resulting in the fixed group of pictures (GOP) length of 15 frames. The GOP size setting was derived from the frame rate as is standard for video intended for transmission through a noisy communication channel when an unreliable protocol is used for data transfer. The GOP length actually defines the interval between two consecutive keyframes—intracoded (I) or predicted (P). Considering this fact, it seems logical that the loss of image information for more than half a second (half the frame rate value) could be unacceptable for real-time applications such as video telephony or live streaming. The number of bidirectional predicted (B) frames was set to 3, as is recommended [23] for H.264 and H.265 codecs. The structure of employed FFmpeg commands was organized as follows from the example

ffmpeg -i Marathon_1920x1080_30fps_420_8bit_YUV.yuv -vcodec libx264 -x264-params keyint=15:min-keyint=15:bframes=3:b-adapt=1:bitrate=1M:vbv-maxrate=1M:vbv-bufsize=1M Marathon_1920x1080_30fps_420_8bit_H264_1M.ts.

Table 3. Description of codecs settings.

Command Example	Command Explanation
ffmpeg	run ffmpeg
-i Marathon_1920x1080_30fps_420_8bit_YUV.yuv	set the input file (name and container/uncompressed mode)
-vcodec libx264	set the codec
-x264-paramas	set the detailed H.264/AVC codec parameters
keyint=15	set the GOP length
min-keyint=15	set the minimum GOP length
bframes=3	set the number of B frames
b-adapt=1	disable the adaptivity of the B frames (default mode)
bitrate=1M	set the target bitrate
vbv-maxrate=1M	set the maxmum bitrate (by variable bitrate mode)
vbv-bufsize=1M	set the maximum buffer size
Marathon_1920x1080_30fps_420_8bit_H264_1M.ts	set the output file (name and container)

Subsequently, a local area network consisting of a streaming server and a streaming client was assembled to serve as a transmission channel for the previously encoded sequences. To simulate a lossy network connection, Clumsy [24] software was installed on the receiving computer, which assumed the role of the streaming client. By employing this tool, a certain percentage of UDP packets from every compressed video sequence was artificially intercepted. To approximate the actual operation of IP networks, clustered packet losses were generated pseudorandomly during video transmission. Both the broadcast server and the client were represented by standard computers—laptops with gigabit LAN ports with the Windows operating system installed. The interconnection of both devices was point-to-point type by using a cat6 metallic network cable. To minimize the occurrence of errors, the Wi-Fi adapters were disabled on both computers and the Windows firewall was also disabled. As for the stream itself, the compressed data was encapsulated in a TS container. The reason for using TS is that it is still currently the most used container for multimedia stream distribution. It is not only in Ethernet networks, but also for distribution in DVB-T/DVB-T2 systems. The transport stream allows multiplexing of streams (PES and PS) that do not necessarily share a common time base for transmission in a noisy environment. Thus, its biggest advantage is robustness to transmission errors. The image data was transmitted as UDP segments by using the RTP transport protocol. RTP was chosen based on the nature of the data. The real-time data and UDP segments are usually used for transmission in broadcast networking, because there is no possibility of retransmitting failed packets. It is not like a video-on-demand service where, in contrast, TCP segments are used. The resulting packet loss rate (PLR) ranged from 0 to 1%. A block diagram illustrating the combined process of video encoding and packet loss simulation is shown in Figure 3.

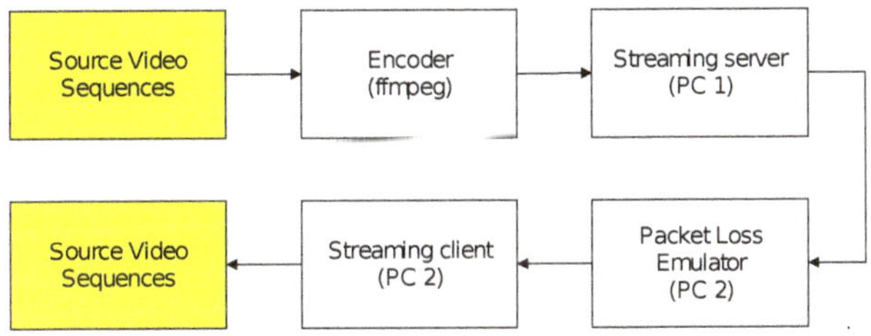

Figure 3. Process of video encoding and packet loss simulation.

Applying all possible combinations of compression parameters and PLR listed in Table 4, 1120 processed video sequences were created. To ensure higher statistical accuracy of the quality-assessment results, each video sequence was transmitted ten times over the assembled network, increasing the total number of test sequences to 11,200.

Table 4. Variable parameters of test sequences.

Parameter	Value
Codec	H.264 AVC, H.265 HEVC
Resolution	1920 × 1080 Full HD, 3840 × 2160 Ultra HD
Bitrate [Mbps]	1, 3, 5, 10, 15
Packet Loss Rate [%]	0, 0.1, 0.2, 0.3, 0.5, 0.75, 0.1

4. Objective Quality Evaluation

Objective quality measurement of the prepared video sequences was performed by using the MSU Video Quality Measurement Tool [25]. Two well-known objective methods, PSNR [26] and SSIM [27], were used for the evaluation; however, only the results of the SSIM metrics are presented in the paper. These showed a slightly higher correlation with the results of the subsequent subjective evaluation. The SSIM is a full-reference metric which focuses on measuring distortions in the structure of the image arising from changes in brightness, contrast, blurring of the image and so forth. This objective measure is grounded on the premise, that the human visual system is better suited to detect structural changes in the image than to identify specific errors. This can also explain the higher correlation of the SSIM quality ratings with the results of subjective quality assessment. The core principle of the SSIM consists of dividing the frames of two video sequences into areas of several pixels (depending on the size of the sliding window), in which the three components of the signal—brightness (l), contrast (c), and structure (s)—are sequentially compared. These components can be calculated by using the following formulas,

$$l(x,y) = (2\mu_x \mu_y + c_1)/(\mu_x^2 + \mu_y^2 + c_1) \quad (4)$$

$$c(x,y) = (2\sigma_x \sigma_y + c_2)/(\sigma_x^2 + \sigma_y^2 + c_2), \quad (5)$$

$$s(x,y) = (\sigma_x y + c_3)/(\sigma_x \sigma_y + c_3), \quad (6)$$

where μ_x and μ_y denote the average values of the two nonnegative signals x and y, σ_x and σ_y are their standard deviations and $\sigma_x y$ expresses the covariance of these signals.

The variables c_1, c_2 and c_3, defined by the following formulas, are included to stabilize the results if the denominator is close to zero,

$$c_1 = (K_1\, L)^2, \tag{7}$$

$$c_2 = (K_2\, L)^2, \tag{8}$$

$$c_3 = c_2/2, \tag{9}$$

where L denotes the dynamic range of pixel values, and K_1 and K_2 are low value constants.

The resulting SSIM value is determined by the weighted product of the brightness, contrast, and structure components as

$$SSIM(x,y) = [l(x,y)]^{\alpha}[l(x,y)]^{\beta}[l(x,y)]^{\gamma}. \tag{10}$$

By setting the weights α, β and γ to 1, the final calculation can be simplified to

$$SSIM(x,y) = \frac{(2\mu_x\mu_y + c_1)(2\sigma_{xy} + c_2)}{(\mu_x^2 + \mu_y^2 + c_1)(\sigma_x^2 + \sigma_y^2 + c_2)}. \tag{11}$$

The results obtained by using the SSIM metrics fall within the interval [0, 1], where 1 expresses the best quality, which can be achieved only if all the compared frames are identical. It is a symmetric method, which means that the order of the reference and degraded video sequence is insignificant. Given that the SSIM is a full-reference metrics, it would seem logical to choose an uncompressed video sequence unimpaired by transmission over the IP network as a reference. The aim of our research was to investigate the effect of PLR on quality of video in different resolutions with various compression parameters. It was not directly about the combination of PLR and effect of compression. This is a reason why the corresponding sequence with identical bit rate undegraded by network transmission was always chosen as the reference sequence for the metrics. In this way, a total of, 9600 objective measurements were performed by using the SSIM method.

Analysis of the Results

In view of the fact that the purpose of this particular research is not examining the impact of the scene content on video quality, graphs contained in the paper present only the average quality ratings. It is the same over all types of sequences, with differing content for each codec and resolution evaluated. Furthermore, it should be noted that the results of the evaluation of each sequence that was streamed 10 times were also averaged. Figure 4 outlines the development of measured video quality affected by the gradual increase in PLR, which is plotted on the x-axes of the graphs. The y-axes represent the measure of video quality evaluated by the SSIM metrics.

Because the video sequences unaffected by PLR encoded to both H.264 and H.265 in full HD and ultra HD, each bit rate of interest was used as a reference for the objective metrics. It is understandable that the quality of every sequence with 0% PLR was rated with an SSIM value of 1. Another common feature of Figure 4 is the tendency to objectively measure video quality at all bit rates to decline with the increasing PLR, as was anticipated. A full-reference objective metric such as SSIM always evaluates a video that lacks more information as of lower quality. An interesting discovery is that increasing PLR had a greater negative effect on the quality of full HD sequences than those in ultra HD resolution. Moreover, videos encoded to H.265 were rated worse on average than H.264-encoded sequences. The last and most unexpected finding is a faster decrease in the quality of sequences with a higher bit rate, which is clearly observable with all combinations of codec and resolution used. This fact, although counterintuitive because the quality of the compressed sequences with 0% PLR grows with the increasing bitrate, may have a logical explanation. For illustration, file sizes of the H.265 ultra HD Marathon video sequence at all examined bit rates are presented in Table 5. It is self-explanatory that by applying the same PLR, more bytes of data will be lost from larger files. Although the loss probability remains unchanged, it will most likely not affect the same part of the compressed sequences. It also will not affect the same number of parts of the video compressed at different bitrates, which is more important. This is due to the fact that the average size of the captured UDP packets is equal for each streamed sequence, which means that at a higher bit rate, the number of lost packets must also be larger. Therefore, most individual losses occur when video at 15 Mbps is transmitted. As the research in [28] proves, video quality degrades more significantly when several individual packets are lost than in a case when loss of the same number of consecutive packets occurs, which can be attributed to the temporal propagation of the errors. Regarding our investigation, it follows that the higher the bit rate used, the more individual or clustered packet losses will occur. Although at higher bit rates the losses are likely to affect a smaller portion of the frames, they may continue to propagate in time and space, causing more substantial video quality degradation.

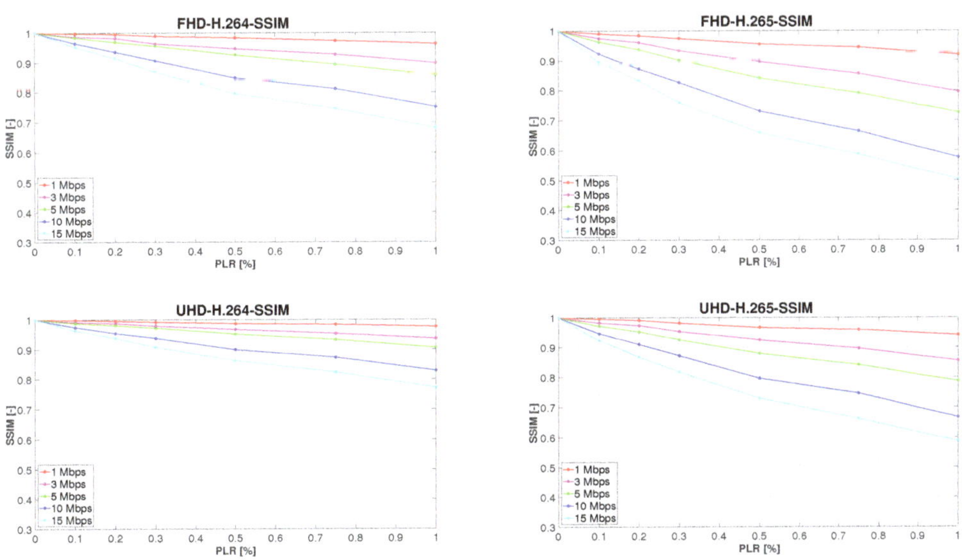

Figure 4. Development of objectively measured video quality with increasing PLR.

Table 5. File sizes of the H.265 UHD Marathon video sequence at 1–15 Mbps.

Test Sequence Name	File Size of the Test Sequence (in Bytes)
Marathon_UHD_30fps_420_8bit_H265_1M	1,436,508
Marathon_UHD_30fps_420_8bit_H265_3M	4,052,716
Marathon_UHD_30fps_420_8bit_H265_5M	6,784,544
Marathon_UHD_30fps_420_8bit_H265_10M	13,445,008
Marathon_UHD_30fps_420_8bit_H265_15M	20,205,864

5. Subjective Quality Evaluation

A subjective video-quality assessment was conducted to verify the findings that emerged from the analysis of the objective quality evaluation results. In addition to determining the strength of the correlation between subjective and objective measurement methods, the second reason for performing subjective tests was to monitor the combined impact of bit rate and PLR on respondents' views of video quality. To ensure sufficient diversity of visual content while limiting the time requirements of test sessions, the four most diverse video sequences were selected for subjective quality assessment according to the SI-TI diagram, namely Campfire Party, Construction Field, Tall Buildings, and Wood. Considering that each compressed video was streamed and objectively evaluated 10 times, the sequence that achieved the median SSIM value was always chosen for the subjective tests. Bearing in mind all possible combinations of the given resolution, codec, bit rate, and packet loss rate, 560 test sequences were thus selected. The popular absolute category rating (ACR) was adopted for subjective quality assessment of the prepared video sequences. This single-stimulus method was chosen on the grounds that the test conditions are analogous to those under which a regular user watches multimedia content. The course of ACR assessment, described in ITU recommendation [21] is faster and less complicated compared to other subjective evaluation methods. In addition, the results of the assessment are reproducible quite accurately even by different groups of respondents, as proven in [29]. The test presentation comprised degraded video sequences lasting approximately 10 seconds, alternating with a medium gray color displayed on the screen during the evaluation period of five seconds. Respondents were instructed to rate the video quality according to the following scale divided into five levels:

- 5—Excellent
- 4—Good
- 3—Fair
- 2—Poor
- 1—Bad.

A total of 36 laymen with no visual impairment participated in the experiment, of whom 14 were women and 22 were men. The age bracket of respondents ranged from 16 to 62 years, and the average age was approximately 30 years. After the completion of the experiment, the elimination of outliers was necessary. For this purpose, a Pearson linear correlation coefficient (PLCC) was calculated between the evaluation results of all video sequences from one respondent and the average rating of each sequence, following the equation

$$PLCC = \frac{\sum_{i=1}^{N}(X_i - \bar{X})(Y_i - \bar{Y})}{\sqrt{\sum_{i=1}^{N}(X_i - \bar{X})^2}\sqrt{\sum_{i=1}^{N}(Y_i - \bar{Y})^2}}, \tag{12}$$

where N is a number of evaluated sequences, X_i, Y_i are individual indexed ratings and $\bar{X}$, $\bar{Y}$ are the arithmetic means of all ratings. The threshold value of PLCC for removing a respondent's results from further processing was set at 0.75. The correlation was examined separately on four sets of video sequences, assorted according to resolution and codec used. The number of detected outliers for each set ranged from 1 to 3 out of all 36 respondents, leaving a sufficient number of results for subsequent statistical analysis.

5.1. Analysis of the Results

The first step in processing the obtained subjective data was the calculation of the mean opinion score (MOS) of each evaluated video sequence. Statistical analysis of the first half of the evaluation results revealed that it is quite sufficient to inquire about the users' opinion on the quality of video with a maximum PLR of 0.5% because after exceeding this value, the subjective quality was rated as poor at the utmost on the ACR scale. Between 0.5% and 1% PLR, the perceived quality further decreased on average by 0.28 and by a maximum of 0.45 MOS points. On that account, and for greater clarity, graphs included in this section show the development of subjectively perceived video quality with packet loss rates ranging from 0 to 0.5%.

5.2. Correlation with the Objective Evaluation Results

In order to calculate and investigate the correlation between subjective and objective evaluation results, the MOS values were shifted in a way such that the highest score of the given sequence was always 5 (Figure 5), and thus the video unaffected by PLR was considered as a hidden reference. According to [30] the Pearson linear correlation coefficient was once more used to calculate the correlation.

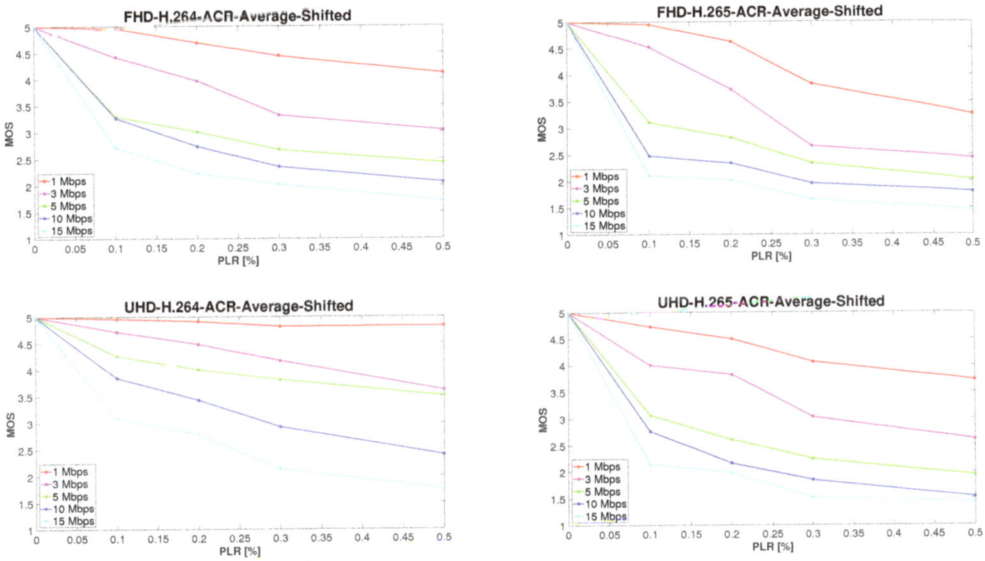

Figure 5. Development of subjectively perceived video quality with increasing PLR. MOS values were shifted in order to regard each video sequence unaffected by PLR as a reference.

Tables 6 and 7 show the correlation between the results of subjective and objective quality assessment. The graphical analysis indicated that the main difference between these results is in the steepness of the decrease in video quality at different bit rates between 0 and 0.1% PLR. Therefore, the correlation of subjectively and objectively measured quality of only those video sequences that were affected by PLR is illustrated in Table 7. By removing sequences impaired solely by compression from the equation, the correlation of the results increased significantly. Nevertheless, it is apparent from Figure 5 and Tables 6 and 7 that the correlation between subjective and objective assessment results is strong for all combinations of compression parameters and resolutions, with the lowest value of correlation coefficient of 0.83, calculated for H.265 ultra HD sequences at 15 Mbps. The observable differences are due to the fact that, unlike the algorithm, a person cannot separate the influence of PLR from the impact of bit rate on the visual quality of the evaluated video sequence. For this reason, a more detailed analysis of the combined effect of PLR and bit rate on the subjectively perceived video quality is presented in the following section.

Table 6. Correlation between the results of subjective and objective video quality evaluation.

	PLCC—Lossless Video Sequences Included				
	1 Mbps	**3 Mbps**	**5 Mbps**	**10 Mbps**	**15 Mbps**
FHD—H.264	0.99	0.97	0.86	0.89	0.84
FHD—H.265	0.97	0.95	0.88	0.84	0.84
UHD—H.264	0.92	1.00	0.94	0.97	0.93
UHD—H.265	0.98	0.97	0.87	0.88	0.83

Table 7. Correlation between the results of subjective and objective video quality evaluation after removing lossless sequences.

	PLCC—Lossless Video Sequences Excluded				
	1 Mbps	**3 Mbps**	**5 Mbps**	**10 Mbps**	**15 Mbps**
FHD—H.264	0.99	0.96	0.97	0.95	0.96
FHD—H.265	0.97	0.92	0.98	0.95	0.98
UHD—H.264	0.88	1.00	0.98	0.99	0.98
UHD—H.265	0.99	0.98	0.96	0.94	0.93

5.3. Combined Impact of PLR and Compression on Subjective Video Quality

Although this paper focuses mainly on the impact of PLR, the subjective method used made it possible to further examine the mixed impact of PLR and compression on subjectively perceived video quality. To illustrate this combined effect, it was sufficient to leave the MOS values unchanged in the graphical representation of the results. This can be seen in Figure 6, which shows the overall picture of the development of perceived video quality affected by alteration of the compression parameters and the gradual increase in PLR. Each graph displays the average MOS values over all four types of sequences with different content. On the x-axes, PLR within the narrowed limits 0–0.5% is plotted, and the y-axes represent the measure of video quality on the five-level MOS scale.

A common aspect of Figure 6a–d is the inclination of perceived video quality at all bit rates to decline with the increasing PLR in correlation with the SSIM ratings. Another anticipated outcome is the increase of the MOS values of video sequences unaffected by packet loss as the bit rate rises. This trend is distinctly observable in the order of the individual curves representing the different bit rates at 0% PLR. However, it is apparent from the graphs that this arrangement changes substantially even if only 0.1% of packets are lost during video transmission. In fact, it follows that the subjective quality of video sequences affected by PLR also decreases with a steeper slope when a higher bit rate is transmitted. This was predicted from the found high correlation between subjective and objective evaluation results and can be seen in better clarity in Figure 5. Figure 6a,b reveal that the results of subjective quality evaluation of full HD sequences are highly consistent, regardless of the codec used. The main common characteristic of the assessment results of all videos at this resolution is the overall MOS distribution of the individual bit rates. It is worth noting that starting at 0.3% PLR, the quality ratings are exactly in the reverse order to the ratings of sequences unimpaired by packet loss.

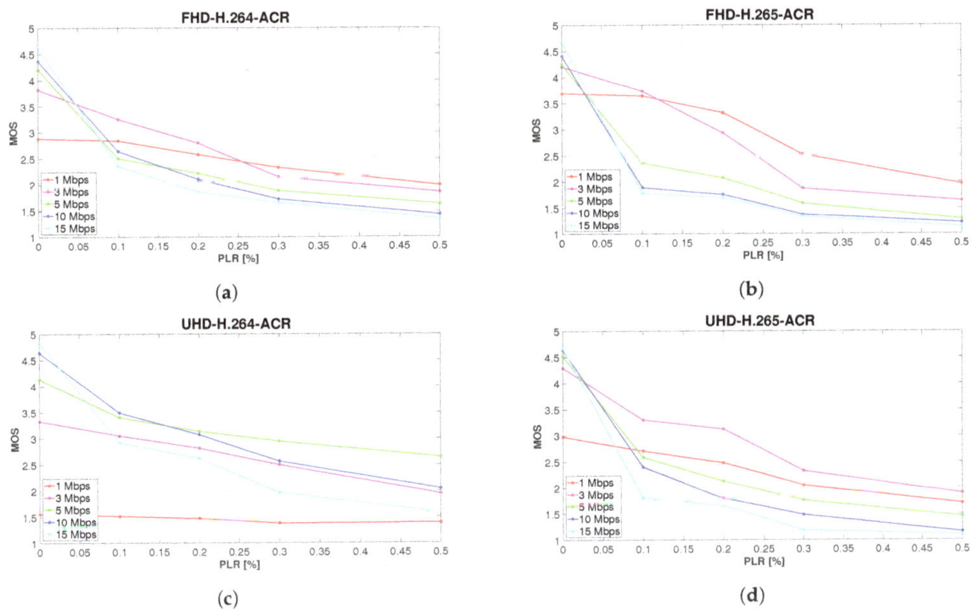

Figure 6. Development of subjectively perceived video quality with increasing PLR.

The essential distinction between the evaluation results of H.264 and H.265-encoded videos is the significantly more pronounced gap between the rating of lower bit rates (1 and 3 Mbps) compared to videos encoded at higher bit rates (5, 10, and 15 Mbps), which is evident on H.265 full HD sequences. This phenomenon is most eloquent when the simulated PLR reaches values from 0.1 to 0.2%. Another difference is a steeper decline in H.265 video quality, observable primarily at 10 and 15 Mbps. Unlike full HD videos, ratings of H.264-encoded ultra HD sequences diverged significantly from those at H.265 format. The only common aspects of Figure 6c,d are those that are universal for all four charts. As can be seen in Figure 6c, the average MOS of H.264 sequences shows that respondents considered 10 Mbps videos to retain the highest quality, when 0.1% packets were discarded during transmission. At higher PLR, 5 Mbps video sequences achieved the best subjective score. Although 1 Mbps sequences were rated higher than expected in the majority of the cases (as shown in Figure 6a,b,d), the quality of each H.264 ultra HD video at this bit rate was considered poor or bad, regardless of the PLR. It is therefore clear that in this particular case, the effect of compression on video quality was more intrusive than the loss of visual information. There is a straightforward explanation for this phenomenon, supported by Figure 7. The quality of H.264 ultra HD sequences at low bit rates is worse than the quality of the corresponding Full HD videos because four times the amount of information is discarded during compression due to the higher resolution. This effect is less pronounced when employing the H.265 codec, as HEVC compression is approximately 40% more efficient than older AVC [31].

Figure 6d indicates that the subjective MOS of H.265 videos affected to any extent by packet loss was unquestionably the highest when using a 3 Mbps bitrate.

Figure 7. Preview of 1 Mbps Wood sequence in all examined compression formats and resolutions.

5.4. Recommended Compression Parameters Under Various Network Conditions

If a certain packet loss rate in an error-prone network is anticipated, it may be convenient to get a better overview of how subjective quality changes as the bit rate increases or when other compression parameters are altered. For this reason, another four graphs were generated (Figure 8), with bit rate plotted on the *x*-axes. Each individual color curve represents a specific PLR. It is evident from these charts that subjective MOS values rise only until a certain value of bit rate is reached, and typically only at low PLR.

Based on the subjective data analysis, it is viable to determine the most suitable combination of resolution, bit rate, and codec for use under different conditions. If the video content is transmitted over a reliable network, the only variable is the adverse effect of compression. In such a case, deploying the higher bit rate below a certain threshold will most certainly increase the perceived video quality. However, after exceeding this limit, respondents may be unable to distinguish between the quality of two video sequences at consecutive data rates, which is why the differences in ratings are narrowing with increasing bit rate. The highest average MOS of 4.833 was achieved by H.264 ultra HD sequences at 15 Mbps. Video sequences at both resolutions and codecs were considered better than good (4 on the ACR scale) at 5, 10, and 15Mbps. By using the H.265 coding format, even 3 Mbps videos surpassed this value. Our research shows that if the assumed PLR in the network is less than 0.1%, it is best to encode the video to H.265 at full HD resolution at 3 Mbps. The average perceived quality of such sequences reached the MOS value of 3.733. The advantage is that the transmission of video with these specifications requires almost the narrowest bandwidth, and is therefore also the fastest and most economical. Although the average MOS of 1 Mbps videos with the same parameters is only slightly lower with a value of 3.642, employing this bitrate is not advisable due to the substantial disparity between ratings of sequences with different content, which we observed during the experiments. For ultra HD video, the highest average MOS of 3.492 was reached by H.264 sequences at 10 Mbps, closely followed by MOS 3.402 of videos at 5 Mbps. If the connection is extremely unreliable with a possible PLR of 0.2% or more, we recommend using an H.264 coding format, ultra HD resolution and 5 Mbps data rate, as the development of video quality with such parameters is the most constant, starting at MOS 3.136 and ending at 2.646 when packet loss reaches 0.5%. For full HD video, the H.265 codec at 1 Mbps seems to be the most advantageous, as the MOS starts at 3.317, although it drops sharply afterward to 1.958 at 0.5% PLR. For greater clarity, Table 8 lists the recommended bit rates for use with different combinations of compression parameters and expected PLR, based on subjective MOS.

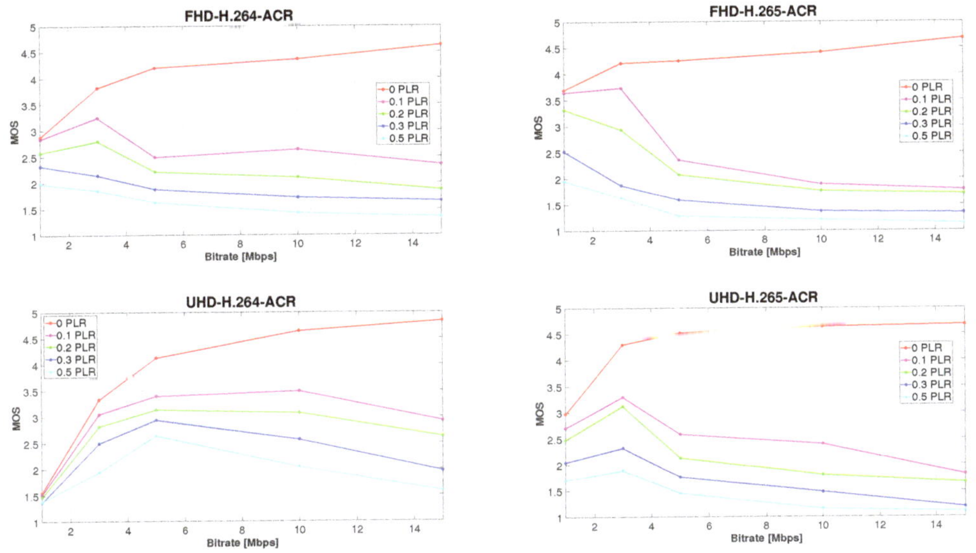

Figure 8. Development of subjectively perceived video quality with increasing bit rate.

Table 8. Bit rate values at which subjective video quality is highest.

Expected PLR [%]		**0**	**0.1**	**0.2**	**0.3**	**0.5**
				Recommended Bitrate Based on MOS [Mbps]		
Full HD	H.264	15	3	3	1	1
	H.265	15	3	1	1	1
Ultra HD	H.264	15	10	5	5	5
	H.265	15	3	3	3	3

6. Conclusions and Further Discussion

The aim of this paper was to analyze the impact of PLR on quality of video in full HD and ultra HD resolution encoded to H.264 and H.265 format at a bit rate ranging from 1 to 15 Mbps. First, a dataset of 11,200 test video sequences with eight different types of content combining all the abovementioned parameters and simulated PLR in range of 0 to 1% was compiled. Section 3 described the selection of sequences and their encoding. Subsequently, objective video quality evaluation was conducted by employing PSNR and SSIM metrics. Subjective quality assessment of 560 selected video sequences was performed by 36 laymen, using the popular ACR method. Evaluation of the video and analysis of the results were described in Section 4 by objective evaluation methods and in Section 5 by subjective evaluation methods. A high correlation was found between the results of evaluation methods; however, the subjective MOS revealed more about the mixed effect of PLR and compression.

Based on the analysis of the results, several conclusions were drawn. As was expected, the video quality declined along with the increasing PLR, irrespective of resolution, codec, or bitrate used. In case of video unaffected by packet loss, both measured and perceived quality improved with rising bit rate. However, this did not apply to video sequences impaired by PLR. In fact, it appears that in the presence of PLR, video quality decreases with a steeper slope when a higher data rate is transmitted. This is caused by the fact that use of a higher bit rate leads to a larger file size of the video sequence to be streamed over the error-prone network, thus splitting the image information into more UDP packets. At the same PLR, more of these packets are discarded during transmission which results in more individual or clustered losses that can propagate in time and space, causing further video quality degradation. It stemmed from the further analysis of subjective MOS that even 0.1% loss generally has more prominent negative effect on perceived video quality than H.264 or H.265 compression artifacts, which means that in most cases, increasing the bit rate over a given threshold value results in greater video-quality degradation. Based on this observation, the paper also includes a recommendation of compression parameters for use when various levels of PLR are expected. Packet losses may generate errors in different types of data. It depends on which part of the packet is affected by the loss. If the header of the packet is affected, which is the worst case, it is not possible to decode the whole packet. However, our effort was not to analyze in detail which part of the packet is influenced by the loss because in real traffic it is not able to control which part of the packet is affected by the loss. Our effort was to take an average in the loss rate, which we achieved by multiple transmission sequences.

Author Contributions: Methodology, M.U.; Validation, J.B.; Formal analysis, A.H.; Resources, J.B.; Data curation, J.B.; Writing—original draft, A.H.; Writing—review & editing, J.B., M.U., L.S. and A.H.; Visualization, M.U. and L.S.; Funding acquisition, L.S. All authors have read and agreed to the published version of the manuscript.

Funding: This work has been supported by the project of Operational Programme Integrated Infrastructure: Independent research and development of technological kits based on wearable electronics products, as tools for raising hygienic standards in a society exposed to the virus causing the COVID-19 disease, ITMS2014+ code 313011ASK8. This work has been partially supported by the Slovak VEGA grant agency, Project No. 1/0588/22 "Research of a location-aware system for achievement of QoE in 5G and B5G networks.

Institutional Review Board Statement: Not applicable.

Informed Consent Statement: Not applicable.

Data Availability Statement: Data available on request from the authors.

Conflicts of Interest: The authors declare no conflict of interest.

References

1. Long, C.; Cao, Y.; Jiang, T.; Zhang, Q. Edge Computing Framework for Cooperative Video Processing in Multimedia IoT Systems. *IEEE Trans. Multimed.* **2018**, *20*, 1126–1139. [CrossRef]
2. Li, M.; Chen, H.L. Energy-Efficient Traffic Regulation and Scheduling for Video Streaming Services Over LTE-A Networks. *IEEE Trans. Mob. Comput.* **2019**, *18*, 334–347. [CrossRef]
3. Duanmu, Z.; Ma, K.; Wang, Z. Quality-of-Experience for Adaptive Streaming Videos: An Expectation Confirmation Theory Motivated Approach. *IEEE Trans. Image Process.* **2018**, *27*, 6135–6146. [CrossRef] [PubMed]
4. Ohm, J.R.; Sullivan, G.J.; Schwarz, H.; Tan, T.K.; Wiegand, T. Comparison of the Coding Efficiency of Video Coding Standards—Including High Efficiency Video Coding (HEVC). *IEEE Trans. Circuits Syst. Video Technol.* **2012**, *22*, 1669–1684. [CrossRef]
5. Nightingale, J.; Wang, Q.; Grecos, C.; Goma, S. The impact of network impairment on quality of experience (QoE) in H.265/HEVC video streaming. *IEEE Trans. Consum. Electron.* **2014**, *60*, 242–250. [CrossRef]
6. Grajek, T.; Stankowski, J.; Karwowski, D.; Klimaszewski, K.; Stankiewicz, O.; Wegner, K. Analysis of Video Quality Losses in Homogeneous HEVC Video Transcoding. *IEEE Access* **2019**, *7*, 96764–96774. [CrossRef]
7. Song, J.; Yang, F.; Zhou, Y.; Gao, S. Parametric Planning Model for Video Quality Evaluation of IPTV Services Combining Channel and Video Characteristics. *IEEE Trans. Multimed.* **2017**, *19*, 1015–1029. [CrossRef]

8. Karthikeyan, V.; Allan, B.; Nauck, D.D.; Rio, M. Benchmarking Video Service Quality: Quantifying the Viewer Impact of Loss-Related Impairments. *IEEE Trans. Netw. Serv. Manag.* **2020**, *17*, 1640–1652. [CrossRef]
9. Kazemi, M.; Ghanbari, M.; Shirmohammadi, S. The Performance of Quality Metrics in Assessing Error-Concealed Video Quality. *IEEE Trans. Image Process.* **2020**, *29*, 5937–5952. [CrossRef]
10. Diaz, C.; Perez, P.; Cabrera, J.; Ruiz, J.J.; Garcia, N. XLR (piXel Loss Rate): A Lightweight Indicator to Measure Video QoE in IP Networks. *IEEE Trans. Netw. Serv. Manag.* **2020**, *17*, 1096–1109. [CrossRef]
11. Silva, C.A.G.D.; Pedroso, C.M. MAC-Layer Packet Loss Models for Wi-Fi Networks: A Survey. *IEEE Access* **2019**, *7*, 180512–180531. [CrossRef]
12. Jacobsson, M.; Rohner, C. Estimating Packet Delivery Ratio for Arbitrary Packet Sizes Over Wireless Links. *IEEE Commun. Lett.* **2015**, *19*, 609–612. [CrossRef]
13. Liang, Y.; Apostolopoulos, J.; Girod, B. Analysis of Packet Loss for Compressed Video: Effect of Burst Losses and Correlation Between Error Frames. *IEEE Trans. Circuits Syst. Video Technol.* **2008**, *18*, 861–874. [CrossRef]
14. Katsenou, A.V.; Dimitrov, G.; Ma, D.; Bull, D.R. BVI-SynTex: A Synthetic Video Texture Dataset for Video Compression and Quality Assessment. *IEEE Trans. Multimed.* **2021**, *23*, 26–38. [CrossRef]
15. Neves, F.; Soares, S.; Assuncao, P.A.A. Optimal voice packet classification for enhanced VoIP over priority-enabled networks. *J. Commun. Netw.* **2018**, *20*, 554–564. [CrossRef]
16. Sevcik, L.; Voznak, M. Adaptive Reservation of Network Resources According to Video Classification Scenes. *Sensors* **2021**, *21*, 1949. [CrossRef]
17. Taha, M.; Ali, A. Smart algorithm in wireless networks for video streaming based on adaptive quantization. *Concurr. Comput. Pract. Exp.* **2023**, e7633. [CrossRef]
18. Taha, M.; Canovas, A.; Lloret, J.; Ali, A. A QoE adaptive management system for high definition video streaming over wireless networks. *Telecommun. Syst.* **2021**, *77*, 63–81. [CrossRef]
19. Taha, M.; Ali, A.; Lloret, J.; Gondim, P.R.L.; Canovas, A. An automated model for the assessment of QoE of adaptive video streaming over wireless networks. *Multimed. Tools Appl.* **2021**, *80*, 26833–26854. [CrossRef]
20. Song, L.; Tang, X.; Zhang, W.; Yang, X.; Xia, P. The SJTU 4K video sequence dataset. In Proceedings of the 2013 Fifth International Workshop on Quality of Multimedia Experience (QoMEX), Klagenfurt am Wörthersee, Austria, 3–5 July 2013.
21. ITU-T. Recommendation ITU-T P.910—Subjective Video Quality Assessment Methods for Multimedia Applications. 2008. Available online: https://www.itu.int/rec/T-REC-P.910-200804-I/en (accessed on 9 January 2023).
22. FFmpeg. A Complete, Cross-Platform Solution to Record, Convert and Stream Audio and Video. ffmpeg.org. Available online: https://www.ffmpeg.org (accessed on 9 January 2023).
23. AWS. Encoding – Group of Pictures (GOP). aws.amazon.com. Available online: https://docs.aws.amazon.com/elemental-live/latest/ug/vq-gop.html (accessed on 9 January 2023).
24. Github. Clumsy, an Utility for Simulating. Available online: https://jagt.github.io/clumsy/ (accessed on 9 January 2023).
25. MSU Video Quality Measurement Tool. Program for Objective Video Quality Assessment. compression.ru. Available online: http://www.compression.ru/video/quality_measure/video_measurement_tool.html (accessed on 9 January 2023).
26. Wu, H.; Rao, K. (Eds.) *Digital Video Image Quality and Perceptual Coding*; CRC Press: Boca Raton, FL, USA, 2017. [CrossRef]
27. Wang, Z.; Bovik, A.; Sheikh, H.; Simoncelli, E. Image Quality Assessment: From Error Visibility to Structural Similarity. *IEEE Trans. Image Process.* **2004**, *13*, 600–612. [CrossRef]
28. Boulos, F.; Parrein, B.; Le Callet, P.; Hands, D.S. Perceptual Effects of Packet Loss on H.264/AVC Encoded Videos. In Proceedings of the Fourth International Workshop on Video Processing and Quality Metrics for Consumer Electronics VPQM-09, Scottsdale, Arizona, 15–16 January 2009.
29. Huynh-Thu, Q.; Garcia, M.N.; Speranza, F.; Corriveau, P.; Raake, A. Study of Rating Scales for Subjective Quality Assessment of High-Definition Video. *IEEE Trans. Broadcast.* **2011**, *57*, 1–14. [CrossRef]
30. ITU-T. Recommendation ITU-T P.1401—Methods, Metrics and Procedures for Statistical Evaluation, Qualification and Comparison of Objective Quality Prediction Models. 2020. Available online: https://www.itu.int/rec/T-REC-P.1401-202001-I/en (accessed on 9 January 2023).
31. Grois, D.; Marpe, D.; Mulayoff, A.; Itzhaky, B.; Hadar, O. Performance comparison of H.265/MPEG-HEVC, VP9, and H.264/MPEG-AVC encoders. In Proceedings of the 2013 Picture Coding Symposium (PCS), San Jose, CA, USA, 8–11 December 2013. [CrossRef]

sensors

MDPI

Article

A New Deep-Learning Method for Human Activity Recognition

Roberta Vrskova, Patrik Kamencay *, Robert Hudec and Peter Sykora

Department of Multimedia and Information-Communication Technologies, University of Zilina, 010 26 Zilina, Slovakia
* Correspondence: patrik.kamencay@feit.uniza.sk

Abstract: Currently, three-dimensional convolutional neural networks (3DCNNs) are a popular approach in the field of human activity recognition. However, due to the variety of methods used for human activity recognition, we propose a new deep-learning model in this paper. The main objective of our work is to optimize the traditional 3DCNN and propose a new model that combines 3DCNN with Convolutional Long Short-Term Memory (ConvLSTM) layers. Our experimental results, which were obtained using the LoDVP Abnormal Activities dataset, UCF50 dataset, and MOD20 dataset, demonstrate the superiority of the 3DCNN + ConvLSTM combination for recognizing human activities. Furthermore, our proposed model is well-suited for real-time human activity recognition applications and can be further enhanced by incorporating additional sensor data. To provide a comprehensive comparison of our proposed 3DCNN + ConvLSTM architecture, we compared our experimental results on these datasets. We achieved a precision of 89.12% when using the LoDVP Abnormal Activities dataset. Meanwhile, the precision we obtained using the modified UCF50 dataset (UCF50mini) and MOD20 dataset was 83.89% and 87.76%, respectively. Overall, our work demonstrates that the combination of 3DCNN and ConvLSTM layers can improve the accuracy of human activity recognition tasks, and our proposed model shows promise for real-time applications.

Keywords: deep learning; 3DCNN; ConvLSTM; human activity recognition

Citation: Vrskova, R.; Kamencay, P.; Hudec, R.; Sykora, P. A New Deep-Learning Method for Human Activity Recognition. *Sensors* **2023**, *23*, 2816. https://doi.org/10.3390/s23052816

Academic Editor: Antonio Fernández-Caballero

Received: 7 February 2023
Revised: 22 February 2023
Accepted: 1 March 2023
Published: 4 March 2023

1. Introduction

Presently, there is an increased emphasis on safety, and video capture and storage devices are constantly evolving to meet this demand. However, these devices need to be equipped with a system capable of accurately classifying various abnormal incidents and reducing human error. By increasing the effectiveness of surveillance systems, we could not only reduce crime and prevent various incidents but also provide first aid as soon as possible. Currently, video classification research in the field of computer vision has become a popular yet challenging topic. In addition to static information, videos also contain time information. Therefore, it is necessary to take into account the previous and subsequent frames to accurately recognize and classify incidents. This makes incident recognition from video more challenging than image recognition. In recent years, 3DCNN (3D Convolutional Network) and ConvLSTM (Convolutional Long Short-Term Memory) networks have emerged as popular techniques for video classification. These models are capable of capturing both spatial and temporal features, enabling them to accurately classify videos with high accuracy. Overall, with the continued development and improvement of video classification techniques, we can create more effective surveillance systems that can enhance safety and security in various settings.

In [1], the authors employ a combination of ConvLSTM and Conv3D layers for the task of human activity recognition. In [2], the authors use 3DCNN networks for Facial Micro-Expression recognition. Gesture recognition is covered in [3], where the classic 3DCNN network and the 3D Resnet network are used. In [4], the authors propose several approaches for the classification of abnormalities in video, including a deep hierarchical architecture that extracts generic video descriptors using 3DCNN and BICLSTM layers.

As part of this work, they also use a combination of 3DCNN, ConvLSTM, and Conv2D layers. In [5], the authors develop a non-contact method of assessing respiratory rate in sub-optimal light using video recordings, combining Euler scaling and a 3DCNN network to eliminate the need for Region of Interest (ROI). In [6], the authors develop a framework that takes pose-based skeleton joint sequences as input, followed by an LSTM network to learn the temporal evolution of the poses. The obtained results are compared with two fine-tuned deep neural networks, ConvLSTM and 3DCNN, and the Histogram of the Optical Flow (HOF) descriptor achieves the best results when used with the MLP classifier. This approach is applied by the authors in the diagnosis of Autism Spectrum Disorder (ASD) using video recordings, which is also covered in [7]. In [7], the authors use popular neural networks, 3DCNN and ConvLSTM, to detect diseases from video. In [8], the authors propose a hierarchical LSTM Convolutional Neural Network for the classification of farmers' behavior in agriculture. Finally, gesture recognition is also dealt with in [9], where the authors use two types of neural networks, 3DCNN and ConvLSTM network [10–13].

In our previous work, we also focused on the classification and recognition of video incidents. In [14,15], we proposed 3DCNN and ConvLSTM neural network approaches, respectively. For the recognition of abnormal incidents, we created the LoDVP Abnormal Activities database [14]. We trained and tested the proposed ConvLSTM network on this database, achieving an accuracy of 96.19%. The proposed 3DCNN neural network was trained and tested on the UCF YouTube action, UCF50, and UCF101 databases, achieving accuracy values of 87.4%, 80.6%, and 78.5%, respectively. It is worth noting that the proposed ConvLSTM network in [14] was specifically designed for our LoDVP Abnormal Activities database, which may have contributed to its high accuracy. Meanwhile, the proposed 3DCNN network in [15] was tested on several well-known databases, which demonstrated its generalizability to different video recognition tasks.

Based on the results we have obtained thus far, as well as the current state of the field, we have decided to focus on combining ConvLSTM and 3DCNN networks. Specifically, we aim to leverage both ConvLSTM layers and Conv3D layers in our approach. By doing so, we hope to achieve better performance in video recognition tasks, as both types of layers have shown promise in previous work. We believe that this approach has the potential to yield improved accuracy and generalizability in video recognition tasks.

2. Materials and Methods

In our research, we are investigating a combination of 3DCNN and ConvLSTM networks for video classification. ConvLSTM networks use their temporal memory to capture spatiotemporal patterns in videos, while 3DCNN networks leverage the third dimension for classification. Both networks are widely used for video and image classification in various fields including industry and medicine.

2.1. 3DCNN Architecture

The 3DCNN neural network can analyze and identifying different moving 2D objects in images and 3D images, such as in medical imaging. In 3DCNN, the 3D convolution operation is applied to the dataset in three directions (x, y, z) using a three-dimensional filter as is shown in Figure 1. The values in the layer within the three-dimensional filter must be set to be non-negative. The equation below defines the value for each position in the 3D convolution map of features in the layer:

$$v_{ij}^{xyz} = \tanh\langle b_{ij} + \sum_{m}\sum_{p=0}^{P_i-1}\sum_{q=0}^{Q_i-1}\sum_{r=0}^{R_i-1} w_{ijm}^{pqr} v_{\langle i-1\rangle m}^{\langle x+p\rangle\langle y+q\rangle\langle z+r\rangle}, \quad (1)$$

where w_{ijm}^{pqr} expresses the value of the kernel attached to the convolutional feature map in the previous layer, R^i expresses the size of the 3D kernel [15].

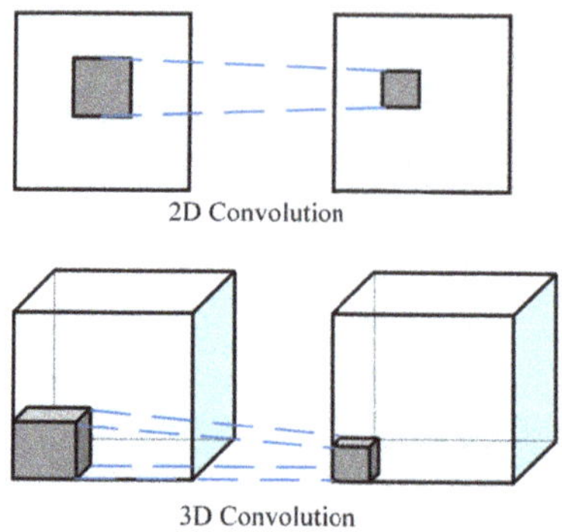

Figure 1. Comparison of the mathematical operation of 2D convolution and 3D convolution [16].

The 3D convolution is created by stacking adjacent layers around the center of the cube, and the convolution maps are interconnected, which captures motion information. However, the convolutional kernel can only extract one type of feature. Generally, 3DCNN is similar to Conv2D (2D Convolutional Neural Network). Combining multiple convolutional layers can improve the results of 3DCNN, similar to 2D convolution. When constructing a 3DCNN, it is crucial to set the number of layers, the number of filters in each layer, and the filter size properly. If pooling is used in the neural network design, the pooling size must have three dimensions to accommodate the 3D data. The output shape from the 3DCNN network is a 3D volume space [16–18].

2.2. ConvLSTM Architecture

The ConvLSTM neural network was developed by combining a Convolutional Neural Network (CNN) and a Long Short-Term Memory (LSTM) network. The ConvLSTM network is similar to an LSTM network in that it is a memory network, but it performs convolution operations on the transitions between layers. The internal design of a ConvLSTM network is illustrated in Figure 2 [19].

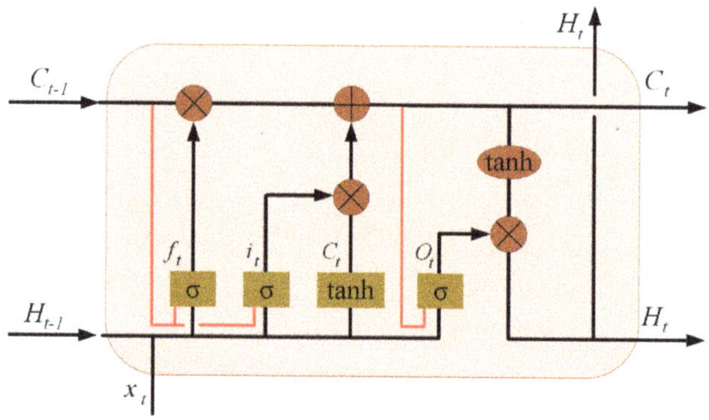

Figure 2. Inner structure of ConvLSTM [19].

The ConvLSTM neural network is commonly used for time-dependent image and video recognition, as it is equipped to capture spatial and temporal correlations. ConvLSTM implements a convolutional operation on the transitions between states and inputs. If we view states as hidden representations of moving objects, a ConvLSTM with a larger transition kernel can capture faster motions, while a network with a smaller kernel can capture slower motions. The key equations of ConvLSTM are derived from LSTM equations by convolutional coupling, as shown below, where "$*$" denotes the convolution operator and "$\circ$" denotes the Hadamard product:

$$i_t = \sigma(W_{xi} * X_t + W_{hi} * H_{t-1} + W_{ci} \circ C_{t-1} + b_i), \tag{2}$$

$$f_t = \sigma(W_{xf} * X_t + W_{hf} * H_{t-1} + W_{cf} \circ C_{t-1} + b_f), \tag{3}$$

$$C_t = f_t \circ C_{t-1} + i_t \circ tanh(W_{xc} * X_t + W_{hc} * H_{t-1} + b_f), \tag{4}$$

$$o_t = \sigma(W_{xo} * X_t + W_{ho} * H_t + W_{co} \circ C_{t-1} + b_o), \tag{5}$$

$$H_t = o_t \circ tanh(C_t), \tag{6}$$

where cell inputs are labeled X_t and cell states are labeled C_t, hidden states are labeled H_t; the gates are i_t, f_t, o_t and σ is a sigmoidal function. We denote the convolution kernels in the equation W_{-} [19].

2.3. Proposed 3DCNN + ConvLSTM Architecture

Our proposed neural network architecture combines Conv3D layers with a ConvLSTM network layer and a Conv2D layer. The architecture, referred to as 3DCNN + ConvLSTM, contains multiple Conv3D layers followed by a single ConvLSTM layer and a single Conv2D layer. The design of the architecture is depicted in Figure 3. This proposed architecture comprises the following layers:

- Conv3D layers: These layers extract spatiotemporal features from the input video data. The number of Conv3D layers can be adjusted based on the complexity of the task. These layers incorporate a three-dimensional filter, which performs convolution by moving in three directions (x, y, z).
- MaxPooling3D layer is a mathematical operation for 3D data (reduction of 3D data).
- ConvLSTM layer: This layer processes the extracted features from the Conv3D layers and captures the temporal dependencies between the frames.
- Conv2D layer is layer, which applies convolution on 2D data. This layer performs the final classification based on the output of the previous layers.
- A flatten layer converts the output matrix to the vector.

The proposed architecture combines the strengths of both Conv3D and ConvLSTM networks. This architecture consists of multiple 3D convolutional layers, a single ConvLSTM layer, and a single 2D convolutional layer, as well as batch normalization, a flattened layer, and a dense layer. The 3D convolutional part of the architecture was adopted from a previous study [15], while the ConvLSTM part was based on another previous research [14]. The hyperparameters of the 3D convolutional layers and MaxPooling, such as the number of filters and kernel size, are determined by mathematical constraints, with the output of the Conv3D layer being constrained to non-negative integer values. The flowchart of the proposed architecture is illustrated in Figure 4.

The proposed 3DCNN + ConvLSTM architecture in this work consists of six 3D convolution layers, four MaxPooling3D layers, and one ConvLSTM layer followed by a single Conv2D layer. The input to the network has dimensions of 100 × 100 × 3 (width, height, and number of channels). The first 3D convolution layer uses 64 filters with a kernel size of 3 × 3 × 3. Following each of these 3D convolution layers are MaxPooling3D layers of size 2 × 2 × 2 with a stride of 2. The next two 3D convolution layers have 128 filters of size 3 × 3 × 3, and after these two layers, there is a MaxPooling3D layer of size 2 × 2 × 2. The last two 3D convolution layers have 256 and 512 filters of size 3 × 3 × 3, respectively, and are used in the 3D convolution part of the network. The entire 3DCNN network has a batch normalization layer after every MaxPooling3D layer to improve the training process. The ConvLSTM network includes one ConvLSTM layer with a size of 3 × 3 and 64 filters. After the ConvLSTM layer, there is a batch normalization layer and a Conv2D layer with 16 filters of size 2 × 2. The output of the Conv2D layer is then passed through a flattened layer that turns a matrix into a vector, and the final dense layer has only one neuron that directly predicts the class of the input. The used optimization algorithm was "Adamax" and the learning rate was set to 0.001. A detailed description of the layers of the proposed 3DCNN + ConvLSTM architecture is shown in Figure 5.

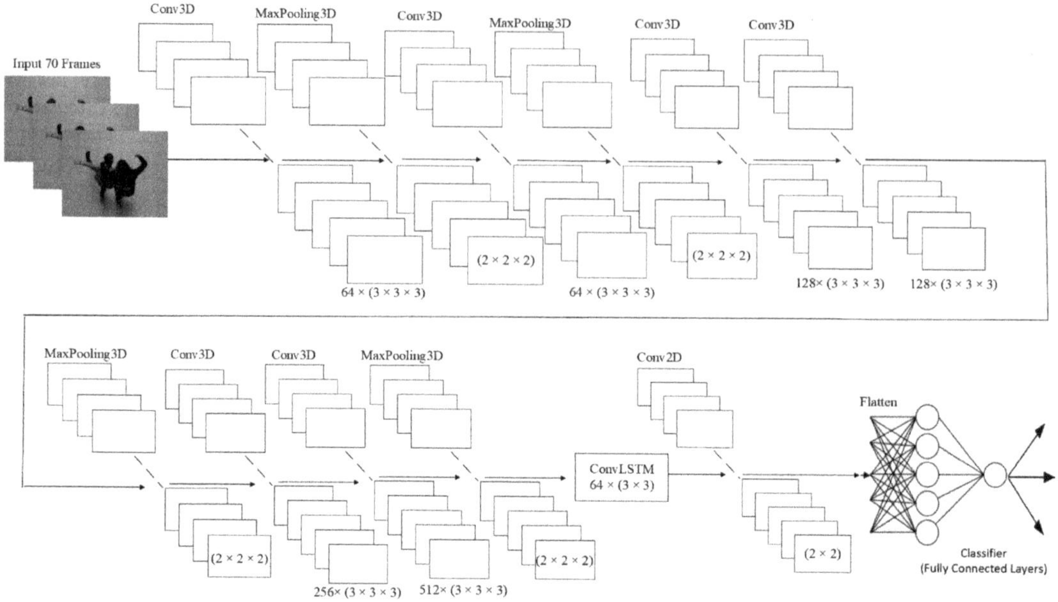

Figure 3. Proposed 3DCNN + ConvLSTM architecture.

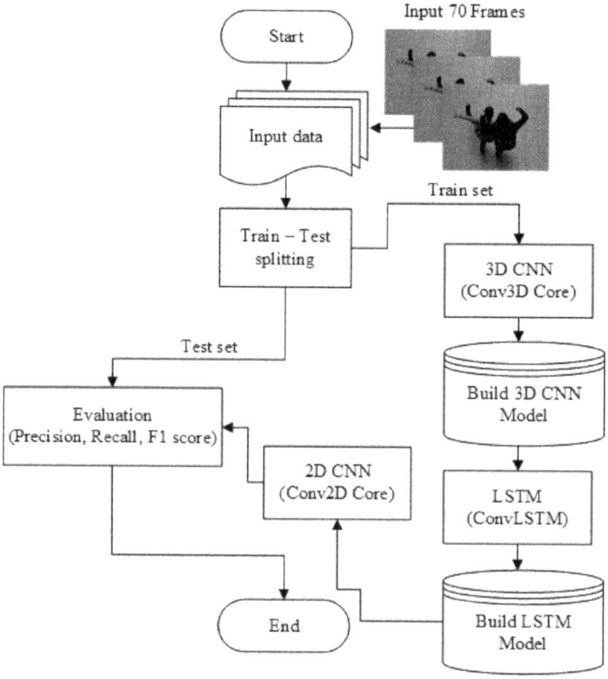

Figure 4. The flowchart of the proposed architecture.

Layer (type)	Output Shape	Param #
conv3d (Conv3D)	(None, 68, 98, 96, 64)	8704
max_pooling3d (MaxPooling3D)	(None, 34, 49, 48, 64)	0
batch_normalization (BatchNo	(None, 34, 49, 48, 64)	256
conv3d_1 (Conv3D)	(None, 32, 47, 46, 64)	110,656
max_pooling3d_1 (MaxPooling3	(None, 16, 23, 23, 64)	0
batch_normalization_1 (Batch	(None, 16, 23, 23, 64)	256
conv3d_2 (Conv3D)	(None, 14, 21, 21, 128)	221,312
conv3d_3 (Conv3D)	(None, 14, 21, 21, 128)	16,512
max_pooling3d_2 (MaxPooling3	(None, 7, 10, 10, 128)	0
batch_normalization_2 (Batch	(None, 7, 10, 10, 128)	512
conv3d_4 (Conv3D)	(None, 7, 10, 10, 256)	33,024
conv3d_5 (Conv3D)	(None, 7, 10, 10, 512)	131,584
max_pooling3d_3 (MaxPooling3	(None, 7, 10, 10, 512)	0
batch_normalization_3 (Batch	(None, 7, 10, 10, 512)	2048
conv_lst_m2d (ConvLSTM2D)	(None, 8, 8, 64)	1,327,360
batch_normalization_4 (Batch	(None, 8, 8, 64)	256
conv2d (Conv2D)	(None, 6, 6, 16)	9232
dropout (Dropout)	(None, 6, 6, 16)	0
dense (Dense)	(None, 6, 6, 256)	4352
dropout_1 (Dropout)	(None, 6, 6, 256)	0
flatten (Flatten)	(None, 9216)	0
dense_1 (Dense)	(None, 50)	460,850

Figure 5. Description of the layers our proposed architecture.

The values in Table 1 show the total number of parameters, as well as the number of trainable and non-trainable parameters.

Table 1. The overall number of parameters of our suggested architecture.

Parameters of the Proposed Architecture	Number of Parameters
Total parameters	2,326,914
Trainable parameters	2,325,250
Non-Trainable parameters	1664

Deep-learning systems are built using Python libraries, such as Keras and TensorFlow, and the experimental results were obtained using the Nvidia CUDA libraries. The input data consisted of images with dimensions of 100 × 100 and 3 channels. For each database, 70% of the data were used for training, 20% for testing, and 10% were reserved for validation (in a 70:20:10 split).

3. Description of the Datasets

In this section, the used datasets will be described. The all experimental results on the LoDVP Abnormal Activities dataset, UCF50 dataset, and MOD20 dataset were obtained.

3.1. UCF50 Dataset

UCF50 dataset consisted of 50 action categories. The dataset included realistic videos from YouTube. The dataset had large variations in camera motion, cluttered backgrounds, illumination conditions, etc.

Videos in the same group may share some common features, such as the same person, similar background, similar viewpoint, etc. The UCF50 dataset can be seen in Figure 6.

The UCF50 dataset consisted of 50 categories. The dataset consists of categories such as Baseball Pitch, Basketball Shooting, Bench Press, Biking, Billiards Shot, Breaststroke, Clean and Jerk, Diving, Drumming, Fencing, Golf Swing, Playing Guitar, High Jump, Horse Race, Horse Riding, Hula Hoop, Javelin Throw, Juggling Balls, Jump Rope, Jumping Jack, Kayaking, Lunges, Military Parade, Mixing Batter, Nun chucks, Playing Piano, Pizza Tossing, Pole Vault, Pommel Horse, Pull Ups, Punch, Push Ups, Rock Climbing Indoor, Rope Climbing, Rowing, Salsa Spins, Skate Boarding, Skiing, Skijet, Soccer Juggling, Swing, Playing Tabla, TaiChi, Tennis Swing, Trampoline Jumping, Playing Violin, Volleyball Spiking, Walking with a Dog, and Yo Yo [20].

Figure 6. Example of the UCF50 dataset [20].

We cut the dataset used into 10 layers and used both datasets in training the network to compare the results. We called the reduced dataset UCF50mini. Both datasets (reduced and total) were divided into three sets (training, test and validation set).

3.2. LoDVP Abnormal Activities Dataset

The LoDVP Abnormal Activity dataset comprises 1069 videos. The incidents in the videos are created by non-professional actors. The videos in the database are created believably. Incidents are reported in the parking lot, in the university campus and in the forest. The scenes were recorded from different angles. The dataset is divided into 11 classes, and each class contains about 100 videos. The length of the video depends on the incident and lasts from 1 s to 30 s. Similar videos belonging to the same class may exhibit common traits, such as a recurring individual, perspective of the camera, and a comparable setting [14].

The LoDVP Abnormal Activity dataset consists of the following classes, which can be seen in Figure 7. For our work, this dataset was also divided into a training test and a validation set in the same ratio as UCF50mini. The division was in the ratio of 70:20:10.

3.3. MOD20 Dataset

The MOD20 dataset consists of 2324 videos, of which six videos were created by a quadrotor UAV and 2318 videos were downloaded from YouTube. All clips are 1:1 aspect ratio. Videos were sampled below 29.97 fps. The videos in the dataset are recorded from both fixed and moving cameras. The videos show realistic scenarios in 20 selected classes (see Figure 8). The dataset consists of classes such as backpacking, cliff jumping, cutting wood, cycling, dancing, fighting, figure-skating, fire-fighting, chainsawing trees, jet skiing, kayaking, motorbiking, football-catching, rock-climbing, running, skateboarding, skiing, standup-padding, surfing, and windsurfing [21].

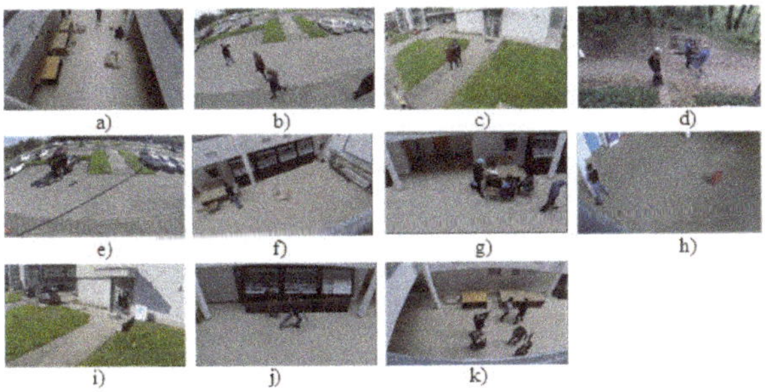

Figure 7. LoDVP Abnormal Activities dataset: (**a**) Begging (**b**) Drunkenness (**c**) Fight (**d**) Harassment (**e**) Hijack (**f**) Knife hazard (**g**) Normal videos (**h**) Pollution (**i**) Property damage (**j**) Robbery (**k**) Terrorism [14].

Figure 8. The example of the MOD20 dataset [21].

4. Experimental Results

This section describes the experimental results achieved on the LoDVP Abnormal Activities dataset, UCF50 dataset, and MOD20 dataset, which demonstrate the superiority of the 3DCNN + ConvLSTM combination in recognizing human activities. To provide a more comprehensive comparison of the proposed architecture, we compared the experimental results achieved on these datasets. In our work, we divided this dataset into a training, testing, and validation set in the same ratio as the UCF50 mini and LoDVP Abnormal Activities datasets, with a data distribution of 70:20:10.

Results

All of the tested datasets, including LoDVP Abnormal Activities, UCF50, and UCF50 mini, were divided into three main parts: training set, testing set, and validation set. In this study, the first step was to classify the LoDVP Abnormal Activities dataset into 11 classes, which included Begging, Drunkenness, Fight, Harassment, Hijack, Knife Hazard, Normal Videos, Pollution, Property Damage, Robbery, and Terrorism. The data underwent preprocessing, which involved resizing each video to 100 × 100 size with 70 frames. The accuracy and loss functions during training are displayed in Figure 9.

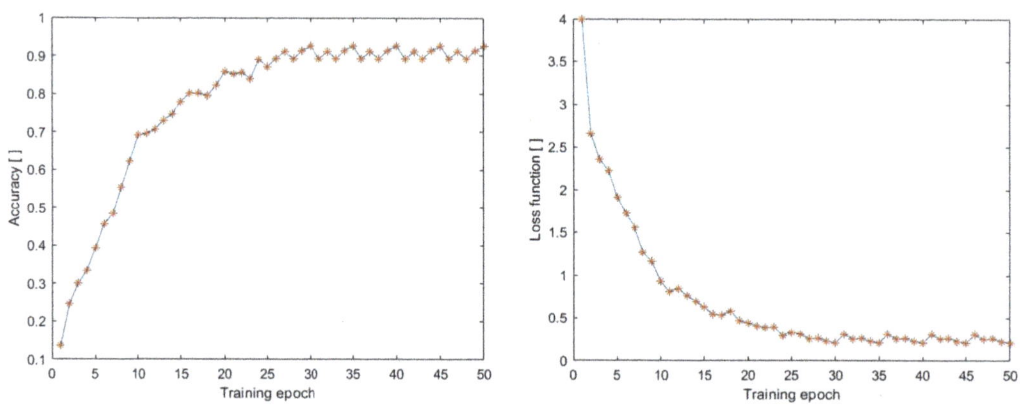

Figure 9. Accuracy during training process on the dataset (LoDVP Abnormal Activities).

In the early epochs, we observe a gradual increase in accuracy, and the highest accuracy achieved during training is 92.5%. At the same time, there is a decrease in the loss function, which is directly proportional to the increase in accuracy during the training process. The accuracy and loss function have a directly proportional relationship, and the lowest achieved loss function value during the training process was 0.2106.

To provide a comparison, we also monitored the accuracy and loss function achieved on the UCF50mini dataset during the training process. The UCF50mini dataset includes 10 classes: Baseball Pitch, Basketball Shooting, Bench Press, Biking, Billiards Shot, Breaststroke, Clean and Jerk, Diving, Drumming, and Fencing. The data were preprocessed in the same way as in the previous case, with a size of 100 × 100 and 70 frames. The accuracy and loss function can be seen in Figure 10, which displays an increase in accuracy and a decrease in loss function similar to the previous case. The highest accuracy attained was 0.9668, and the lowest loss function value during training was achieved by the neural network architecture with a value of 0.1042.

We also monitored the training process for comparison with the MOD20 dataset and focused on accuracy and loss function (see Figure 11). The MOD20 dataset consists of 20 classes (with backpacking, cliff jumping, cutting wood, cycling, dancing, fighting, figure-skating, fire-fighting, chainsawing trees, jet skiing, kayaking, motorbiking, football-catching, rock-climbing, running, skateboarding, skiing, standup-padding, surfing, and windsurfing). The data were preprocessed in the same way as in the previous cases, with a size of 100 × 100 and 70 frames. Figure 11 shows a linear increase in accuracy and a decrease in loss function, as observed previously. The highest accuracy value achieved was 0.8630, while the smallest value of the loss function was 0.4223. However, compared to the other datasets, we achieved the worst results during the training process on this dataset.

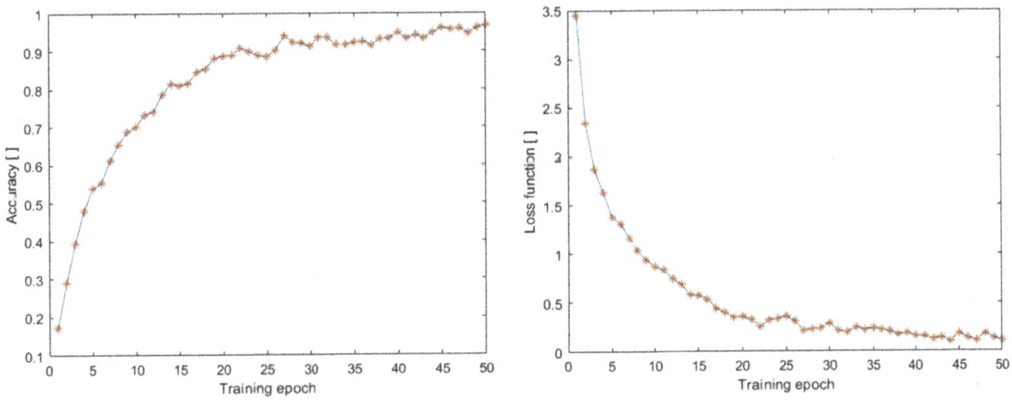

Figure 10. Accuracy during training process on the dataset (UCF50mini).

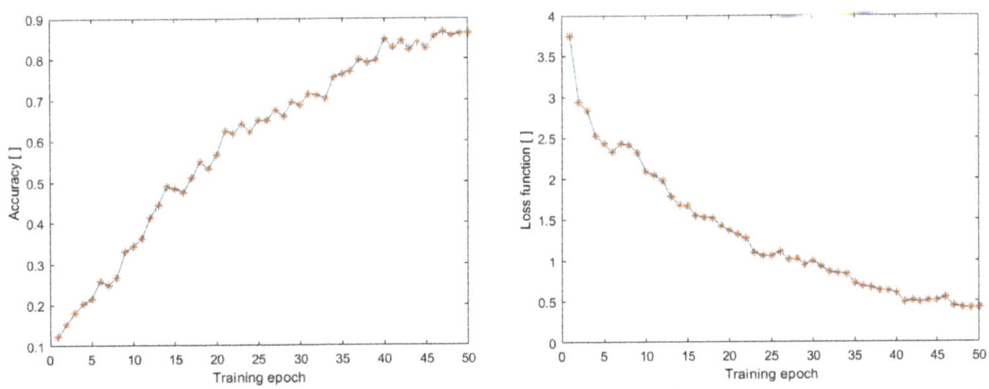

Figure 11. Accuracy during training process on the dataset (MOD20).

To make a clearer comparison, we created Table 2, which shows the training and testing results of our proposed model on all three datasets, namely LoDVP Abnormal Activities, UCF50, and UCF50mini. The table presents the values of training loss and training accuracy achieved during the training process, as well as the test loss and test accuracy obtained on a separate set of data. We computed the test results by evaluating the trained model on the test set, which was not used for training or validation. The Table 2 provides a comprehensive comparison of the performance of our proposed model on all three datasets, allowing us to evaluate its generalization ability and robustness to different activity categories.

Table 2. The accuracy and loss function of the model over 50 epochs, during both training and testing phases.

Evaluation Metrics	MOD20	UCF50mini	LoDVP Abnormal Activities
Train loss	0.4223	0.2106	0.1042
Train accuracy	86.30%	92.50%	96.68%
Test loss	0.5614	0.3568	0.3982
Test accuracy	78.21%	87.78%	83.12%

In both cases, we evaluated the confusion matrix for the LoDVP Abnormal Activities and UCF50mini datasets. The first confusion matrix is shown in Table 3, which expresses the results of the neural network architecture tests on the LoDVP Abnormal Activities dataset. All classes were divided into the following categories: 1. Begging, 2. Drunkenness, 3. Fight, 4. Harassment, 5. Hijack, 6. Knife hazard, 7. Normal videos, 8. Pollution, 9. Property damage, 10. Robbery, 11. Terrorism. The confusion matrix shows how the tested videos were correctly and incorrectly classified into the given categories. For most classes, the proposed architecture did not have a classification problem. However, for the third class (Fight), we observe an increased error rate, where four videos were misclassified into the first class (Begging) and six videos into the second class (Drunkenness). These errors may have occurred due to the similarity of the videos. For example, sometimes a small fight can be confused with begging, and begging can turn into a fight. The similarity between a fight and drunkenness is also quite high, as drunk people can push each other violently, which can lead to a fight.

Table 3. The example of the confusion matrix for LoDVP Abnormal Activities.

Targeted/ Predicted	1	2	3	4	5	6	7	8	9	10	11
1	16	0	0	0	0	0	0	0	0	0	0
2	0	14	2	0	0	0	0	0	0	0	0
3	4	6	18	0	0	0	0	0	0	0	0
4	0	0	0	16	0	0	4	0	0	0	0
5	0	0	0	0	20	0	0	0	0	0	0
6	0	2	0	0	0	8	0	0	0	0	0
7	0	0	0	2	0	0	18	2	0	0	0
8	0	0	0	6	0	0	2	12	0	0	0
9	0	0	0	0	0	0	0	0	26	0	0
10	2	0	0	0	0	0	0	0	0	4	0
11	0	0	0	0	0	0	0	0	0	0	14

Furthermore, we created a confusion matrix to display the results of testing the proposed neural network architecture on the UCF50mini dataset (see Table 4). The classes in the confusion matrix are divided into the following categories: 1. Baseball Pitch, 2. Basketball Shooting, 3. Bench Press, 4. Biking, 5. Billiards Shot, 6. Breaststroke, 7. Clean and Jerk, 8. Diving, 9. Drumming, 10. Fencing. The confusion matrix shows the increased accuracy during classification within the testing process. Upon observing the confusion matrix, we can assess that the biggest problem in the classification occurred in category two Basketball Shooting, where four videos were incorrectly classified into category four Biking. The neural network architecture also had a problem classifying class seven Clean and Jerk, where it misclassified one video into class three Bench Press and two videos into the Biking class.

Similarly, we evaluated the confusion matrix for the UCF50mini dataset, which is shown in Table 4. The dataset consists of ten classes, and the confusion matrix shows how the tested videos were correctly and incorrectly classified into these classes. The proposed architecture performed well for most classes, with only a few misclassifications. However, the model had difficulty distinguishing between two classes, namely Billiards Shot and Drumming. Some videos were misclassified as Billiards Shot when they should have been Drumming, and vice versa. This could be due to the similarity in the movements of the two activities, such as hand-eye coordination and rhythmic movements.

Table 4. The example of the confusion matrix for UCF50mini.

Targeted/ Predicted	1	2	3	4	5	6	7	8	9	10
1	14	0	0	1	0	0	0	0	0	0
2	0	11	0	4	0	0	0	0	0	0
3	0	0	5	0	0	0	0	0	1	0
4	1	1	0	13	0	0	0	1	0	1
5	0	0	0	0	15	0	0	0	0	0
6	0	0	0	0	0	8	0	0	0	0
7	0	0	1	2	0	0	9	0	0	0
8	0	0	0	0	1	0	0	12	0	0
9	0	0	1	0	0	0	0	0	17	0
10	0	0	0	0	0	0	1	0	1	11

We also created a confusion matrix to display the results of testing the neural network architecture on the MOD20 dataset (see Table 5). In the confusion matrix, the classes are assigned numbers as follows: 1. tourism, 2. cliff jumping, 3. chopping wood, 4. cycling, 5. dancing, 6. fighting, 7. figure-skating, 8. Fire-fighting, 9. motorized saw-trees, 10. jet ski, 11. kayak, 12. motorcycle, 13. FOOTBALL-catching, 14. rock-climbing, 15. running, 16. skateboarding, 17. skiing, 18. stand-up paddling, 19. surfing, and 20. windsurfing. The confusion matrix enables us to monitor the accuracy of classification within the testing process. By observing the confusion matrix, we can evaluate that there is not one class with the biggest problem with classification, but the problems within the classification are evenly distributed. The biggest mistakes made in placing a video into a class were two videos. However, we can state that the best results were achieved by categories 13 (FOOTBALL-catching), 18 (stand-up paddling), and 20 (windsurfing), where the neural network architecture had no problem with classification. The unproblematic classification of the mentioned classes may be due to the clarity of the activity occurring in the video.

Table 5. The example of the confusion matrix for MOD20.

Targeted/ Predicted	1	2	3	4	5	6	7	8	9	10	11	12	13	14	15	16	17	18	19	20
1	9	0	0	1	0	0	0	0	0	2	0	0	0	1	2	0	0	0	0	1
2	0	9	0	2	0	0	1	0	0	0	0	0	1	0	0	0	0	0	0	0
3	1	0	10	0	0	0	0	0	0	1	0	0	0	0	1	0	0	1	0	0
4	0	0	0	7	1	0	0	1	0	0	0	1	0	0	0	0	0	0	0	0
5	2	0	0	0	6	0	2	0	0	1	0	0	0	2	0	0	0	0	0	1
6	1	0	0	2	0	6	0	0	0	0	0	1	0	0	0	0	1	0	0	0
7	0	0	0	0	0	0	14	0	0	0	0	0	1	0	0	0	0	0	0	0
8	0	0	0	0	0	0	0	14	0	0	0	0	0	0	0	0	0	2	0	0
9	0	0	0	1	0	2	0	0	7	0	0	0	0	0	1	0	0	0	0	1
10	1	0	1	0	0	0	0	0	1	9	0	0	0	1	0	0	0	0	0	0
11	0	0	0	0	0	0	0	0	0	0	11	0	0	0	0	1	0	1	0	0
12	0	0	0	0	0	0	0	0	0	0	0	12	2	0	0	0	0	0	0	0
13	0	0	0	0	0	0	0	0	0	0	0	0	11	0	0	0	0	0	0	0
14	1	0	0	0	0	0	0	0	0	0	0	0	1	9	0	0	0	0	1	0
15	1	0	0	0	0	0	0	0	0	0	0	0	0	0	10	0	0	0	0	0
16	0	0	0	0	0	0	0	0	0	0	1	0	0	0	0	11	0	0	0	0
17	0	0	2	0	0	0	0	0	0	0	0	0	0	0	0	0	9	0	0	0
18	0	0	0	0	0	0	0	0	0	0	0	0	0	0	0	0	0	14	0	0
19	0	0	0	0	0	0	0	0	0	0	0	0	0	1	0	0	0	1	12	0
20	0	0	0	0	0	0	0	0	0	0	0	0	0	0	0	0	0	0	0	11

In general, we have evaluated the results of our tests in several tables. First, we assessed the resulting values of metrics such as F1 score, Precision, and Recall across all three datasets (see Table 6). We can observe that the best results using the proposed architecture were achieved on the UCF50mini dataset. However, the metric results were roughly similar across all three datasets.

Table 6. The evaluation criterion of the proposed neural network architectures using different datasets.

Evaluation Metrics	MOD20	UCF50mini	LoDVP Abnormal Activities
Precision (P)	83.89%	87.76%	89.12%
Recall (R)	81.09%	88.63%	87.69%
F1 score (F1)	81.57%	87.84%	89.32%

Moreover, we compared the performance of our proposed architecture with other available architectures. We applied the architectures to the given datasets and compared their accuracy values. In Table 7, we can observe the resulting accuracy value after testing on the LoDVP Abnormal Activities dataset. We tested ConvLSTM architectures published in the article [21] and 3D Resnet networks such as 3D Resnet50, 3D Resnet101, and 3D Resnet152 [22] on the same dataset. From these results, we can see that our proposed architecture, which combines 3DCNN with ConvLSTM, has the third-best value. Therefore, we can evaluate that our architecture can classify various abnormal incidents such as harassment, fight, etc. very well compared to other architectures. However, compared to classical ConvLSTM, it did not achieve the best results.

Table 7. Accuracy comparison of the proposed architecture with various neural network architectures using the LoDVP dataset for detecting abnormal activities.

Video Recognition Architectures	Accuracy [%]
Proposed architecture	93.41
ConvLSTM [21]	92.38
3D Resnet50 [22]	36.19
3D Resnet101 [22]	61.90
3D Resnet152 [22]	90.48

In Table 8, we can observe the accuracy results of the same neural network architectures as in the previous case, i.e., ConvLSTM and 3D Resnet50, 101, 152, tested on the UCF50mini dataset. In this case, our proposed architecture achieved significantly better results than the other architectures after the testing process. Our 3DCNN + ConvLSTM architecture achieved an accuracy of 87.7%.

Table 8. Accuracy comparison of the proposed architecture with various neural network architectures using the UCF50mini dataset.

Video Recognition Architectures	Accuracy [%]
Proposed architecture	87.78
ConvLSTM [21]	80.38
3D Resnet50 [22]	71.53
3D Resnet101 [22]	75.91
3D Resnet152 [22]	83.39

Furthermore, we trained and tested the MOD20 dataset with our proposed architecture and compared the results with the Kernelized Rank-Based Pooling (KRBP) and Feature Subspace-Based Kernelized Rank Pooling (KRP-FS) approaches used by the authors [21]. The best results were achieved using the 3DCNN + ConvLSTM architecture, with an accuracy of 78.21%. The accuracy results of our proposed neural network architecture and the aforementioned approaches are shown in Table 9.

Table 9. Accuracy comparison of the proposed architecture with various neural network architectures using the MOD20 dataset.

Video Recognition Architectures	Accuracy [%]
Proposed architecture	78.21
BKRP [21]	66.55
KRP-FS [21]	74.00

Based on the results obtained on the three datasets (LoDVP Abnormal Activities, UCF50mini, and MOD20), we can conclude that combining 3DCNN and ConvLSTM layers can lead to a neural network architecture whose results are comparable to or better than other available approaches. In terms of experimental results, we compared the performance of our proposed architecture on the LoDVP Abnormal Activities and UCF50mini datasets with architectures such as ConvLSTM and 3D Resnet50,101,152. In addition, we compared the accuracy values obtained on the MOD20 dataset with the BKRP and KRP-FS approaches. Our proposed architecture achieved an accuracy of 89.41% on the LoDVP Abnormal Activity dataset, 87.78% on the UCF50mini dataset, and 78.21% on the MOD20 dataset.

5. Conclusions and Future Work

This paper proposes a mixed-architecture neural network for classifying human activities from videos. The architecture combines a 3DCNN network layer and a ConvLSTM layer. We trained and tested our network on three databases: UCF50mini (where we selected the first 10 classes), MOD20, and LoDVP Abnormal Activity. To ensure fair comparison between the datasets, we reduced the UCF50 database to match the number of classes in the other two datasets. For UCF50mini, we performed classification on 10 classes: Baseball pitch, Basketball shooting, Bench press, Cycling, Billiard shooting, Breaststroke, Pure movement, Diving, Drumming, and Fencing. For LoDVP Abnormal Activities, we classified 11 classes: begging, drunkenness, fighting, harassment, kidnapping, knife danger, common videos, pollution, property damage, robbery, and terrorism. MOD20 is the largest dataset with 20 classes: tourism, cliff jumping, logging, cycling, dancing, fighting, figure-skating, fire-fighting, chainsawing trees, water skiing, kayaking, motorcycle, football-catching, climbing, running, skateboarding, skiing, standup-paddling, surfing, and windsurfing. The input videos in all datasets were cropped to 100 × 100 × 3 RGB, and we used 70 frames as input for the neural network architecture.

The results showed that the combined 3DCNN + ConvLSTM neural network was effective in classifying video data containing various human activities. The training on the UCF50mini dataset resulted in a decrease in the loss function to 0.2106 and an increase in accuracy to 92.50%. For the LoDVP Abnormal Activities dataset, the loss function decreased to 0.1042 and accuracy increased to 96.68% during training. On the MOD20 dataset, the loss function during training was 0.4223 and accuracy increased to 86.30%. When comparing the results on the datasets during testing, the combined architecture coped well with the problem of temporal continuity between images. The confusion matrix across all three datasets showed that the classification process was successful, with minimal errors in the average of each class. The overall accuracy of the UCF50 mini dataset test was 87.78%, with precision of 87.76% and recall of 88.63%. The F1 score was 87.84%. For the LoDVP Abnormal Activity dataset, the overall accuracy was 93.41%, with precision of 89.12%, recall of 87.69%, and F1 score of 89.32%. On the MOD20 dataset, the overall accuracy was 78.21%, with precision of 83.89%, recall of 81.09%, and F1 score of 81.57%. The results showed that the 3DCNN + ConvLSTM neural network is capable of classifying video data containing various human activities, with high accuracy and minimal errors in the average of each class. The proposed architecture achieved good results when compared to other existing networks designed for video-based human behavior classification. Overall, the results demonstrate the success in creating a neural network architecture combining 3DCNN and ConvLSTM layers for classifying human behavior in videos.

However, we aim to continue our work and improve the classification results of human activities captured in videos. Accurate classification of human behavior by neural networks can significantly enhance their practical applications. To provide a more comprehensive evaluation of our network's performance, we compared it to other available neural network architectures, such as 3D ResNet 50,101,152, ConvLSTM, KRBP, and KRP-FS approaches, using three different datasets. Our contribution focuses primarily on recognizing and classifying non-standard human behavior in public spaces, which has a significant impact on the scientific community. The proposed 3DCNN + ConvLSTM architecture has wide-ranging applications in fields such as security and medicine and is comparable to existing networks designed for video-based human behavior classification. However, monitoring and detecting unusual behavior in public places such as city parks and squares is still a challenging task, and our proposed combination of 3DCNN and ConvLSTM has some limitations, including:

- Limited interpretability: 3DCNN with ConvLSTM is a deep-learning architecture, and like most deep-learning models, it is not transparent in how it makes predictions (understanding how the model arrives at a particular decision can be challenging).
- Limited availability of training data: The training of 3DCNN with ConvLSTM requires a large amount of high-quality data to produce good results. This can be a significant limitation in many applications where such data are not readily available.
- Difficulty in tuning hyperparameters: 3DCNN with ConvLSTM involves several hyperparameters that need to be tuned correctly to achieve optimal performance. Tuning these hyperparameters can be time-consuming and requires a significant amount of expertise and experimentation.
- Sensitivity to noise and missing data: The combination of 3DCNN and ConvLSTM relies on the temporal coherence of data for accurate predictions. Therefore, the model can be sensitive to noise and missing data in the input, which can significantly affect the model's performance.

In summary, the proposed combination of 3DCNN with ConvLSTM is a powerful deep-learning architecture with several limitations, which can impact its scalability, interpretability, data requirements, hyperparameter tuning, and sensitivity to noise and missing data.

In the future work, we plan to explore the incorporation of additional sensor data, such as depth cameras and inertial measurement units, to enhance the performance of the proposed model. We also plan to investigate the use of transfer learning techniques to adapt the model to different domains and environments. Furthermore, we aim to investigate the use of the proposed model for other related tasks such as anomaly detection. Finally, we will also investigate the possibility of deploying the model on edge devices for real-time monitoring. We believe that incorporating additional sensor data could further improve the performance of our model and we look forward to exploring this direction in future work.

Author Contributions: Conceptualization, R.V., P.K. and R.H.; methodology, R.V.; software, R.V. and P.S.; validation, P.K., R.V. and R.H.; formal analysis, P.K., R.H. and P.S.; data curation, R.V.; writing—original draft preparation, R.V. and P.K.; writing—review and editing, R.V. and P.K.; visualization, R.V., P.K. and P.S.; supervision, R.H.; project administration, R.H. and P.K.; funding acquisition, R.H. All authors have read and agreed to the published version of the manuscript.

Funding: This work was supported by the Slovak Research and Development Agency under the contracts no. APVV-21-0502 BrainWatch.

Institutional Review Board Statement: Ethical review and approval were waived for this study, due to the experiment did no harm to all subjects.

Informed Consent Statement: Patient consent was waived due to the fact that we used existing datasets for testing (UCF50, MOD20, LoDVP Abnormal Activities datasets). These datasets were cited in the text of the article text of the article.

Data Availability Statement: The data presented in this study are available on request from the corresponding author. This is according to the laboratory rules.

Conflicts of Interest: The authors declare no conflict of interest.

Abbreviations

ConvLSTM	Convolutional Long Short-Term Memory
3DCNN	3D Convolutional Network
KRP-FS	Feature Subspace-Based Kernelized Rank Pooling
BKRP	Kernelized Rank-Based Pooling
BILSTM	Bidirectional Long Short-Term Memory
Conv2D	2D Convolutional Neural Network
Conv3D	3D Convolutional Neural Network
ROI	Region of Interest
MLP	Multilayer perceptron
HOF	Histogram of oriented gradients
ASD	Autism Spectrum Disorder

References

1. Wang, T.; Li, J.; Zhang, M.; Zhu, A.; Snoussi, H.; Choi, C. An enhanced 3DCNN-ConvLSTM for spatiotemporal multimedia data analysis. *Concurr. Comput. Pract. Exp.* **2021**, *33*, e5302. [CrossRef]
2. Islam, M.S.; Gao, Y.; Ji, Z.; Lv, J.; Mohammed, A.A.Q.; Sang, Y. 3DCNN Backed Conv-LSTM Auto Encoder for Micro Facial Expression Video Recognition. *Mach. Learn. Intell. Commun.* **2021**, *438*, 90–105.
3. Zhu, G.; Zhang, L.; Shen, P.; Song, J.; Shah, S.A.A. Continuous Gesture Segmentation and Recognition using 3DCNN and Convolutional LSTM. *IEEE Trans. Multimed.* **2019**, *21*, 1011–1021. [CrossRef]
4. Krishna, N.S.; Bhattu, S.N.; Somayajulu, D.V.L.N.; Kumar, N.V.N.; Reddy, K.J.S. GssMILP for anomaly classification in surveillance videos. *IEEE Expert Syst. Appl.* **2022**, *203*, 117451. [CrossRef]
5. Pediaditis, M.; Farmaki, C.; Schiza, S.; Tzanakis, N.; Galanakis, E.; Sakkalis, V. Contactless respiratory rate estimation from video in a real-life clinical environment using eulerian magnification and 3D CNNs. In Proceedings of the IEEE International Conference on Imaging Systems and Techniques, Kaohsiung, Taiwan, 21–23 June 2022.
6. Negin, F.; Ozyer, B.; Agahian, S.; Kacdioglu, S.; Ozyer, G.T. Vision-assisted recognition of stereotype behaviors for early diagnosis of Autism Spectrum Disorders. *Neurocomputing* **2022**, *446*, 145–155. [CrossRef]
7. Kaçdioglu, S.; Özyer, B.; Özyer, G.T. Recognizing Self-Stimulatory Behaviours for Autism Spectrum Disorders. In Proceedings of the Signal Processing and Communications Applications Conference, Gaziantep, Turkey, 5–7 October 2020; Volume 28, pp. 1–4.
8. Zhao, W.; Xu, J.; Li, X.; Chen, Z.; Chen, X. Recognition of Farmers' Working Based on HC-LSTM Model. *Neurocomputing* **2022**, *813*, 77–86.
9. Zhang, L.; Zhu, G.; Shen, P.; Song, J.; Shah, S. A.; Bennamoun, M. Learning Spatiotemporal Features Using 3DCNN and Convolutional LSTM for Gesture Recognition. In Proceedings of the IEEE International Conference on Computer Vision Workshops, Venice, Italy, 22–29 October 2017; pp. 3120–3128.
10. Xu, C.; Chai, D.; He, J.; Zhang, X.; Duan, S. InnoHAR: A Deep Neural Network for Complex Human Activity Recognition. *IEEE Access* **2019**, *7*, 9893–9902. [CrossRef]
11. Almabdy, S.; Elrefaei, L. Deep Convolutional Neural Network-Based Approaches for Face Recognition. *Appl. Sci.* **2019**, *9*, 4397. [CrossRef]
12. Zheng, W.; Yin, L.; Chen, X.; Ma, Z.; Liu, S.; Yang, B. Knowledge Base Graph Embedding Module Design for Visual Question Answering Model. *Pattern Recognit.* **2021**, *120*, 108153. [CrossRef]
13. Mutegeki, R.; Han, D.S. A CNN-LSTM Approach to Human Activity Recognition. In Proceedings of the 2020 International Conference on Artificial Intelligence in Information and Communication (ICAIIC), Fukuoka, Japan, 19–21 February 2020; IEEE: Piscataway, NJ, USA, 2020; pp. 362–366.
14. Vrskova, R.; Hudec,R.; Kamencay,P.; Sykora, P. A New Approach for Abnormal Human Activities Recognition Based on ConvLSTM Architecture. *Sensors* **2022**, *22*, 2946. [CrossRef] [PubMed]
15. Vrskova, R.; Hudec, R.; Kamencay, P.; Sykora, P. Human Activity Classification Using the 3DCNN Architecture. *Appl. Sci.* **2022**, *12*, 931. [CrossRef]
16. Chengping, R.; Yang, L. 3D Convolutional Neural Networks for Human Action Recognition. *Comput. Mater. Sci.* **2013**, *35*, 221–231.

17. Partila, P.; Tovarek, J.; Ilk, H.G.; Rozhon, J.; Voznak, M. Deep learning serves voice cloning: How vulnerable are automatic speaker verification systems to spooting trial. *IEEE Commun. Mag.* **2020**, *58*, 100–105. [CrossRef]
18. Ji, S.; Xu, W.; Yang, M.; Yu, K. Three-dimensional convolutional neural network (3D-CNN) for heterogeneous material homogenization. *IEEE Trans. Pattern Anal. Mach. Intell.* **2013**, *184*, 221–231. [CrossRef] [PubMed]
19. Yuan, Z.; Zhou X.; Yang, T. Hetero-ConvLSTM: A Deep Learning Approach to Traffic Accident Prediction on Heterogeneous Spatio-Temporal Data. In Proceedings of the 24th ACM SIGKDD International Conference on Knowledge Discovery & Data Mining, London, UK, 19–23 August 2018.
20. Reddy, K.K.; Shah, M. Recognizing 50 Human Action Categories of Web Videos. *Mach. Vis. Appl. J. (MVAP)* **2013**, *24*, 971–981. [CrossRef]
21. Perera, A.G.; Law, Y.W.; Ogunwa, T.T.; Chahl, J. A Multiviewpoint Outdoor Dataset for Human Action Recognition. *IEEE Trans.-Hum.-Mach. Syst.* **2020**, *50*, 405–413. [CrossRef]
22. Ghodhbani, E.; Kaanich, M.; Benazza-Benyahia, A. An Effective 3D ResNet Architecture for Stereo Image Retrieval. In Proceedings of the 16th International Joint Conference on Computer Vision, Imaging and Computer Graphics Theory and Applications (VISIGRAPP 2021), Virtual Event, 8–10 February 2021; pp. 160580—160595.

sensors

MDPI

Article

A Development of an IoT-Based Connected University System: Progress Report †

Slavomir Matuska *, Juraj Machaj, Miroslav Hutar and Peter Brida

Faculty of Electrical Engineering and Information Technology, University of Zilina, 010 26 Zilina, Slovakia
* Correspondence: slavomir.matuska@uniza.sk; Tel.: +421-41-513-2215

† This paper is an extended version of our paper published in S. Matuska, and R. Hudec. A Functional IoT-based System Design of the Connected University. In Proceeding of the conference ELEKTRO 2022, Krakow, Poland, 23–26 May 2022.

Abstract: In this paper, a report on the development of an Internet of Things (IoT)-based connected university system is presented. There have been multiple smart solutions developed at the university over recent years. However, the user base of these systems is limited. The IoT-based connected university system allows for integration of multiple subsystems without the need to implement all of them in the same environment, thus enabling end-users to access multiple solutions through a single common interface. The implementation is based on microservice architecture, with the focus mainly on system robustness, scalability, and universality. In the system design, four subsystems are currently implemented, i.e., the subsystem for indoor navigation, the subsystem for parking assistants, the subsystem for smart classrooms or offices, and the subsystem for news aggregation from university life. The principles of all implemented subsystems, as well as the implementation of the system as a web interface and a mobile application, are presented in the paper. Moreover, the implementation of the indoor navigation subsystem that uses signals from Bluetooth beacons is described in detail. The paper also presents results proving the concept of the Bluetooth-based indoor navigation, taking into account different placements of nodes. The tests were performed in a real-world environment to evaluate the feasibility of the navigation module that utilizes deterministic fingerprinting algorithms to estimate the positions of users' devices.

Keywords: IoT; smart systems; indoor navigation; mobile application

Citation: Matuska, S.; Machaj, J.; Hutar, M.; Brida, P. A Development of an IoT-Based Connected University System: Progress Report. *Sensors* **2023**, *23*, 2875. https://doi.org/10.3390/s23062875

Academic Editors: Alberto Gotta and Alessandra Rizzardi

Received: 30 January 2023
Revised: 20 February 2023
Accepted: 6 March 2023
Published: 7 March 2023

1. Introduction and Motivation

In recent years, Internet of Things (IoT) technology has experienced a global boom. It is currently used in various industrial applications as well as in the daily lives of ordinary people. The fast and efficient development and deployment of IoT applications are enabled by miniaturizing and reducing hardware costs and the existence of various software-oriented services.

In recent years, several independent intelligent systems were developed at the University of Zilina. These systems were either used independently with a limited number of users or were not used at all and have fallen into oblivion. The main motivation for the development of an IoT-based connected university system was to create a versatile and robust system built on microservices so that it will be possible to implement individual solutions as sub-systems within that system.

Taking into account that the individual services may not have anything in common and may be built in completely different architectures and programming languages, the decision was made to build the proposed system using a microservice architecture. In this architecture, the individual microservices can be created in various programming languages, since each service operates in its own container with its own environment variables. Moreover, the advantage of the microservice architecture is that the possible malfunction of a single service does not affect the other services. The proposed system is open for the implementation of novel sub-systems in the future, due to the use of the microservices architecture.

The individual services implemented within the system are intended to assist or facilitate the normal daily routine within the university. As part of the system design, four sub-systems were proposed, while each service will provide benefits for users at the university in a different way. For example, the news aggregation subsystem is designed to cause it to be easier to bring together news from different sources of university life and display it in one place. This subsystem also retrieves current and historical weather information from around the university. Moreover, the subsystem for assisted parking is intended to speed up finding a free parking space, saving time but also saving the environment. On top of that, the smart classroom subsystem is to assist students and staff in following the principles of proper chair seating and, in this way, help to prevent health problems. Furthermore, the indoor navigation subsystem is useful for junior students and helps them navigate the university campus and find the right classroom.

The principles within each subsystem are based on previously published works and thus do not provide novel approaches. However, the main contribution of this paper is the design and implementation of a robust IoT-based connected university system, which connects the different subsystems and causes them to be accessible through a single common interface, i.e., web interface or mobile application. The system, therefore, enables the creation of novel services for a user taking into account data from multiple subsystems.

The remainder of the paper is organized as follows, Section 2 provides a review of the related work on IoT system development, Section 3 describes the proposed system design and architecture, Section 4 presents the subsystems and their implementation, Section 5 describes the system implementation from the web interface and mobile application point of view, Section 6 presents the proof of concept of implemented BLE localization, and Section 7 concludes the paper.

2. Related Work

The development of IoT systems and applications is currently of interest to many research groups. IoT systems and applications are widely used in multiple areas. In general, IoT systems and applications collect information from sensors and use the captured data to optimize processes or create insights into the data that can be useful.

A smart air quality monitoring IoT-based infrastructure for industrial environments was presented in [1]. The authors presented a complete air quality monitoring infrastructure based on the IoT paradigm that is fully integrable with current industrial systems. The monitoring infrastructure includes highly precise compact devices to facilitate the real-time monitoring of particulate matter concentrations and polluting gases in the air. The Big Data collected by this system are processed using machine learning techniques to predict whether safety levels might be surpassed.

Anumala et al. [2] described an IoT-based air quality monitoring system to continuously observe the air quality inside a room. The system collects information about the temperature, humidity, dust, and gas level. All the data are stored in the database and also displayed in real time on the LCD in the room. When the measured values exceed threshold values, a warning message is sent to the house owner's mobile device using the Global System for Mobile communication (GSM) module.

A long-range outdoor air quality monitoring system based on the LoRaWAN was proposed in [3]. The system consists of multiple sensors (NO_2, SO_2, CO_2, CO, PM2.5, temperature, and humidity), an Arduino microcontroller, a LoRa shield, a LoRaWAN gateway, and The Thing Network (TTN) IoT platform. The IoT nodes are powered by a rechargeable battery with a photovoltaic solar panel. The measured data are stored in the cloud and system users can easily access them on the web-based dashboard. The authors validated the collected data against the high-technology Aeroqual air quality monitoring devices and proved that their system can reliably monitor various air quality indicators.

The monitoring system applied to an aeroponic greenhouse based on an IoT architecture was presented in [4]. The system provides information on the status of the climatic variables and the appearance of the crop in addition to managing the irrigation timing and the frequency of the visual inspections using an Android application called Aeroponics Monitor. The authors use the Thingspeak cloud for data analysis and Firebase servers for data storage.

The challenges in agriculture were addressed in [5]. The proposed IoT-based system was developed for monitoring environmental parameters such as soil moisture, temperature, and humidity. The AgriWealth farming system utilizing the IoT sensors and the Android application that helps farmers control and manage the farm was proposed in [6]. The machine learning model is used to predict suitable crops in accordance with varying weather parameters.

Another flexible IoT agriculture system for irrigation control based on Software Services was described in [7]. The presented paper adopts a software-centric perspective to model and design IoT systems in a flexible manner and provided a simple and novel view of the design of IoT systems in agriculture from the software perspective.

Ahmed et al. [8] designed and implemented a flexible IoT-based platform for the remote monitoring of agriculture farms of different scales, enabling continuous data collection from various IoT devices. IoT nodes are based on the LoRaWAN technology. The authors undertook an experimental validation and showed that the platform can be used to obtain valuable analytics of real-time monitoring that enable decisions and actions such as, for example, controlling the irrigation system or generating alarms.

Security is an important aspect of all information systems, especially in IoT. Vashishtha et al. [9] presented security and detection mechanisms in IoT-based cloud computing using a hybrid approach. They combine RSA and RC4 algorithms for the generation of the key to obtain the superior security method. Security vulnerabilities continue to exist, and security incidents continue to increase. Kim and Yoo [10] classified and analyzed the most common security vulnerabilities for IoT devices and identified the essential vulnerabilities of IoT devices that should be considered for security when producing IoT devices.

IoT technology is also used in smart homes. Vishwakarma et al. [11] proposed a smart energy-efficient home automation system that can remotely access and control home equipment. The solution is based on NodeMCU, Adafruit sensors, and the Arduino platform. The design and implementation of a Cloud-IoT-Based Home Energy Management System were presented in [12]. The developed system allows for collecting and storing energy consumption data from appliances and the main load of the home. Two scenarios were designed and an implemented AWS cloud was used to store all the collected data.

Wu et al. [13] proposed a wearable sensor network system for health applications. The proposed network incorporates multiple wearable sensors to monitor the environmental and physiological parameters of individuals. Based on the sensor specification, the implemented sensors communicate using the LoRa network or Bluetooth. The system includes a warning mechanism that is triggered when a harmful environment is detected. Onasanya and Elshakankiri [14] described a Smart integrated IoT healthcare system for cancer care. They proposed the implementation of an IoT-enabled medical system for the enhanced treatment, diagnosis, detection, and monitoring of cancer patients based on cancer care services. They also created business analytics for insights creation, decision-making, data transmission, and reporting.

The authors in [15] described an IoT-connected e-textile wearable for neonatal medical monitoring called NeoWear. The proposed chest belt wearable should monitor the respiration rate and detect apnea events in babies. The NeoWear is a wearable system consisting of a sensor belt, a wearable embedded system, and an edge computing device. The IoT device is connected to the edge computing device and sends the data over the MQ Telemetry Transport (MQTT) protocol. Their findings show an average error of 0.89 BrPM in respiration rate measurement and 97% accuracy in apnea detection on the programmable baby mannequin.

Other important fields, where IoT systems are widely used are smart cities and smart transportation. Zhu et al. [16] presented a vision as well as work on the integration of artificial intelligence and the intelligent transportation system (ITS) to create an enhanced intelligence of IoT-enabled ITS. A design of an IoT-based smart city model on Raspberry Pi was proposed in [17]. Kumar et al. created an urban IoT system utilizing Raspberry Pi to help smart cities solve domestic difficulties. One of the major smart cities issues is energy optimization.

The authors in [18] proposed a model that can be used to optimize energy consumption in smart homes and smart cities alike. They equipped all smart city electric appliances with the sensors, such as street lighting, building and street billboards, smart homes, and smart parking. The suggested model was evaluated using mathematical modeling and the findings indicated that the proposed model may assist in improving energy usage in smart cities. Smart cities also include people who can use their mobile devices to participate in sensing various environmental variables. This is called crowd-sensing.

The authors in [19] proposed privacy-preserving hybrid sensing for smart cities. They explored an integrated paradigm called "hybrid sensing" that harnesses both IoT-sensing and crowd-sensing in a complementary manner. The authors presented their hybrid sensing on the smart system parking, by which users can inquire and find the available parking spaces in outdoor parking lots. IoT solutions are widely implemented as parking sensing systems that deploy robust outdoor vehicle localization and recognition methodology.

In the [20], the authors proposed a new low-cost sensor system allowing for real-time parking occupancy monitoring along with parking payment without the requirement of any user/driver interaction. A two-stage hybrid approach to help drivers find a parking space was presented in [21]. The proposed solution should decrease the time and energy consumed in finding the parking lot. The first stage focuses on car parks with at least one free parking space located near the target address. The most suitable parking space is searched for and presented in the second stage. The authors also proposed a dynamic cloud-based parking lot reservation system.

A smart parking solution based on the integration of NB-IoT radio communication technology and the core IoT platform was proposed by authors in [22]. Their solution for smart parking benefits from the usage of narrow-band Internet of Things (NB-IoT) technology. The authors created a mobile application for the real-time checking of parking lot availability. A parking space can be reserved via a smartphone application, which will help drivers to find and reserve spots, park their vehicles, and pay.

Sobeslav and Horacek [23] presented A Smart Parking System Based on a Mini PC Platform and Mobile Application for Parking Space Detection. The solution relies on an MPU–9250 magnetometer sensor connected to an Arduino mini microprocessor to detect the vehicle in the parking lot. To send data to the Raspberry Pi server, the IQRF, which is a wireless mesh technology in sub-GHz ISM radio bands, was used. It is possible for a user to create parking place reservations via the mobile application and then the application navigates the user to the reserved spot. The problem with this solution is that it needs a lot of IoT nodes to cover larger parking lots. The camera-based solution is more widespread since it is much easier to implement.

The solutions proposed in [24] and ref. [25] describe the real-time parking lot occupancy detection systems based on a visual sensor network. In both papers, a camera-based solution is used and the achieved results proved that this type of system is suitable for parking occupancy detection.

Indoor localization and navigation is still a hot research topic and an active field of development for many researchers. The most accurate indoor positioning systems are based on UWB technology [26]. A comprehensive survey for indoor positioning systems for IoT-based applications was presented in [27]. Their final finding was that the Global

Positioning System (GPS) and the indoor positioning system (IPS) can act as complements to offer a comprehensive location-based service for both indoor and outdoor environments.

Schroeer [28] presented a real-time UWB multi-channel indoor positioning system for industrial scenarios. The author uses four transceivers instead of one to increase the system robustness for each base station. The root mean square error has been reduced by 0.1 m in comparison with the single-channel indoor positioning system. However, it is necessary to install a based station and equip users or objects with the active tag in order to use the UWB technology for positioning and indoor navigation.

Another option is to use smartphones and their sensors for localization and navigation. Bluetooth localization technology, its principles, applications, and future trends are summarized by the authors in [29]. They reviewed the applications and existing commercial solutions, revealing the possible directions for the industrialization of Bluetooth localization. The Bluetooth Low Energy (BLE)-based indoor localization systems are limited to using only three advertisement channels. The work presented in [30] analyzed the impact of channel diversity on the accuracy of BLE-based indoor localization. The experiments conducted in a 100 m^2 office area show that using signal strength measurements in 40 channels improves the average localization accuracy by approximately 50%.

The challenges related to indoor positioning using Wi-Fi signals were summarized in our previous work [31,32]. An improved Wi-Fi location fingerprint positioning algorithm for robot indoor positioning and navigation was described in [33]. In order to eliminate the location fingerprints that degrade the localization accuracy, Ye and Peng integrated an improved adaptive K-value WKNN algorithm at the end of the localization algorithm. The experimental results show that the probability of the improved algorithm's positioning error within 0.4 m is 49%, which is a 35% improvement over the conventional algorithm. A smartphone-based indoor navigation system was presented in [34]. They present enhancing the particle filter by utilizing a map constraint and k-means clustering and integrating Bluetooth low energy along with pedestrian dead reckoning for positioning. For the overall performance of PFMK, a mean error of <1.5 m in the test environments was achieved.

3. IoT-Based Connected University System Design

The IoT-based connected university system is a huge IoT system that allows users to access various services provided at the campus of the University of Zilina. The system includes multiple different subsystems implemented as micro-services while providing the end-users information from the system via a single common user interface (mobile application and web). The proposed subsystems are as follows:

- Subsystem for news aggregation;
- Subsystem for smart assistant parking;
- Subsystem for smart classroom or office;
- Subsystem for indoor navigation.

The design of the system is focused on the robustness, scalability, and universality. The proposed design is open, so it is possible to implement new services in the future. The principle of the system design is illustrated in Figure 1.

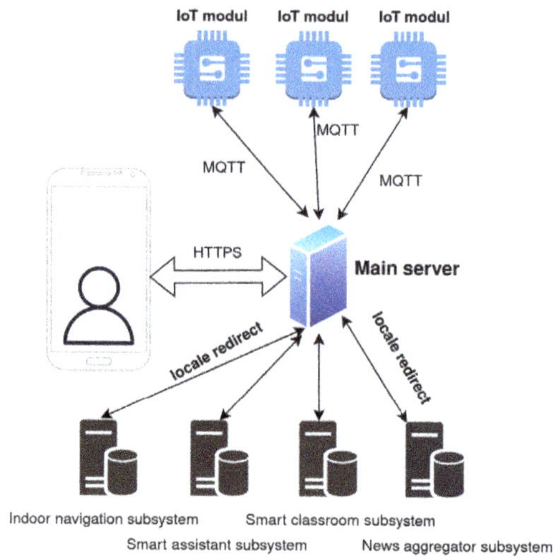

Figure 1. The principle of the system design of the IoT-based connected university.

The main unit of the proposed IoT-based connected university system is a server placed on the premises. The users communicate with the server exclusively through HTTPS requests to ensure communication security. Overall, the system provides two levels of users—non-authorized users and authorized users. Some of the services implemented in the system are considered to be free services, which are accessible also to non-authorized users and are available on the home page of the system. However, some subsystems, including indoor navigation and smart classroom subsystems, are only available to authorized users. The system architecture design is based on microservices. Microservices, also known as microservice architecture, is an architectural style that structures an application as a collection of services. The advantage of microservices is that individual services can be independently deployed, are highly maintainable, scalable, and robust, and can be organized around business capabilities. The microservice architecture implemented in the system is shown in Figure 2.

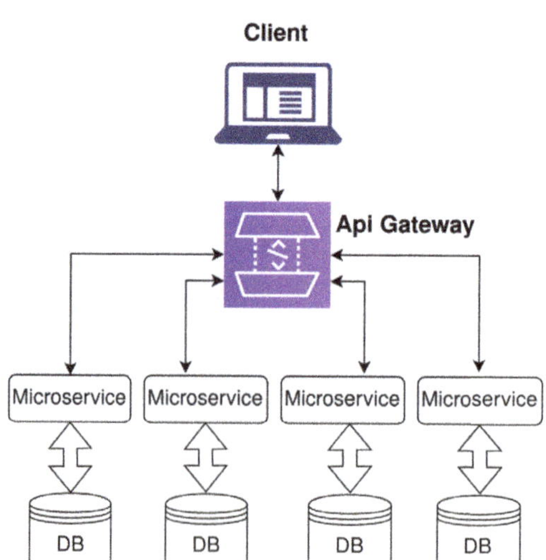

Figure 2. Microservice architecture.

The microservice architecture also includes Application Programming Interface (API) gateway and management/orchestration tools. The API gateway represents an entry point for the clients and decouples the clients from the services. In the proposed solution, the Nginx server is used as an API gateway. The Nginx server is responsible for routing the incoming request to the particular service based on the request URL, as can be seen in Figure 3.

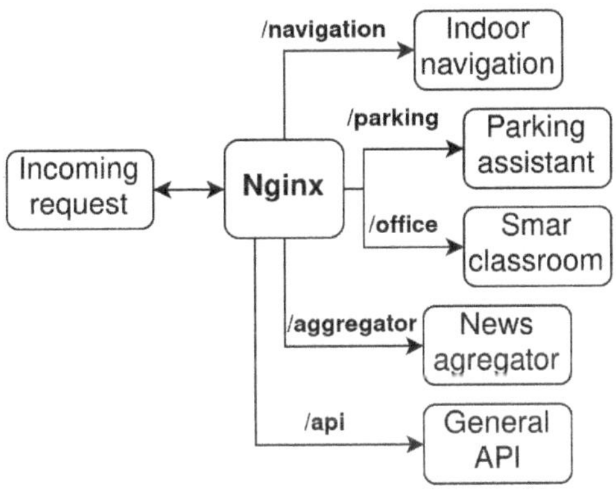

Figure 3. URL-based request routing flow diagram.

The services can also communicate with each other. For this purpose, internal APIs are specified, and the Nginx server is responsible for routing the internal API between the services within the virtual network. Management/orchestration tools are responsible for placing services on nodes, identifying failures, re-balancing services across nodes, and so forth. In the proposed system, Docker and Docker-compose are used for microservices management. The whole system is designed to be ready and compatible for deployment into off-the-shelf technologies, such as Kubernetes. Moreover, it is also possible to use Docker in the swarm mode to boost the scalability of microservices.

4. Subsystems Specification

Currently, there are four subsystems designed in the IoT-based connected university system. This section is dedicated to subsystem specification.

4.1. News Aggregator

The news aggregator subsystem is a free service, it collects information from multiple sources and provides a single interface to display all of them. The main advantage of this subsystem is that students can find all information in one place. The News aggregator collects the news feeds from multiple university web pages. Many of these are created by a Content Management System (CMS) such as WordPress. The WordPress-based web page provides predefined XML files as an RSS news feed that are easily accessible. However, some of the pages are currently built using a special Joomla CMS plugin and do not provide a news feed. Therefore, it is necessary to parse the page content and search for news. The subsystem also gathers information from the university canteen as well as the university library. In this subsystem, the weather information from local weather stations at the university is also gathered. The source of the weather information is our previously developed system [35] for collecting meteorological data from numerous small weather stations. The users can see weather information from various locations as well as the historical data. The design of the proposed subsystem is shown in Figure 4.

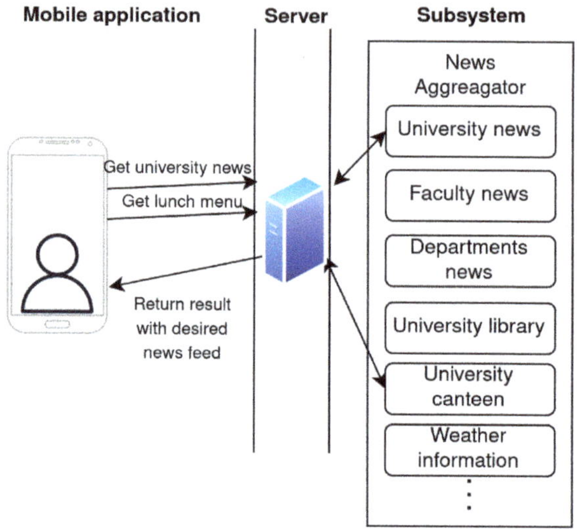

Figure 4. Proposed design of subsystem for news aggregator.

4.2. *Smart Assistant Parking*

The subsystem for smart assistant parking is one of the free services implemented in the IoT-based connected university system. The goal is to provide information about parking occupancy and navigate the user to the nearest free parking spot. There are two ways how to evaluate parking occupancy, i.e., camera- and sensor-based solutions. In the proposed design, the camera system deployed on the university campus is used to collect information about parking occupancy. The concept of the proposed subsystem is presented in Figure 5. The parking detection is based on a convolutional classifier with a residual architecture [36] that achieved an average accuracy of 98.42% on the PKLot database.

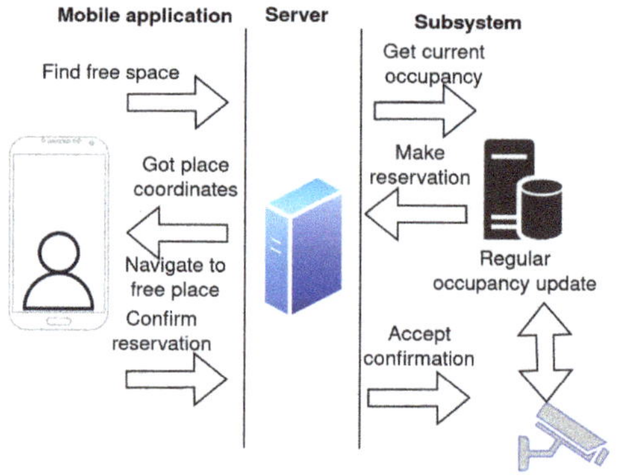

Figure 5. Proposed subsystem for smart assistant parking.

The application automatically detects that the user has approached the parking area and sends a request to find the free parking place with the current position of the user. Since the university campus covers a wide area, the user is allowed to set their preferred parking place area (e.g., based on the building they are going to). The server redirects the request to the subsystem for the smart parking assistant to receive the current parking lot occupancy. Based on the request and parking occupancy data, the subsystem creates a parking place reservation and sends it in a response message to the client. On the client side, the application receives data and navigates the user to the reserved parking place. The user is allowed to accept or decline the reservation. On the subsystem side, the parking occupancy information is updated on a regular basis.

4.3. Smart Classroom or Office

The primary motivation for smart classroom subsystems is to help users who pay attention to their health and proper sitting posture during work or classroom lectures. Moreover, the system is able to estimate the occupancy of offices and lecture rooms based on information from smart chairs. Both students and lecturers spend many hours seated on chairs or stools. Therefore, adopting the correct sitting position is essential for maintaining good posture and a healthy back and spine. There are multiple ways how to monitor the seating posture of individuals. An overview of systems on sitting posture monitoring can be found in [37]. Basically, there are three main approaches used to obtain data about sitting posture:

- Computer image processing;
- Wearable clothing with sensors;
- Measuring the load distribution on some form of substrate.

As part of previous work, a smart system for sitting posture detection was developed. The system is based on force sensors implemented in a chair [38]. The basic concept of the proposed system is illustrated in Figure 6.

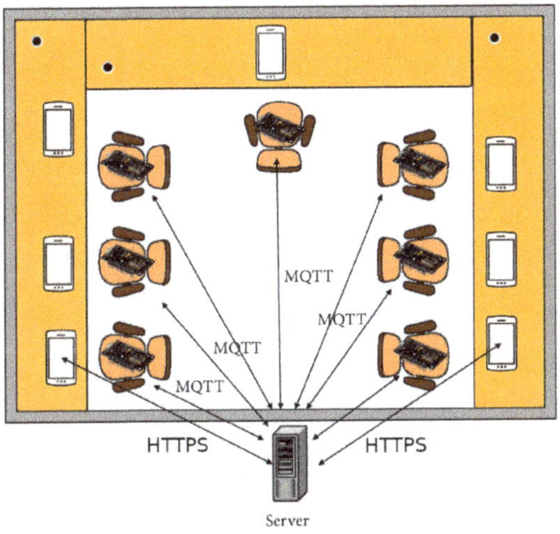

Figure 6. Proposed subsystem for smart classroom.

In each smart chair, six flexible force sensors FSR402 are implemented. Two sensors are placed in the backrest and four in the bottom seat. A NodeMCU board collects data about changes in the resistance of individual sensors and sends these data to the server using the MQTT protocol. There can be a variable number of chairs in one classroom. On the server side, the system uses the Node-RED application for the evaluation and processing of the data. The user can access information about sitting posture correctness as well as detailed information using the mobile application.

Based on data from the experiments, an algorithm for the evaluation of correct sitting posture was proposed. The algorithm has minimal computation power requirements. The data collected during the testing phase were divided into three groups based on the correctness of the sitting posture. During the evaluation of the results, threshold values were defined for each of the three groups. The typical routine for the individuals in the smart classroom should be as follows:

- The individual chooses a free chair in the classroom;
- Using a mobile application, authorized users scan the QR code of their chair;
- The application provides information about the sitting posture to the user as regular updates.

The design of the smart system for sitting posture detection is easy to implement as a subsystem in the IOT-based connected university system. Therefore, authorized users have access to the data from smart chairs and can keep track of the correctness of their sitting postures.

4.4. *Indoor Navigation System Design*

Bluetooth beacons will be deployed in the buildings of the University of Zilina to collect environmental data. Moreover, the signals from these beacons will be used for the implementation of the navigation subsystem. Since BLE beacons will be implemented in the indoor environment with harsh signal propagation conditions, the implemented localization microservice relies on fingerprint positioning.

The main advantage of fingerprinting localization is that there is no need for distance estimation using a signal propagation model. The fingerprinting localization is based on a comparison of measured data with a radio map database. Since the database was collected in the same environment, fingerprinting localization seems to be immune to multipath signal propagation. However, the drawback of fingerprinting localization is related to the time complexity of radio map measurements.

Since position estimation is implemented in the application created for mobile devices, simple deterministic localization algorithms have to be used. These algorithms are based on an assumption that the received signal strength (RSS) from multiple tags depends on the position of the mobile device. Therefore, the position of the mobile device can be estimated by selecting points with the lowest Euclidean distance between the RSS samples collected during localization and the RSS samples stored in the radio map. More details on algorithms tested during the initial trial will be provided in Section 5.

The primary goal of the indoor navigation subsystem is to provide position estimates based on available Bluetooth signals and deploy an indoor navigation system to help the students with orientation inside the university campus. This is a crucial service especially for new students, as they often find it difficult to navigate through the tangle of corridors and arrive in time for lectures or laboratory exercises.

Therefore, the use of the mobile application with data from this subsystem can help students with finding the right route to the desired lecture room. The subsystem is implemented as an independent service in the mobile application. The application requires permission to use a Bluetooth device on the smartphone in order to be able to obtain the position estimates.

To implement indoor navigation in a complex building on a university campus, we decided to split buildings into sub-regions, in order to reduce computational complexity related to position estimation. The example of the one-floor splitting is shown in Figure 7. The particular sub-regions are highlighted in different colors.

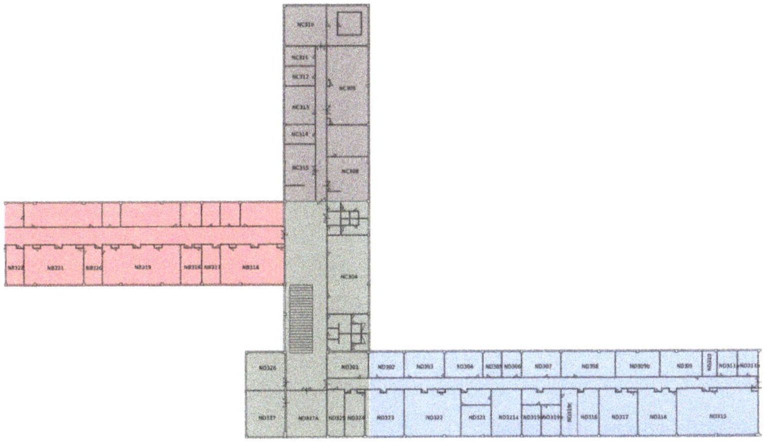

Figure 7. The example of floor divided into the sub-regions.

The positioning is performed within these small sub-regions using the fingerprinting method. When the application service on the mobile device for localization is activated, the application estimates an approximate position on the campus. This process is explained in Figure 8.

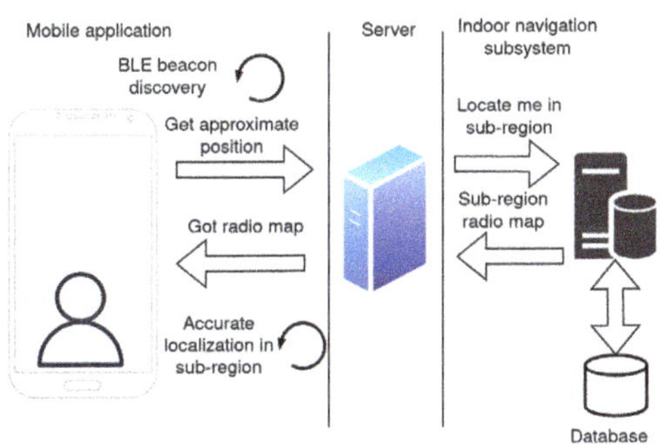

Figure 8. Subsystem for indoor navigation initial localization.

As a first step, the application will scan for signals from the BLE beacons. The data about BLE beacons in the area are sent in the request to the server for the approximate position in the term of sub-region. The server estimates the sub-region based on the signals from the surrounding beacons. The server response contains information about the sub-region and its radio map. The application then uses a positioning algorithm for accurate localization within the sub-region.

Once the position of the mobile device is estimated, the user can search for the destination room by choosing a room label from the list. The actual list of rooms is provided by the server. Once the room is selected, the application sends the request to find the route to the server. The server response contains data about the navigation route. The mobile application receives the response and initializes the navigation within a single sub-region. Once the user moves, the application periodically updates the position.

When the user approaches a neighboring sub-region, the application sends a request to the server to download data from the next sub-region for localization and navigation. This is repeated until the user reaches the destination sub-region, where they will be pointed to the desired office. The navigation process is depicted in Figure 9.

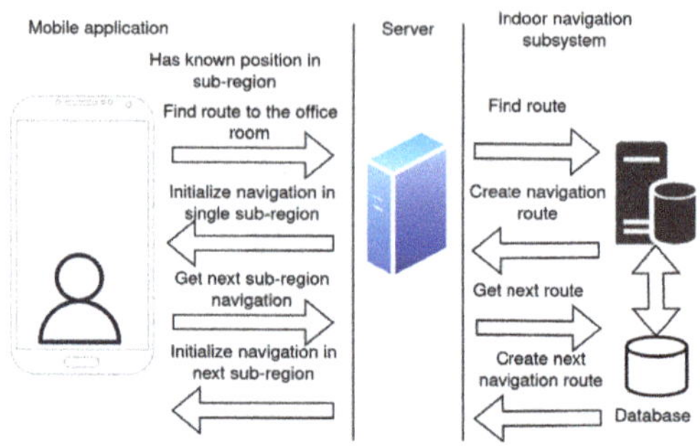

Figure 9. The process of indoor navigation in the mobile application.

5. System Implementation

In this section, the implementation of the microservices of the IoT-based connected university system will be described in detail. To develop the proposed system, the Docker platform was used. Docker is a platform that allows developers to easily create, deploy, and run applications in containers. Containers represent lightweight, portable, and self-sufficient executable packages that include everything needed to run an application, including the code, runtime, system tools, libraries, and settings.

In the proposed architecture, multiple containers are used. The Docker Compose tool was utilized to handle the containers, networks, and services required for the proposed solution. This tool enables the specification and execution of multi-container Docker applications. It offers a convenient way to manage the entire process of a multi-container application, from development to production. Therefore, it is easy to run the same application in different environments. It is an excellent resource for development, testing, and deployment. This tool is using the docker-compose.yml file for the system setup. Used docker-compose.yml file contains the following services:

- Nginx—entry point to the system;
- MongoApi—serves as storage for the API service;
- API—backend for setting and authentication usage;
- Web—holds React web application;
- Aggregator—implementing logic for news aggregator subsystem;
- MongoAggregator—serves as storage for the aggregator service;
- Localization—implementing logic for indoor navigation subsystem;
- MongoLocalization—serves as storage for the localization service;
- Mongo-express—for development purposes only.

The system starts with the single command "docker-compose up" or "docker-compose up-d" to run the system in the background. During the first run, a single administration account is created with a default username and password, which should be changed after the initial login. The administrator has to set up other services to ensure they work properly. After the initialization process, the IoT-based connected university system is ready to be deployed.

5.1. Web-Based User Interface

All communication from the Internet is routed via port 3050. The communication then passes through the Nginx server. In the Nginx server, URL base routing is implemented, see Figure 3. The User Interface (UI) was implemented using React—a JavaScript library with the utilization of Material UI (MUI) components. The home page of the IoT-based connected university system for the non-authorized user is shown in Figure 10.

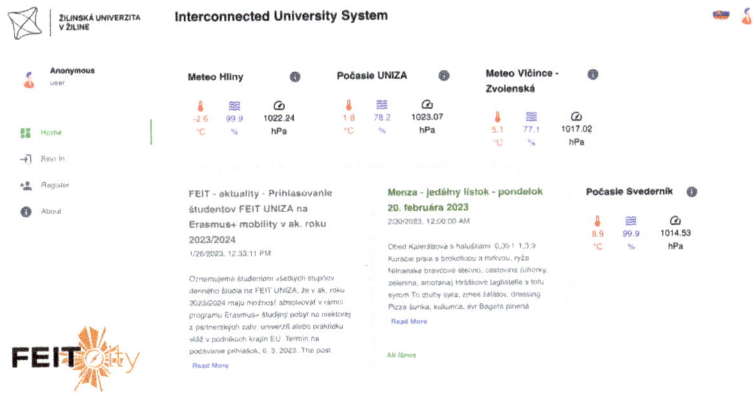

Figure 10. The IoT system for connected university home page for the non−authorized user.

The non-authorized user can only access free services, such as feeds from news aggregators. The user can manage the displayed services. For the non-authorized user, all settings are stored in the browser's local storage. Detailed information for each item can be displayed by clicking on it. The detailed news feed is shown in Figure 11 and a detailed view of the meteorological data is shown in Figure 12.

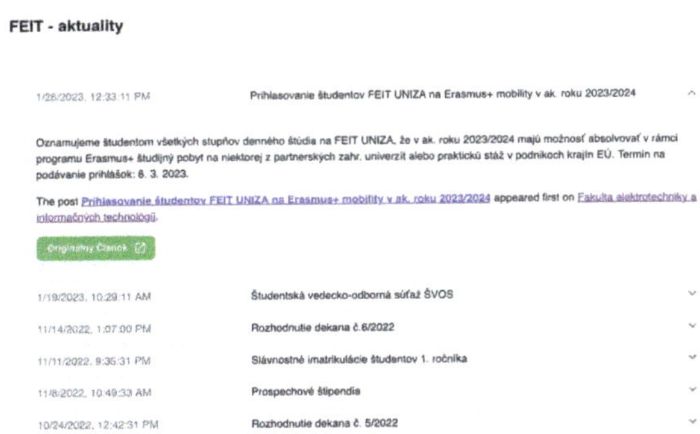

Figure 11. Detailed information for item−type news from aggregator service.

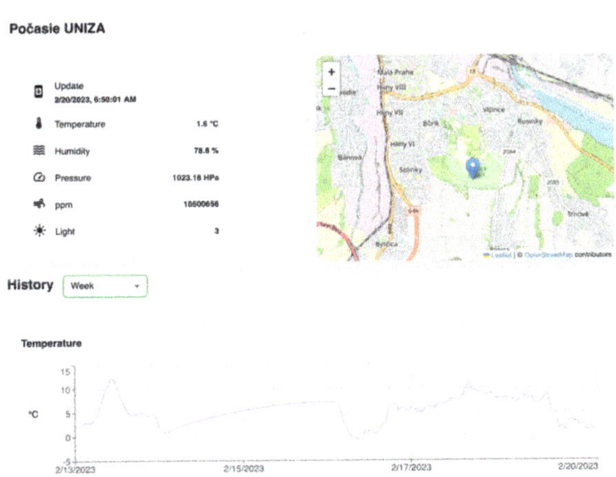

Figure 12. Detailed view of meteorological data.

The system also provides an administration section, where the system administrator can manage all the services, users, and notices. The view on the administration sections with settings of the individual services is shown in Figure 13.

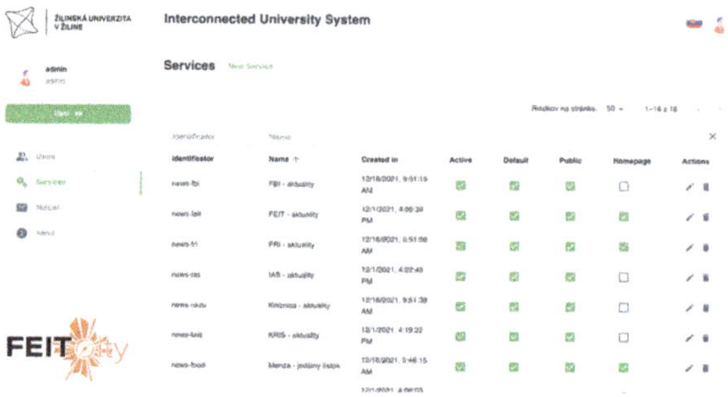

Figure 13. System administration section.

As can be seen from the figure, the administrator is able to set different flags for each service, such as an active flag, default flag, homepage flag, or public flag. Each service has a custom identification assigned, in order to identify the service in the system architecture. The name of the service is displayed as the service name on the user's home screen. Some of the services may not be visible in a web application. An example of such a service is indoor navigation since the native mobile application is required to collect the data required for the proper function of the service.

5.2. Mobile Application

A mobile Connected UNIZA application was developed in order to improve the user experience and provide services with added value. The application was developed using the React Native framework. React Native is a JavaScript framework for building mobile applications using React. It allows developers to build mobile apps for iOS and Android platforms using a single codebase, leveraging the power of React and its component-based architecture. React Native uses native components rather than web components as building blocks and allows developers to access the device's native APIs, such as the camera, GPS, and more. The framework is open-source and actively maintained by Facebook and the community.

The screens of the developed application are shown in Figure 14. The application is available only for authenticated users. The first screen after the application startup is a login screen, as can be seen in Figure 14a. The user must fill in the credentials in order to login into the application. The users are also able to create an account using the register screen. The home screen for the authorized user is shown in Figure 14d. By default, all services with the "homepage" flag are shown for the user who is logged in for the first time. Then, the user can alter which services will be displayed in the feeds. The authorized users are allowed to change and save the layout of the feed for the next login. The user interface of the application is implemented in two languages, Slovak and English. Users can change the language and password in the settings screen, see Figure 14b. The navigation between the screens is implemented via a drawer navigator, shown in Figure 14c. An example of an indoor navigation screen is shown in Figure 14e. The user can choose the destination room number and the application will navigate the user toward the desired destination.

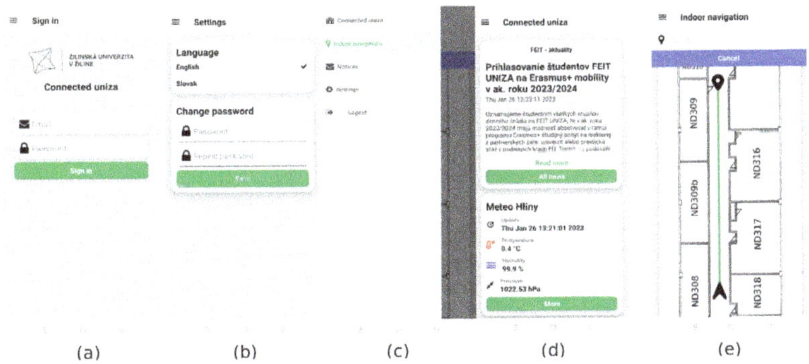

Figure 14. The mobile application of the connected university system, (**a**) Login screen. (**b**) Settings screen for authorized users. (**c**) Navigation drawer. (**d**) Connected university home screen showing news feeds from the aggregator subsystem. (**e**) Indoor navigation screen.

6. Proof of Concept for Indoor Localization

In order to evaluate the feasibility of BLE-based localization for indoor navigation purposes, the preliminary tests were performed at a typical corridor in one of the buildings at the University of Zilina. The corridor has dimensions of 2 × 69 m. The reference points used for radio map measurements were placed in a grid with a spacing of 1 m, resulting in a radio map with measurements recorded at 134 points. The corridor was covered by five BLE beacons. The Holyiot nRF51822 Bluetooth 4.0 beacon BLE module was used. For testing purposes, the position was estimated at 21 positions. The positions of reference as well as the testing points were estimated using a laser distance measurement tool.

The tests were performed in two scenarios, and the positions of the BLE beacons in these scenarios are presented by red dots in Figure 15, together with a view of the corridor. In both scenarios, the BLE beacons were attached to the ceiling, so the signal should not be obstructed by other users present in the area.

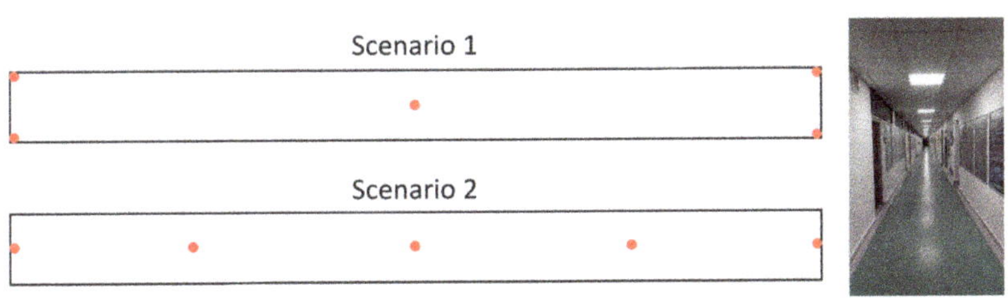

Figure 15. Placement of BLE beacons in testing scenarios.

For testing purposes, the positioning was performed using NN, KNN, and WKNN algorithms. In all these algorithms, the position estimate $\overline{x}$ is provided by:

$$\overline{x} = \frac{\sum_{i=1}^{M} \omega_i p_i}{\sum_{j=1}^{M} \omega_j}, \quad (1)$$

where p_i is the position of the i-th reference point in the radio map, while ω_i and ω_j are weights assigned to the i-th and j-th reference points and M is the number of reference points in the radio map. The weights are calculated as the inverse value of the Euclidean distance between RSS measurements from the mobile device and radio map.

The Nearest Neighbour (NN) algorithm takes into account only the reference point with the highest weight, i.e., the smallest Euclidean distance between RSS samples. The k-Nearest Neighbour (KNN) algorithm uses the k-highest weights and sets them to 1, while the other weights are set to 0. Therefore, KNN estimates the position as the center of gravity of the selected reference points. Similarly, the Weighted k-Nearest Neighbor (WKNN) considers the k reference points with the highest weights; however, the weights are considered in the final estimate, i.e., the estimate is calculated as the weighted average of the selected reference points.

In the first step, the impact of a number of the reference points k in both KNN and WKNN algorithms was evaluated for both scenarios. The achieved results are shown in Figure 16.

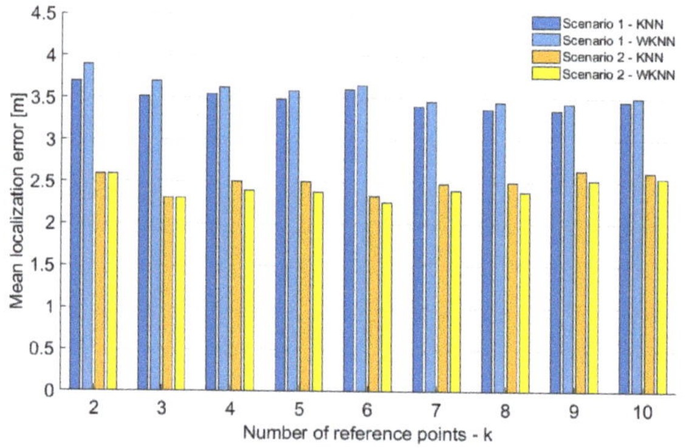

Figure 16. Localization Impact of a number of the refernece points on accuracy of KNN and WKNN algorithms.

From the figure, it is clear that, for scenario 2, the best accuracy was achieved when the number of reference points used for position estimation $k = 3$. On the other hand, in scenario 1, the best results were achieved with $k = 9$. However, it is important to note that the impact of a number of reference points used for position estimation in both scenarios is relatively small. Moreover, the results achieved in scenario 1 were in all cases significantly worse than in scenario 2. The difference in the mean error between the scenarios was between 0.7 and 1.4 m. That means the placement of BLE nodes in scenario 2 helped to improve the accuracy by 21–37% compared to scenario 1. Based on these results, the number of reference points considered in the KNN and WKNN algorithms was set as $k = 3$ for further investigation. Moreover, it was already proven that the number of reference points $k = 3$ provides reasonably good results using data from Wi-Fi networks [39].

The results achieved during the testing of the localization concept in scenarios 1 and 2 are shown in Figure 17 and Figure 18, respectively. In the figures, the median is shown as a line inside the box, the 5% confidence interval is represented by the shaded area, the lower and upper quartiles define the edge of the box, the minimum and maximum values that are not outliers are presented by whiskers, and circles represent the outliers in the data.

From the achieved results, it can be seen that the localization errors achieved in the second scenario were significantly smaller, although there were some outliers. This might be caused by the fact that the BLE nodes in the second scenario are distributed more evenly across the area, therefore providing stronger signals from multiple sources.

It can also be concluded that the NN algorithm achieved the lowest performance, i.e., the highest localization error, in both cases. This is due to the fact that part of the test points, used for the evaluation of the positioning algorithms, were placed at different positions as reference points in the radio map.

The best results were achieved using the KNN localization algorithm; in the second test scenario, therefore, this configuration will be considered in the further development and evaluation of the system. Based on these preliminary results, it can be concluded that BLE-based positioning has the potential to provide the accuracy that should be sufficient for navigation in corridors along the university campus.

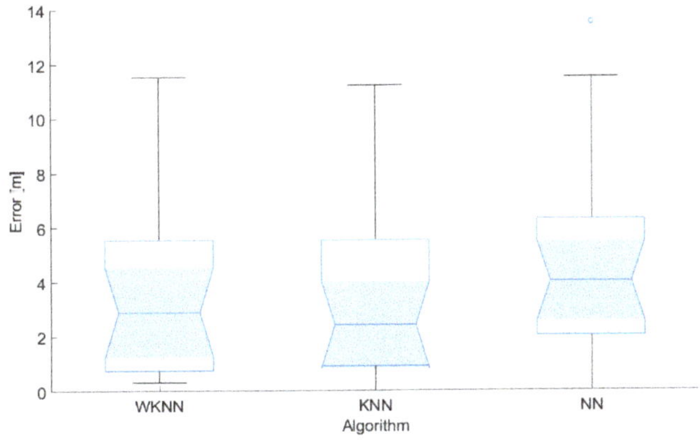

Figure 17. Localization errors achieved in scenario 1.

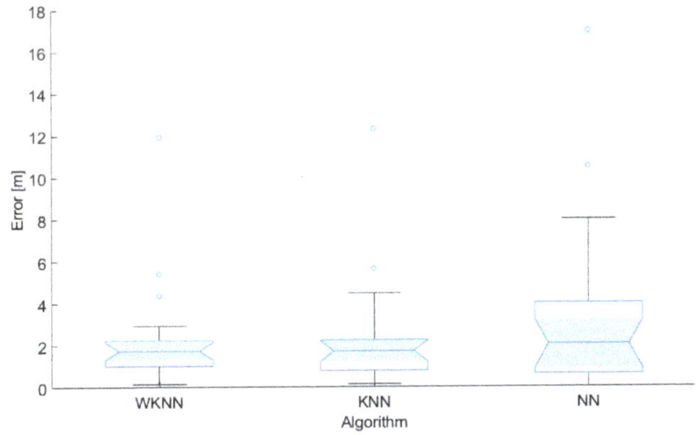

Figure 18. Localization errors achieved in scenario 2.

7. Conclusions and Future Work

The paper presented the design and implementation of a functional IoT-based connected university system—a robust and massive IoT system for integrating various services as subsystems. In the pilot design, we proposed four subsystems, i.e., the subsystem for indoor navigation, the subsystem for parking assistance, the subsystem for smart classrooms or offices, and the subsystem for news aggregation from the university. In the initial system implementation, the basic system functionality, the subsystem for news aggregation, and the subsystem for indoor navigation were developed. The subsystems are implemented using the microservice architecture and, thus, are reliable, scalable, highly maintainable, and independently deployable. The system design is open to implementing new services in the future.

In this paper, the implementation of the system was described in detail. Moreover, the performance of the BLE localization microsystem was tested to prove the proposed system design. Based on the achieved results, it can be concluded that the positioning system can achieve reasonable localization accuracy with the correct placement of the BLE beacons and the system can provide position estimates with an accuracy of 2 m. In this paper, different placements of BLE beacons and their impacts on the accuracy of the positioning algorithms were investigated. From the results, it can be concluded that the positions of the BLE beacons have a significant impact on the achieved accuracy. The localization error was decreased by 31% by a change in the BLE node placement. Therefore, it can be concluded that, with the correct placement of BLE beacons, the implemented positioning solution is suitable for navigation in university buildings using low-cost beacons and off-the-shelf smartphones.

In the future, the system will be tested by a limited number of users to evaluate its functionality in realistic daily use. The sensors with BLE communication modules will be implemented in university buildings to gather environmental data and support positioning services in all corridors. The system will also be extended with novel services based on the analysis of data from connected sensors.

Author Contributions: Conceptualization, S.M., J.M. and P.B.; methodology, J.M. and P.B.; software, S.M. and M.H.; validation, S.M., J.M., M.H. and P.B.; formal analysis, S.M., J.M. and P.B.; investigation, S.M. and J.M.; resources, M.H. and P.B.; data curation, S.M., J.M., M.H. and P.B.; writing—original draft preparation, S.M. and J.M.; writing—review and editing, S.M. and J.M.; visualization, S.M. and J.M.; supervision, P.B.; project administration, P.B. and J.M.; funding acquisition, P.B. and J.M. All authors have read and agreed to the published version of the manuscript.

Funding: This work has been supported by the Slovak VEGA grant agency, Project No. 1/0588/22 "Research of a location-aware system for achievement of QoE in 5G and B5G networks".

Institutional Review Board Statement: Not applicable.

Informed Consent Statement: Not applicable.

Data Availability Statement: The data presented in this study are available on request from the corresponding author.

Conflicts of Interest: The authors declare no conflict of interest.

Abbreviations

The following abbreviations are used in this manuscript:

IoT	Internet of Things.
AWS	Amazon Web Services.
MQTT	MQ Telemetry Transport.
GPS	Global Positioning System.
UWB	Ultra-Wide Band.
BLE	Bluetooth Low Energy.
HTTPS	Hypertext Transfer Protocol Secure.
API	Application Programming Interface.
URL	Uniform Resource Identifier.
CMS	Content Management System.
RSS	Received Signal Strength.
UI	User Interface.
MUI	Material User Interface.
NN	Nearest Neighbors.
KNN	K-Nearest Neighbors.
WKNN	Weighted k-Nearest Neighbor.
DOAJ	Directory of Open Access Journals.
TLA	Three Letter Acronym.
LD	Linear Dichroism.

References

1. García, L.; Garcia-Sanchez, A.J.; Asorey-Cacheda, R.; Garcia-Haro, J.; Zúñiga-Cañón, C.L. Smart Air Quality Monitoring IoT-Based Infrastructure for Industrial Environments. *Sensors* **2022**, *22*, 9221. [CrossRef] [PubMed]
2. Anumala, H.; Addepalli, S.; Kodali, T.; Pravallika, K.; Anuradha, T. Air Monitoring System Using IOT. In *Advances in Information Communication Technology and Computing*; Springer Nature: Singapore, 2022; pp. 327–333. [CrossRef]
3. Jabbar, W.A.; Subramaniam, T.; Ong, A.E.; Shu'Ib, M.I.; Wu, W.; de Oliveira, M.A. LoRaWAN-Based IoT System Implementation for Long-Range Outdoor Air Quality Monitoring. *Internet Things* **2022**, *19*, 100540. [CrossRef]
4. Méndez-Guzmán, H.A.; Padilla-Medina, J.A.; Martínez-Nolasco, C.; Martinez-Nolasco, J.J.; Barranco-Gutiérrez, A.I.; Contreras-Medina, L.M.; Leon-Rodriguez, M. IoT-Based Monitoring System Applied to Aeroponics Greenhouse. *Sensors* **2022**, *22*, 5646. [CrossRef]

5. Prema, P.; Sivasankari, B.; Kalpana, M.; Vasanthi, R. Smart Agriculture Monitoring System using IoT. *Int. J. Pure Appl. Biosci.* **2019**, *7*, 160–165. [CrossRef]
6. Sarpal, D.; Sinha, R.; Jha, M.; TN, P. AgriWealth: IoT based farming system. *Microprocess. Microsyst.* **2022**, *89*, 104447. [CrossRef]
7. Palomar-Cosín, E.; García-Valls, M. Flexible IoT Agriculture Systems for Irrigation Control Based on Software Services. *Sensors* **2022**, *22*, 9999. [CrossRef]
8. Ahmed, M.A.; Gallardo, J.L.; Zuniga, M.D.; Pedraza, M.A.; Carvajal, G.; Jara, N.; Carvajal, R. LoRa Based IoT Platform for Remote Monitoring of Large Scale Agriculture Farms in Chile. *Sensors* **2022**, *22*, 2824. [CrossRef]
9. Vashishtha, M.; Chouksey, P.; Rajput, D.S.; Reddy, S.R.; Reddy, M.P.K.; Reddy, G.T.; Patel, H. Security and detection mechanism in IoT-based cloud computing using hybrid approach. *Int. J. Internet Technol. Secur. Trans.* **2021**, *11*, 436. [CrossRef]
10. Kim, H.H.; Yoo, J. Analysis of Security Vulnerabilities for IoT Devices. *J. Inf. Process. Syst.* **2022**, *18*, 489–499. [CrossRef]
11. Vishwakarma, S.K.; Upadhyaya, P.; Kumari, B.; Mishra, A.K. Smart Energy Efficient Home Automation System Using IoT. In Proceedings of the 2019 4th International Conference on Internet of Things: Smart Innovation and Usages (IoT-SIU), Ghaziabad, India, 18–19 April 2019; IEEE: New York, NY, USA, 2019. [CrossRef]
12. Condon, F.; Martínez, J.M.; Eltamaly, A.M.; Kim, Y.C.; Ahmed, M.A. Design and Implementation of a Cloud-IoT-Based Home Energy Management System. *Sensors* **2022**, *23*, 176. [CrossRef]
13. Wu, F.; Wu, T.; Yuce, M.R. Design and Implementation of a Wearable Sensor Network System for IoT-Connected Safety and Health Applications. In Proceedings of the 2019 IEEE 5th World Forum on Internet of Things (WF-IoT), Limerick, Ireland, 15–18 April 2019; IEEE: New York, NY, USA, 2019. [CrossRef]
14. Onasanya, A.; Elshakankiri, M. Smart integrated IoT healthcare system for cancer care. *Wirel. Netw.* **2019**, *27*, 4297–4312. [CrossRef]
15. Cay, G.; Solanki, D.; Rumon, M.A.A.; Ravichandran, V.; Hoffman, L.; Laptook, A.; Padbury, J.; Salisbury, A.L.; Mankodiya, K. NeoWear: An IoT-connected e-textile wearable for neonatal medical monitoring. *Pervasive Mob. Comput.* **2022**, *86*, 101679. [CrossRef]
16. Zhu, F.; Lv, Y.; Chen, Y.; Wang, X.; Xiong, G.; Wang, F.Y. Parallel Transportation Systems: Toward IoT-Enabled Smart Urban Traffic Control and Management. *IEEE Trans. Intell. Transp. Syst.* **2020**, *21*, 4063–4071. [CrossRef]
17. Kumar, A.; Akhtar, M.A.K.; Pandey, A. Design of Internet of Things (IoT) System Based Smart City Model on Raspberry Pi. *IETE J. Res.* **2022**, *1*, 1–8.
18. Humayun, M.; Alsaqer, M.S.; Jhanjhi, N. Energy Optimization for Smart Cities Using IoT. *Appl. Artif. Intell.* **2022**, *36*, 2426–2443.
19. Zhu, H.; Chau, S.C.K.; Guarddin, G.; Liang, W. Integrating IoT-Sensing and Crowdsensing with Privacy: Privacy-Preserving Hybrid Sensing for Smart Cities. *ACM Trans. Internet Things* **2022**, *3*, 1–30.
20. Kanan, R.; Arbess, H. An IoT-Based Intelligent System for Real-Time Parking Monitoring and Automatic Billing. In Proceedings of the 2020 IEEE International Conference on Informatics, IoT, and Enabling Technologies (ICIoT), Doha, Qatar, 2–5 February 2020; IEEE: New York, NY, USA, 2020. [CrossRef]
21. Canli, H.; Toklu, S. AVL Based Settlement Algorithm and Reservation System for Smart Parking Systems in IoT-based Smart Cities. *Int. Arab J. Inf. Technol.* **2022**, *19*, 793–801.
22. Kadusic, E.; Zivic, N.; Ruland, C.; Hadzajlic, N. A Smart Parking Solution by Integrating NB-IoT Radio Communication Technology into the Core IoT Platform. *Future Internet* **2022**, *14*, 219. [CrossRef]
23. Sobeslav, V.; Horalek, J. A Smart Parking System Based on Mini PC Platform and Mobile Application for Parking Space Detection. *Mob. Inf. Syst.* **2020**, *2020*, 8875301. [CrossRef]
24. Dhuri, V.; Khan, A.; Kamtekar, Y.; Patel, D.; Jaiswal, I. Real-Time Parking Lot Occupancy Detection System with VGG16 Deep Neural Network using Decentralized Processing for Public, Private Parking Facilities. In Proceedings of the 2021 International Conference on Advances in Electrical, Computing, Communication and Sustainable Technologies (ICAECT), Bhilai, India, 19–20 February 2021; IEEE: New York, NY, USA, 2021. [CrossRef]
25. Baroffio, L.; Bondi, L.; Cesana, M.; Redondi, A.E.; Tagliasacchi, M. A visual sensor network for parking lot occupancy detection in Smart Cities. In Proceedings of the 2015 IEEE 2nd World Forum on Internet of Things (WF-IoT), Milan, Italy, 14–16 December 2015; IEEE: New York, NY, USA, 2015. [CrossRef]
26. Cheng, Y.; Zhou, T. UWB Indoor Positioning Algorithm Based on TDOA Technology. In Proceedings of the 2019 10th International Conference on Information Technology in Medicine and Education (ITME), Qingdao, China, 23–25 August 2019; IEEE: New York, NY, USA, 2019. [CrossRef]
27. Farahsari, P.S.; Farahzadi, A.; Rezazadeh, J.; Bagheri, A. A Survey on Indoor Positioning Systems for IoT-Based Applications. *IEEE Internet Things J.* **2022**, *9*, 7680–7699. [CrossRef]
28. Schroeer, G. A Real-Time UWB Multi-Channel Indoor Positioning System for Industrial Scenarios. In Proceedings of the 2018 International Conference on Indoor Positioning and Indoor Navigation (IPIN), Nantes, France, 24–27 September 2018; IEEE: New York, NY, USA, 2018. [CrossRef]
29. Zhuang, Y.; Zhang, C.; Huai, J.; Li, Y.; Chen, L.; Chen, R. Bluetooth Localization Technology: Principles, Applications, and Future Trends. *IEEE Internet Things J.* **2022**, *9*, 23506–23524. [CrossRef]
30. Nikodem, M.; Szelinski, P. Channel Diversity for Indoor Localization Using Bluetooth Low Energy and Extended Advertisements. *IEEE Access* **2021**, *9*, 169261–169269. [CrossRef]

31. Machaj, J.; Brida, P.; Majer, N. Challenges introduced by heterogeneous devices for Wi-Fi–based indoor localization. *Concurr. Comput. Pract. Exp.* **2019**, *32*, e5198. [CrossRef]
32. Machaj, J.; Brida, P. Impact of optimization algorithms on hybrid indoor positioning based on GSM and Wi-Fi signals. *Concurr. Comput. Pract. Exp.* **2016**, *29*, e3911. [CrossRef]
33. Ye, H.; Peng, J. Robot Indoor Positioning and Navigation Based on Improved WiFi Location Fingerprint Positioning Algorithm. *Wirel. Commun. Mob. Comput.* **2022**, *2022*, 8274455. [CrossRef]
34. Junoh, S.A.; Subedi, S.; Pyun, J.Y. Floor Map-Aware Particle Filtering Based Indoor Navigation System. *IEEE Access* **2021**, *9*, 114179–114191. [CrossRef]
35. Vestenicky, M.; Matuska, S.; Hudec, R.; Kamencay, P. Sensor network proposal based on IoT for a prediction system of the power output from photovoltaic panels. In Proceedings of the 2018 28th International Conference Radioelektronika (RADIOELEKTRONIKA), Prague, Czech Republic, 19–20 April 2018; IEEE: New York, NY, USA, 2018. [CrossRef]
36. Gregor, M.; Pirník, R.; Nemec, D. Transfer Learning for Classification of Parking Spots using Residual Networks. *Transp. Res. Procedia* **2019**, *40*, 1327–1334.
37. Tlili, F.; Haddad, R.; Ouakrim, Y.; Bouallegue, R.; Mezghani, N. A Survey on sitting posture monitoring systems. In Proceedings of the 2018 9th International Symposium on Signal, Image, Video and Communications (ISIVC), Rabat, Morocco, 27–30 November 2018; pp. 185–190. [CrossRef]
38. Matuska, S.; Paralic, M.; Hudec, R. A Smart System for Sitting Posture Detection Based on Force Sensors and Mobile Application. *Mob. Inf. Syst.* **2020**, *2020*, e6625797. [CrossRef]
39. Machaj, J.; Brida, P. Wireless Positioning as a Cloud Based Service. In *Intelligent Information and Database Systems*; Springer International Publishing: New York, NY, USA, 2015; pp. 430–439. [CrossRef]

sensors

MDPI

Article

Comparison of the Usability of Apple M2 and M1 Processors for Various Machine Learning Tasks

David Kasperek *, Pawel Antonowicz, Marek Baranowski, Marta Sokolowska and Michal Podpora

Department of Computer Science, Opole University of Technology, Proszkowska 76, 45-758 Opole, Poland
* Correspondence: d.kasperek@student.po.edu.pl

Abstract: This paper compares the usability of various Apple MacBook Pro laptops were tested for basic machine learning research applications, including text-based, vision-based, and tabular data. Four tests/benchmarks were conducted using four different MacBook Pro models—M1, M1 Pro, M2, and M2 Pro. A script written in Swift was used to train and evaluate four machine learning models using the Create ML framework, and the process was repeated three times. The script also measured performance metrics, including time results. The results were presented in tables, allowing for a comparison of the performance of each device and the impact of their hardware architectures.

Keywords: machine learning; deep learning; neural processing unit; neural processing cores; NPU benchmark; processor architectures; Apple M1; Apple M2; CoreML; neural engine

1. Introduction

In the current age, artificial intelligence algorithms are becoming more and more omnipresent, not only in robotic applications, but also in a wide range of application areas [1–3]. From ads and video recommendations [4–6], through text auto completion [7,8], to algorithms capable of producing award-winning art [9,10], an increasing number of people are using deep learning (DL) models in their work [11–14]. The production cycle of a deep learning model is time consuming and preferably requires understanding of complex concepts such as deep neural networks, neural network topology, training, and validation [15–17], as well as the application field, e.g., computer vision or natural language processing. Having that knowledge, training a production-capable model requires a lot of data [15,18,19] (or, alternatively, the use of transfer learning [18–20], which requires the knowledge of where to find such a model, and which one to use). Moreover, it is essential to have proficiency in using DL frameworks such as TensorFlow [21] or PyTorch [22]. Learning to create a good model [23] requires an investment of a considerable amount of time (preferably introduced at an early stage of education [24]) and knowledge of the basics of model preparation.

The authors have noticed that the process of learning and experimenting with machine learning for many researchers, students, or professionals is often preceded or accompanied by a difficult question—which hardware platform to choose [25–28]. This multi-factor optimization always includes an economical aspect [12,29], but the computational capabilities are not inessential [30,31]. Proper evaluation of the 'money-to-value' assessment is actively hindered by a 'marketing fog' [32,33], which tries to make the choice emotion-based instead of being based on any measurable factors.

Choosing a purposeful notebook for both everyday work and DL-oriented research is difficult. The general rule-of-thumb (better CPU, more RAM memory, modest graphics card) might still be a valid intuition-based choice; however, the evolution of CPUs brought a new player to the game: ARM-based (ARM—Advanced RISC Machine) 'Apple M1' chip (and its newer versions) [34–37], equipped with specialized GPU cores and NPU (Neural Processing Unit) cores. The presence of NPU cores sounds especially promising; however, not much

Citation: Kasperek, D.; Antonowicz, P.; Baranowski, M.; Sokolowska, M.; Podpora, M. Comparison of the Usability of Apple M2 and M1 Processors for Various Machine Learning Tasks. *Sensors* **2023**, *23*, 5424. https://doi.org/10.3390/s23125424

Academic Editors: Róbert Hudec, Patrik Kamencay and Peter Hockicko

Received: 5 May 2023
Revised: 6 June 2023
Accepted: 6 June 2023
Published: 8 June 2023

evidence of the actual computational benefits is available. For this reason, the authors have decided to design and conduct a series of typical DL-related models/tasks, based upon readily available datasets, to evaluate and compare the new processors.

In this article, the authors verify the validity of Apple's Deep Learning framework for some of the common DL challenges— image classification and regression using the Animals dataset (available on Kaggle's webpage [38], consisting of over 29,000 images), image classification using a custom-made mini-dataset of 24 photos, a tabular dataset (the Kaggle's Payment Fraud Detection Dataset [39]), and text-based use case—the Kaggle's Steam Reviews [40] dataset.

1.1. Motivation

Deep learning researchers and enthusiasts worldwide are keen to obtain knowledge concerning new hardware that is affordable and could potentially speed up their experimentation with models. The marketing language is often not specific enough to explain the performance of the product. While, for most people, Neural Processing Unit performance is not the most important aspect of the laptop, for those who intend to work on deep learning, it could be a deciding matter. Moreover, the knowledge of the performance of particular hardware models at specific price points could be beneficial in terms of deciding whether to buy the more expensive chip or not. Having clear information about current hardware capabilities, especially the newest ones, may be of great interest for researchers, who have to decide on their next project, its budget, and its scope. Although it is feasible to carry out basic ML experiments on contemporary computers, the authors aim to delve into and juxtapose the "ML usability" of the aforementioned hardware platforms. In this context, "usability" is comprehended and examined as a quantitative assessment derived from employing systematic research methodologies for time-based evaluation of specific hardware platforms in preliminary machine learning experiments. The primary criterion under scrutiny is the computation time; nonetheless, it is advisable for the reader to expand the benefits of reading this paper by also taking into consideration the current prices of respective models.

This study is intended to deliver reliable information and arguments to scientists and enthusiasts who are interested in purchasing a new notebook equipped with hardware capable of accelerating neural network computations.

1.2. Scope and Limitations

The intended readership of this study comprises scientists and practitioners of deep learning who are considering purchasing an Apple laptop for their research, rather than investing in specialized High-Performance Computing (HPC) or Workstation equipment. Apple's processors with a Neural Processing Unit (NPU) are marketed as ones that are indeed capable of accelerating model computations. Therefore, it is reasonable to compare only laptops that are equipped with Apple's NPU.

Authors of this study compared Apple's M-chip CPU family, including M1, M1 Pro, M2, and M2 Pro (details regarding exact hardware specifications are included in Section 2.4). Research was conducted using the same operating system (latest available to date, which was macOS Ventura 13.2) to introduce as little potential interference as possible. Additionally, Section 3.2 presents an alternative comparison–three different macOS versions whilst using the same hardware. All tests were designed to be used with the same code, environment, framework, and libraries (see Section 2.4) for all processors tested.

The comparison was performed using the Create ML framework [41] developed by Apple, designed and implemented with full compatibility and maximum efficiency of the CPU/hardware. Comparison of other DL frameworks may also be interesting, but, since it is strongly dependent on the availability of hardware support for the Apple M1/M2 chip as well as a proper implementation of the CPU extensions within a particular framework, a fair comparison is not yet possible.

While it would be interesting to measure the performance differences using statistical analysis, this work focuses primarily on the processing time of the datasets, as well as training and evaluation time of models created by Create ML.

1.3. Performance Measurements

The performance measurements were conducted with the usage of specialized software that stress-tested the hardware's computational capabilities. The authors utilized common deep learning problems such as computer vision (classification) and regression, while using popular real-world datasets to test the viability of Apple's chips in the tasks presented in Section 2.3. The proper analyses, as well as the correct interpretation of results, are key after collecting enough measurements on a particular hardware platform. The results wre converted and visualized in an easy-to-read and understandable form.

2. Materials and Methods

To ensure the repeatability and the ease of implementation of the models and datasets used within the research, the authors opted for the most native choices for the macOS-based platforms, which were readily available with fairly low entry threshold. The analyses and comparisons were implemented using the Swift programming language [42], Xcode Integrated Development Environment (IDE) [43], Xcode Playground (part of the Xcode IDE, introduced in 2014 [44], especially useful for rapid prototyping), and Create ML [41] (merged into a unified Apple ecosystem for creating, managing, and using machine learning models, with full support of the available hardware acceleration [45], as well as the ability to deploy the models onto mobile platforms).

Swift is a high-level programming language developed by Apple, released in 2014 as a replacement for Objective-C. It is commonly used as the first-choice language for applications built for Apple platforms [42].

Xcode is an integrated development environment (IDE) designed by Apple for developing software for macOS, iOS, iPadOS, watchOS, and tvOS . It includes a suite of tools for developing software, and also provides access to a wide range of Software Development Kits (SDKs) and Application Programming Interfaces (APIs) that are required for building applications for Apple's platforms [43].

Xcode Playground is an interactive programming environment that allows developers to experiment with Swift code in an interactive way. It provides a lightweight environment for writing and running Swift code with live code execution. Xcode Playground is integrated within the Xcode IDE [44].

Create ML [41] is a framework and a collection of components and tools intended for easy preparation of machine learning models as well as their easy integration into custom applications. It features a GUI-based application for creating and training a model, which can be later distributed and used on other devices, including mobile applications. Create ML implements the Core ML framework to be able to benefit from the hardware it targets.

Core ML is Apple's machine learning framework [45], designed specifically to benefit from the hardware acceleration capabilities of processors used in Apple devices. The Core ML framework is said to enable optimization of on-device performance by also using the GPU, NPU (named Neural Engine), and optimization of memory usage and power consumption [45].

2.1. Model Creation

The measurements were conducted using the 'Benchmarker.playground' script, available (open-source) in [46]. For a detailed insight into how the experiment was carried out, please refer to [37]. Within the 'Benchmarker.playground' script, each model was created by the use of custom functions written in the Swift programming language. The appropriate datasets were passed as arguments, and the trained models were returned. Finally, the models were tested using custom testing functions. The execution time of each stage and function was measured, which allowed the comparison of devices on which the script was running to be made. All the results were logged into the console output. The process of training and evaluation of all models was repeated three times.

2.2. Model Export as .mlmodel

Models created with the use of Create ML, (whether implemented in the Playground sandbox or in an actual application), can be easily exported as a file [47]. This is carried out using Apple's Core ML framework file format—an '*.mlmodel' file [48].

The .mlmodel file contains the prediction methods of a machine learning model, including all of its configuration data and metadata [49]. These parameters were previously extracted from its training environment and then processed to be optimized for Apple device performance [50].

The '.mlmodel' file includes the following sections [51]:

1. Metadata—Defines the model's metadata;
2. Interface—Defines the input and output features;
3. Architecture—Encodes the model's architecture;
4. Parameters—Stores all values extracted during the model training.

To correctly interpret the input data and produce valid output predictions, features need to be defined and specified in the .mlmodel [51,52]. This includes "Metadata", such as the author, license, and model version, stored in the form of a dictionary [51,53,54]. The "Model description" information such as the names, data types, and shapes of the features are saved in the "Interface" module [51,52,54].

In the next step, the architecture of the model needs to be defined [51]. This involves the definition of the model's structure, including the number and type of layers, the activation functions, and other operations [50,51].

Then, the model parameters are defined [55]. These are values of variables and coefficients, including weights and biases of each layer [51,55,56].

The model metadata, interface, architecture, and parameters are encoded into a data structure called 'protobuf message definitions' [51,54,55]. The Protocol Buffer syntax allows encoding and decoding information about the Core ML model in any language that supports the Protocol Buffer serialization technology, (including Python, C++, C#, or Java) [54,55]. 'Model.proto' is the essential file of the Core ML Model *protobuff message definitions* [57]. It describes the structure of the model, the type of inputs and outputs it can have, and metadata [54,57]. The file also includes the 'specification version', which determines the versions of Core ML format specification and functionalities that it can provide [54,55,57,58]. Each version of the target device's operating system has its own way of implementing the model, so it is crucial to also include these in the model specification [58].

All of the Model's *protobuff message definitions* are encoded into binary format, which can be deployed on Apple platforms and encoded while loading the model [51,55].

2.3. Datasets Used

The authors used four distinct datasets for their study. These datasets were used for different purposes: two for image classification, one for tabular classification, and one for tabular regression. Of these, one dataset was created by the authors, and the rest were obtained from the Kaggle [59] website.

2.3.1. The Animals Dataset

The Animals dataset [38] was downloaded from Kaggle.com [59]. It contains 29,071 images divided into a training subset and testing subset. It is published under a Creative Commons license. Images are in different shapes, have three colour channels, and animals often are only partially visible in the picture (e.g., only the head of an ostrich), while in other cases, the whole body is portrayed. To set the image size for the network properly, it is essential to know what objects are in the images. One picture could be presenting a large animal (e.g., an elephant), but from a large distance so that it appearz small, while another could be a picture of a shrimp, but taken from a close distance and zoomed in.

Knowing that the depicted objects may vary in size, and the classes can be similar enough that differentiating between them requires a certain level of detail, it becomes justified to increase the input size of the neural network. Hence, it is important to take a good look at the data.

The training set consists of 22,566 images divided into 80 classes, making the problem a multiclass classification. The distribution of data in the training subset is presented in Figure 1. The testing set is made of 6505 images. Figure 2 portrays a random sample of 20 pictures of different classes from the Animals dataset.

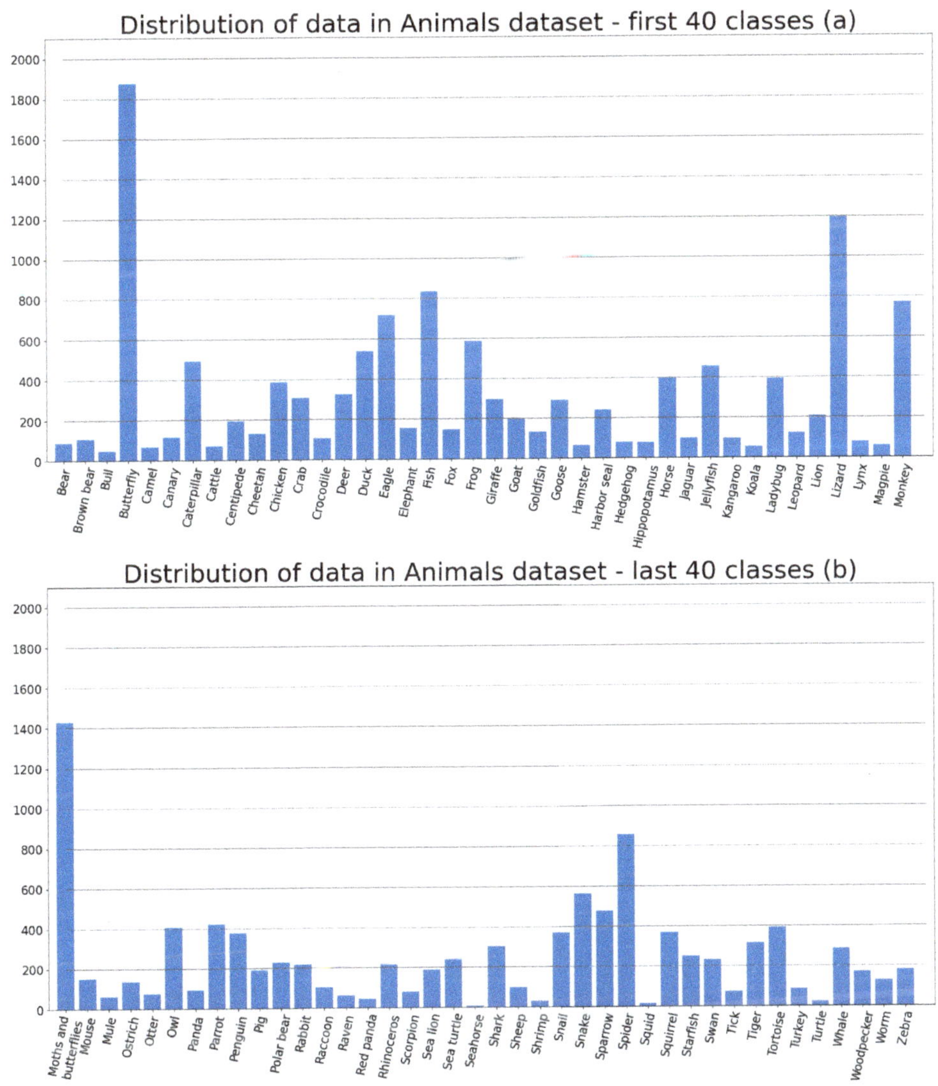

Figure 1. (**a**)—The distribution of data of the first 40 classes, (**b**)—the distribution of the remaining 40 classes.

Figure 2. A sample of 20 random images from the Animals dataset [38] subset.

2.3.2. The Payment Fraud Detection Dataset

The Online Payments Fraud Detection Dataset was published on Kaggle under Creative Commons Attribution-NonCommercial-ShareAlike 4.0 International license. The dataset consists of 5,080,807 entries in a ".csv" file, which translates into 493.5 MB. The data are divided into two classes (fraud or non-fraud), making it a binary classification problem. Every entry has nine features, as explained on the Kaggle's dataset webpage [39].

2.3.3. The Steam Reviews Dataset

The Steam Reviews Dataset 2021, obtained from Kaggle, represents the most recent dataset used in this study. It comprises approximately 21 million user reviews pertaining to approximately 300 games on the Steam platform. The dataset is available under the GNU GPL 2 license. To prepare the dataset for utilization in "Create ML", the authors conducted a data cleaning process using the Python language, along with the "Pandas" library. The cleaning procedure involved removing specific columns such as "comment text", "author", "creation date", and others. The resulting cleaned dataset was saved as "SteamReviewsCleaned.csv", resulting in a reduction in size from 3 GB to 2.15 GB. This dataset was utilized for a tabular regression problem within the context of this study.

2.3.4. The ClassifierData Dataset

This very small custom dataset was created by one of the authors. It was composed of four classes: "IPhone", "MacBook", "Apple Watch", and "AirPods". Every class contained 25 photos—19 in the training subset and 6 in the test subset. Every image was taken from different angles as well as in varyied lighting. Some pictures were taken of objects held in hand, while others were taken while lying on the floor, table, or carpet. Similarly to the Animals dataset, Figure 3 presents a random sample of images from all four classes. The data distribution of both training and test subsets is visualized in Figure 4. The subset was structured in a Create ML-compliant format [47], i.e., as image files placed inside the class-related folders. All images were sampled using a mobile device, photographing the target in various angles and light conditions.

Figure 3. A random sample of 20 images from the ClassifierData training subset.

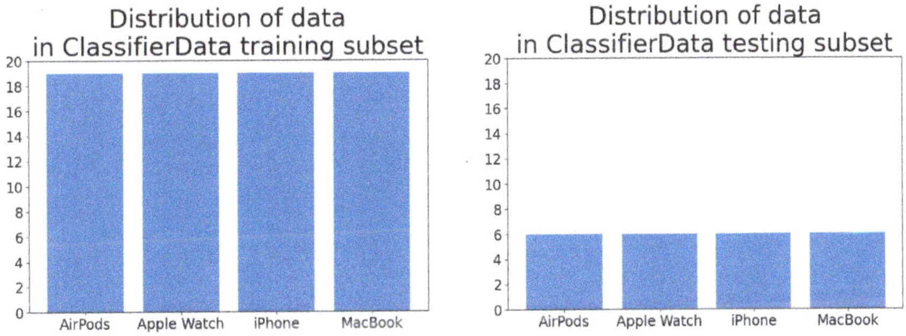

Figure 4. Distribution of data in training (**left**) and testing (**right**) subsets of the ClassifierData dataset.

2.4. Framework and Hardware Used in the Trials

The study was conducted on four notebooks, each one running the macOS Ventura operating system version 13.2 [60], with the Xcode Integrated Development Environment, version 14.2 [61].

The primary objective of the research was to investigate the usability of modern laptops equipped with ARM-based M1/M2-series CPUs in popular machine learning tasks. Since the manufacturer of M1- and M2-equipped laptops declares that the presence of the NPU cores in the CPUs makes them useful and interesting in machine learning applications, and since there are many researchers willing to buy a suitable portable platform for everyday work, the authors have decided that it may be worthwhile and interesting to put the eligibility of the NPU-equipped CPUs in DL tasks to a test.

All examined machines were ARM-based Apple MacBook Pro notebooks.

The first tested computer was the 2020 M1 MacBook Pro. The model started Apple's transition from Intel to ARM architecture [34]. It was equipped with the first version of the Apple M1 chip. The CPU included four high-performance and four energy-efficient cores. The chip was also equipped with an eight-core GPU. The first version of Neural Engine—a 16-core neural processing unit (NPU) —was also included in this chip [34]. The device had 8 GB of RAM. It is referred to as 'M1' in this work.

Another device from the M1-series was the MacBook Pro 2021. It included the strengthened version of the M1 Pro chip, which also included an eight-core CPU, but the allocation of the cores differed: six were high-performance and two were energy-saving.

The M2 series processor is an upgraded version of the previous M1 processor, boasting a speed increase of about 40%. It features eight cores, which are designed to be four performance cores and four efficiency cores [35]. Additionally, the processor includes 10 GPU cores and 16 Neural Engine cores. The M2-equipped laptop used in the research had 16 GB of RAM.

The fourth laptop used for the research was a MacBook Pro equipped with an Apple M2 Pro processor and 16 GB of RAM. This processor was composed of 10 cores, with 6 of them being performance cores and 4 being efficiency cores [36]. The processor also included 16 GPU cores and 16 Neural Engine cores.

3. Results

The most important result presented within this paper is the comparison of the computational performance of the M1 and M2 processors in ML tasks, presented in Section 3.1 and discussed in Sections 4 and 5. The comparison is made based on the performance (processing time) of ML models included within the 'Benchmarker.playground' project.

Section 3.2 includes an additional, also interesting, analysis of a possible impact of the versions of macOS and Xcode on the ML tasks' processing time.

3.1. *Measurement of the Impact of the Processor Model on the Model Creation Time*

During the research, three measurements of model training and testing time were performed. The computer was consistently connected to a power source throughout the script execution to ensure uninterrupted performance and avoid any potential limitations caused by energy-saving features. This allowed us to benchmark the efficiency of M-series processors in machine learning tasks using Apple's ecosystem.

3.1.1. Running the Benchmark

The measurements were performed on four computers. The 'Benchmarker.playground' file was copied to the hard drive of each machine. Then, the file was opened in the Xcode environment and executed. During the tests, each computer was left without any additional tasks. When the running of the script had finished, the console output was saved to a '.txt' file.

A screenshot of the 'M1 Pro.txt' file with the console output log from the 'Benchmarker' playground executed on the M1 Pro Mac is presented in Figure 5.

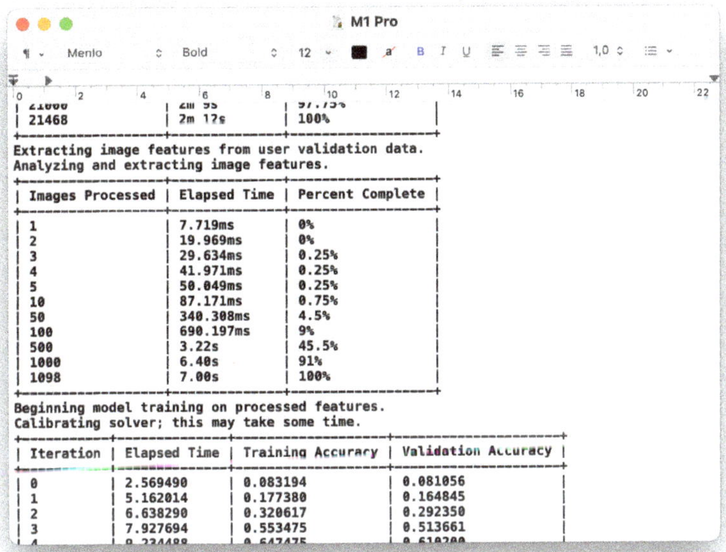

Figure 5. A part of the benchmark report of the 'M1 Pro' computer.

3.1.2. The Results of the Benchmark

The results of the benchmark performed on the ClassifierData dataset were similar in terms of overall time, except for the M1 Pro, which was over two times slower than the other processors. Each training took, on average, a different number of iterations (epochs); the means spanned from 11 to 14.667. The M1 achieved the best average result by a slight margin, outperforming the second-fastest (which was the M2 Pro) by 383 ms. The third average result was achieved by M2, which lost about 70 ms to the Pro variant. The worst performance on the ClassifierData dataset was the M1 Pro; despite taking the second-lowest average number of iterations (11.333), it scored by far the worst time of 8.622. Every model trained on each chip achieved 100% for both training and validation accuracy. This test was the quickest one due to the small size of the dataset.

Table 1 shows the time and accuracy results of the model training and testing process, performed on the 'ClassifierData' dataset using Create ML.

Table 1. Average results for the processor-related test performed using the 'ClassifierData' dataset.

	M1	M1 Pro	M2	M2 Pro
Training data—analysis time (s)	2.543	7.083	3.157	2.987
Validation data—analysis time (s)	0.346	0.912	0.366	0.402
Model training—total time (s)	0.213	0.214	0.261	0.256
Model training—number of iterations	11	11.333	14.667	12.333
Training Accuracy	100%	100%	100%	100%
Validation Accuracy	100%	100%	100%	100%
Total model creation time (s)	3.689	8.622	4.145	4.072
Evaluation data—analysis time (s)	0.824	2.203	0.876	0.949
Evaluation Accuracy	97.22%	98.6%	98.6%	95.8%
Total model evaluation time (s)	1.047	2.421	1.022	1.147

The multiclass classification test was performed by utilizing the Animals dataset of over 29,000 images, split into training and testing subsets. The results of the benchmark are presented in Table 2. The quickest of all tested processors while training the model on the

Animals dataset was M2 Pro. It took the M2 Pro 169.7 s to complete the test. The second-fastest processor, (M2), took 186.689 s to train the model, which is 9% slower than the Pro variant; however, the evaluation time was basically the same, at 40.796 for the Pro and 40.796 for the basic M2. The M1 Pro finished the training process in 193.219 s, which earned it third place. This result is 12% slower than the M2 Pro. The evaluation time was approximately 5 s slower than both M2 and M2 Pro. The slowest one, M1, achieved a result of 236.285 s. It was 28% slower than the fastest processor; simultaneously, it was the only one that completed the training in over 200 s on average. The evaluation also took the longest, exceeding 47 s. All processors achieved similar training and validation accuracy, of about 88% and 86.5%, respectively.

Table 2. Average results for the processor-related test performed using the 'Animals' dataset.

	M1	M1 Pro	M2	M2 Pro
Training data—analysis time (s)	132.333	132.333	116.667	116.333
Validation data—analysis time (s)	7.240	7.043	6.087	6.133
Model training—total time (s)	79.364	39.442	49.746	33.356
Training Accuracy	88.26%	88.24%	88.11%	88.13%
Validation Accuracy	85.98%	86.64%	86.13%	86.52%
Total model creation time (s)	236.285	193.219	186.689	169.777
Evaluation data—analysis time (s)	41.050	40.393	36.033	36.050
Evaluation Accuracy	84.87%	84.84%	84.91%	84.87%
Total model evaluation time (s)	47.284	45.415	40.768	40.796

Upon examining the results of the benchmark conducted on the PaymentFraud dataset displayed in Table 3, it is apparent that the accuracy levels for all tested cases were comparable, with minor disparities emerging during the data analysis phase. During this stage, the M2 Pro processor exhibited the quickest performance, taking only 1.924 s, while the slowest was the M1 at 2.310 s. The M2 processor, on the other hand, completed the data analysis in 2.01 s, and the M1 Pro required 2.161 s.

The most notable differences in processing time were found during the overall model building phase. The M2 processor was the speediest in this regard, finishing the model building task in 102.109 s. The M1 processor took 13% longer, completing the same assignment in 117.546 s, while the Pro version of the M1 took 146.641 s, which was 43% slower than the M2 processor. Interestingly, in this case, the M2 Pro processor proved to be the slowest, taking 151.659 s to complete the task, which was 48% slower than the M2's base version.

Table 3. Average results for the processor-related test performed using the 'PaymentFraud' dataset.

	M1	M1 Pro	M2	M2 Pro
Data processing time (s)	2.310	2.161	2.010	1.924
Training accuracy	99.96%	99.96%	99.96%	99.96%
Validation accuracy	99.96%	99.95%	99.97%	99.96%
Total model creation time (s)	117.546	146.641	102.109	151.659
Evaluation accuracy	99.96%	99.96%	99.96%	99.95%
Total model evaluation time (s)	1.280	1.279	1.139	1.107

Table 4 displays the benchmark outcomes for the SteamReviewCleaned dataset. Noticeably, the table does not present the maximum error and root-mean-square error findings for the training, validation, and test data. These results are excluded due to their consistency across all cases, as was shown in our preceding publication [37].

The M2 Pro processor boasted the swiftest processing time, taking only 7.981 s to complete the task, while the M1 Pro and M2 processors processed the data in nearly the same amount of time, clocking in at 8.143 s and 8.255 s, respectively. Meanwhile, the M1 proved to be the slowest, taking 9.276 s.

During the model building phase, the M2 Pro processor was once again the fastest, completing the task in only 12.545 s. The M2 processor followed closely, requiring 13.395 s to build the model. The M1 Pro took 14.713 s to finish the task, while the M1 took 15.545 s. With the exception of the M1, which took 2.023 s, all processors required less than 1.75 s to evaluate the model. The M2 Pro processor was once again the quickest in this task, taking only 1.596 s, while the M1 Pro and M2 processors achieved similar times of 1.736 and 1.665 s, respectively.

Table 4. Average results for the processor-related test performed using the 'SteamReviewsCleaned' dataset.

	M1	M1 Pro	M2	M2 Pro
Data processing time (s)	9.276	8.143	8.255	7.891
Total model creation time (s)	15.545	14.713	13.395	12.545
Total model evaluation time (s)	2.023	1.736	1.665	1.596

Table 5 displays the execution times of a script on various tested platforms. The MacBook with an M2 processor completed the entire script in the fastest time, taking 1088.989 s, with an average of 362.996 s per iteration. All iterations took a similar amount of time, with the fastest iteration completed in 359.398 s and the slowest in 364.796 s. The MacBook with an M2 Pro processor took 9% longer to execute the script, taking 1186.557 s, with an average of 395.519 s per iteration, and the slowest iteration took 397.401 s. The MacBook with an M1 Pro processor took 1281.761 s, with an average of 427.254 s per iteration, which was 18% longer than the M2 MacBook. The MacBook with M1 took the longest time, taking 1314.943 s, with an average of 438.314 s per iteration. This took 21% longer than the fastest tested MacBook. The fastest iteration took 436.706 s, and the slowest iteration took 439.858 s on the slowest MacBook tested.

Table 5. Average iteration times of the 'Benchmarker_Playground.playground' program.

	M1	M1 Pro	M2	M2 Pro
Average iteration time (s)	438.314	427.254	362.996	395.519
Fastest iteration (s)	436.706	423.018	359.398	394.179
Slowest iteration (s)	439.858	431.328	364.796	397.401
Total measurement time (s)	1314.943	1281.761	1088.989	1186.557

3.2. *Measurement of the Impact of the macOS Version on the Model Creation Time*

An evaluation of the influence of the macOS wersion on the script execution time was carried out on the MacBook Pro referenced as 'M1 Pro'. The measurement was performed on three various versions of macOS and Xcode IDE:

- macOS Monterey 12.4 and Xcode 13.4 ;
- macOS Ventura 13.0.1 and Xcode 14.2;
- macOS Ventura 13.2 and Xcode 14.2.

The computer was not used during each execution of the script. The device remained connected to the power source at all times to prevent any limitations due to energy-saving features.

Table 6 displays the benchmark results for various versions of systems and Xcode for the ClassifierData dataset. The Macbook with macOS 12.4 and Xcode 13.4 installed achieved the fastest training dataset, finishing the task in 6.957 s. A very similar time was recorded on a Macbook with macOS 13.2 and installed Xcode 14.2, at 7.083 s. The slowest was macOS 13.0.1, taking 7.573 s to complete the task. However, when analyzing the validation set, macOS 13.2 Macbook was the fastest, taking merely 0.912 s. The other platforms analyzed the validation set at a similar time, with macOS 12.4 and 13.0.1 completing the task in 0.921 and 0.922 s, respectively. The macOS 13.2 performed the fastest training, with the entire training taking 0.214 s, attaining completion after 11.3 iterations. The training took 0.217 s on a Macbook with macOS 12.4 after 12 iterations. The longest training, lasting 0.263 s, was on macOS 13.0.1, ending after 14 iterations. In all cases, the model achieved 100% accuracy on the training and validation sets. Overall, the entire process of analyzing the sets and building the model was the fastest on a Macbook with macOS 12.4, taking 8.571 s. A few milliseconds longer, the task was completed on a Macbook with macOS 13.2 in 8.662 s. The Macbook with macOS 13.0.1 took the longest at 9.175 s. The test set was analyzed the fastest on a Macbook with macOS 12.4, completing the task in 1.993 s. On a Macbook with macOS 13.0.1, the task was completed in 2.093 s. This task took the longest on a Macbook with macOS 13.2, requiring 2.203 s. The evaluation of the entire model on the test set was the fastest on a Macbook with macOS 12.4, taking 2.138 s, and the previously prepared model on this version achieved 95.83% accuracy. The same accuracy was achieved by the model prepared for macOS 13.0.1, but it took 2.309 s. The best accuracy was achieved on macOS 13.2, at 98.61%, but evaluating the model on the test set required 2.421 s.

Table 6. Average results for the OS-related test performed using the 'ClassifierData' dataset.

	macOS 12.4 Xcode 13.4	**macOS 13.0.1 Xcode 14.2**	**macOS 13.2 Xcode 14.2**
Training data—analysis time (s)	6.957	7.573	7.083
Validation data—analysis time (s)	0.921	0.922	0.912
Model training—total time (s)	0.217	0.263	0.214
Model training—number of iterations	12	14.7	11.3
Training Accuracy	100%	100%	100%
Validation Accuracy	100%	100%	100%
Total model creation time (s)	8.571	9.175	8.622
Evaluation data—analysis time (s)	1.933	2.093	2.203
Evaluation Accuracy	95.83%	95.83%	98.61%
Total model evaluation time (s)	2.138	2.309	2.421

Table 7 presents the development times of the model for the Animals set using different versions of macOS and Xcode. The analysis of the training set took the longest time on macOS 12.4, i.e., 137.667 s. The Macbook with macOS 13.2 analyzed the set for 132.333 s, while macOS 13.0.1 achieved the best result by completing the task in 130.667 s. The validation set was analyzed for 7.16 s on macOS 12.4 and 7.043 s on macOS 13.2, while the shortest time of 6.937 s was achieved on macOS 13.0.1. The model was trained for 40.424 s on a MacBook with macOS 12.4, achieving an accuracy of 87.91% for the training set and 86.12% for the validation set. The entire modeling process took 198.985 s. On macOS 13.0.1, the model was trained for 39.709 s, achieving an accuracy of 87.97% for the training set and 86.95% for the validation set. The entire modelling process took 191.641 s. For macOS 13.2, it took 193.219 s to build the entire model, with 39.442 s dedicated to training. In this case, the model achieved an accuracy of 88.24% for the training set and 86.64% for the validation set. The shortest time to evaluate the model on the test set was on a MacBook with macOS 13.2, taking 45.415 s, with an achieved accuracy of 84.84%. The model was evaluated on the test set in 46.095 s on macOS 13.0.1, with an accuracy of 84.87%. On a MacBook with macOS 12.4, the model was evaluated for the longest time of 46.590 s, achieving an accuracy of 84.94% for the test set.

Table 7. Average results for the OS-related test performed using the 'Animals' dataset.

	macOS 12.4 Xcode 13.4	macOS 13.0.1 Xcode 14.2	macOS 13.2 Xcode 14.2
Training data—analysis time (s)	137.667	130.667	132.333
Validation data—analysis time (s)	7.160	6.937	7.043
Model training—total time (s)	40.424	39.709	39.442
Training Accuracy	87.91%	87.97%	88.24%
Validation Accuracy	86.12%	86.95%	86.64%
Total model creation time (s)	198.985	191.641	193.219
Evaluation data—analysis time (s)	41.977	40.887	40.393
Evaluation Accuracy	84.94%	84.87%	84.84%
Total model evaluation time (s)	46.590	46.095	45.415

Table 8 presents the results acquired for the PaymentFraud dataset. Each system analyzed the training set for a similar amount of time. The macOS 13.0.1 with Xcode 14.2 installed analyzed the fastest set in 2.153 s, and the slowest analysis was on macOS 13.2 with the same Xcode version in 2.161 s. The analysis on macOS 12.4 took 2.158 s. The accuracy of the test set was 99.96% on each tested system, with only minor differences. The highest accuracy of 99.99% was achieved on macOS 13.0.1, and the lowest accuracy of 99.95% was achieved on macOS 13.2. The model trained on macOS 12.4 achieved an accuracy of 99.97% for the same validation set. The fastest model was built on macOS 12.4 in 143.456 s, while the slowest model building process was on macOS 13.2 in 146.641 s. It took 144.639 s to build the entire model on macOS 13.0.1. The accuracy measurement for the test set took 1.298 s on macOS 12.4 and 1.292 s on macOS 13.0.1. The fastest accuracy measurement of the model was on macOS 13.2 in 1.279 s.

Table 8. Average results for the OS-related test performed using the 'PaymentFraud' dataset.

	macOS 12.4 Xcode 13.4	macOS 13.0.1 Xcode 14.2	macOS 13.2 Xcode 14.2
Data processing time (s)	2.158	2.153	2.161
Training accuracy	99.96%	99.96%	99.96%
Validation accuracy	99.97%	99.99%	99.95%
Total model creation time (s)	143.456	144.639	146.641
Evaluation accuracy	99.96%	99.96%	99.96%
Total model evaluation time (s)	1.298	1.292	1.279

The training times for the SteamReviewsCleaned dataset on various versions of Macbook Pro with M1 Pro processor are displayed in Table 9. The quickest model was trained on macOS 13.2 equipped with Xcode 14.2, which took 14.713 s, with 8.143 s spent on analyzing the dataset. The longest training time was documented on macOS 13.0.1, where the whole process took 15.006 s, and dataset analysis took 8.356 s. The model evaluation process on different systems had similar timings, with macOS 13.0.1 having the longest evaluation time of 1.834 s and macOS 12.4 having the shortest evaluation time of 1.732 s. In the case of macOS 13.2, the evaluation process took 1.736 s.

Table 9. Average results for the OS-related test performed using the 'SteamReviewsCleaned' data-set.

	macOS 12.4 Xcode 13.4	macOS 13.0.1 Xcode 14.2	macOS 13.2 Xcode 14.2
Data processing time (s)	8.347	8.356	8.143
Total model creation time (s)	14.924	15.006	14.713
Total model evaluation time (s)	1.732	1.834	1.736

Table 10 provides a summary of the benchmark results for different versions of macOS. The quickest benchmark was completed in 1276.076 s on a MacBook running macOS 13.0.1 with Xcode 14.2 installed. On average, each iteration took 425.359 s. The fastest iteration was completed in 421.602 s, while the slowest iteration took 432.384 s. The benchmark took the longest time to complete on a MacBook running macOS 12.4 with Xcode 13.4 installed, taking 1293.449 s to finish, with one iteration taking 431.150 s. In this case, the slowest iteration took 436.282 s, while the quickest iteration was completed in 431.150 s. On macOS 13.2, the benchmark took 1281.761 s to complete, with an average of 427.254 s needed for each iteration.

Table 10. Cumulative results of all OS-related tests performed.

	macOS 12.4 Xcode 13.4	**macOS 13.0.1 Xcode 14.2**	**macOS 13.2 Xcode 14.2**
Average iteration time (s)	431.150	425.359	427.254
Fastest iteration (s)	428.403	421.602	423.018
Slowest iteration (s)	436.282	432.384	431.328
Total measurement time (s)	1293.449	1276.076	1281.761

4. Discussion

The gathered results present the comparative computational performances of the Apple laptops equipped with four different M-family processors, including the most recent M2 Pro chip. All of the tested hardware (including the previous M1 generation) is perfectly capable of performing ML tasks that do not require processing millions of images or hundreds of gigabytes (or more) of data. Each and every dataset has been successfully analyzed and processed. Every created model had similar efficacy; regardless of the chip it was trained on, the results were satisfactory.

Three of the used datasets are available to the public; as well, the hardware and software specifications provided in this research ensure that the reproducibility and comparability of the results is possible for other researchers.

The chips were tested for machine learning applications with the use of Apple's Create ML. A rather surprising average result is the overall performance of the M2 Pro variant. It was outperformed by the base variant by approximately 9%; however, this was mainly due to poor performance of the more expensive variant on the PaymentFraud dataset. This result may affect someone's decision as to whether it is beneficial to increase their budget to buy the Pro chip or save money and buy the cheaper standard M2. The M1 Pro also had the same difficulties with the same dataset that its newer counterpart had, achieving a much worse time than the base version of the chip.

The research also included an evaluation of the script execution time on various macOS versions. The tasks were performed on the same MacBook Pro laptop, with different versions of the operating system and development environment.

The obtained results showed, that the version of the macOS has an impact on the script execution times. For the 'ClassifierData', almost all times were longer after updating the operating system from the previous 'major version' (macOS 12.4) to the next new 'major version' (macOS 13.0.1). However, the installation of a system update with bug fixes and improvements (macOS 13.2) decreased the execution time to a value which was comparable with the results from the previous 'major version'.

In case of the 'Animals' dataset, the data analysis time also changed with the version of the operating system, with macOS 13.0.1 being the fastest. The model training time was comparable in each test, which suggests that system updates have no impact on the training process.

The tests performed on tabular datasets ('PaymentFraud' and 'SteamReviewsCleaned') showed no remarkable difference between the execution time of the training and evaluation process. This shows that differences are visible only when working with image datasets.

5. Conclusions

Upon conducting an in-depth analysis of the collected results, our study revealed significant findings that may provide insights into the performance of distinct chips when employed for training and testing models using Create ML.

The results presented in the paper demonstrate that the M2 chip may exhibit superior performance compared to the M2 Pro (as shown in Table 5), implying that the M2 chip may be a favorable choice for tasks demanding efficient model creation.

While it is difficult to provide a definitive recommendation for future processors or operating systems due to their evolving nature, in this study, the authors proposed a methodology for evaluating the effectiveness of specific hardware architectures, which can be investigated by the researchers themselves, using the proposed [46] benchmark.

The observations also reveal that the 'Pro' series of respective chips (namely, M1 Pro and M2 Pro) do not meet the anticipated time-related performance efficiency for model creation using the 'Payment Fraud' dataset (see Table 3). Moreover, the processing performance of the 'M1 Pro' chip proved to be below average for the small 'ClassifierData' dataset, whereas the 'M1' chip exhibited surprisingly good performance on the same dataset (see Table 1). These observations indicate that certain characteristics of datasets can indeed impact the performance of specific chip models. The research provided evidence that the expected superiority of the 'Pro' variant should be challenged for model training, even when using Apple's own 'Create ML.'

Lastly, the experimental comparative research resulted in the formulation of additional insights of minor significance: it was confirmed that the multiclass classification performance results were consistent with the CPU-generation-related expectations, and that the operating system version had an impact on the processing time, particularly in the case of image datasets.

The results presented within this study bring theoretical and managerial implications that extend beyond the immediate scope of hardware platform performance evaluation. The insights gained from comparing respective processors and their performance in machine learning tasks using Create ML shed light on the complexities and nuances of hardware-platform-specific characteristics. From a theoretical standpoint, these findings help to understand the impact of particular hardware choices on the efficiency and effectiveness of ML computation time. By examining the performance (and its variations across various chips), researchers can refine their theoretical models and develop more nuanced frameworks for leveraging the benefits of hardware acceleration for typical Machine Learning applications. On a managerial level, the research findings have substantial value for decision-makers who are considering hardware platforms for machine learning researchers and initiatives. The performance disparities observed among the tested chips help to highlight the benefits and importance of careful evaluation of the hardware requirements, based on the specific needs and characteristics of current machine learning projects.

Future Work

The research conducted in this study opens avenues for further exploration in the realm of multi-platform analysis (the comparison of Apple platforms against other, non-Apple, platforms). While this research focused on the performance evaluation of hardware platforms utilizing primarily Create ML, future studies could extend the analysis to encompass a broader range of machine learning frameworks, especially the most popular ones—TensorFlow and PyTorch. By conducting experiments using TensorFlow and PyTorch across different hardware platforms and operating systems, a more comprehensive understanding of the performance variations and platform compatibility can be obtained. Additionally, investigating the impact of different frameworks on the efficiency and effectiveness of model creation would contribute valuable insights to the field. Such multi-platform approaches will provide a more comprehensive assessment of the hardware-platform-specific characteristics and guide researchers and practitioners in making informed decisions regarding the choice of frameworks and hardware configurations for their machine learning tasks.

Author Contributions: Conceptualization, D.K., P.A., M.B. and M.P.; methodology, D.K.; software, D.K.; validation, D.K., M.B. and P.A.; formal analysis, M.P.; investigation, D.K.; resources, D.K.; data curation, D.K.; writing—original draft preparation, D.K., P.A., M.B., M.S. and M.P.; writing—review and editing, D.K., P.A., M.B. and M.P.; visualization, D.K. and P.A.; supervision, M.P. All authors have read and agreed to the published version of the manuscript.

Funding: This research received no external funding.

Institutional Review Board Statement: Not applicable.

Informed Consent Statement: Not applicable.

Data Availability Statement: The 'Benchmarker.playground' project, including all source code, is made available as Open Source on GitHub, at https://github.com/dKasperek/Benchmarker (accessed 3 March 2023). The code is implemented to run three iterations, each one creating four ML models, using the following datasets: 'Animals' (available from [38]), 'PaymentFraud' ([39]), 'SteamReviews' ([40]), and the 'ClassifierData' (available within the above-mentioned GitHub project). The datasets should be imported into the 'Resources' folder inside the Xcode Playground project.

Conflicts of Interest: The authors declare no conflict of interest.

References

1. Angra, S.; Ahuja, S. Machine learning and its applications: A review. In Proceedings of the 2017 International Conference on Big Data Analytics and Computational Intelligence (ICBDAC), Chirala, India, 23–25 March 2017; IEEE: Piscataway, NJ, USA, 2017; pp. 57–60.
2. Shinde, P.P.; Shah, S. A review of machine learning and deep learning applications. In Proceedings of the 2018 Fourth International Conference on Computing Communication Control and Automation (ICCUBEA), Pune, India, 16–18 August 2018; IEEE: Piscataway, NJ, USA, 2018; pp. 1–6.
3. Bertolini, M.; Mezzogori, D.; Neroni, M.; Zammori, F. Machine Learning for industrial applications: A comprehensive literature review. *Expert Syst. Appl.* **2021**, *175*, 114820. [CrossRef]
4. Choi, J.A.; Lim, K. Identifying machine learning techniques for classification of target advertising. *ICT Express* **2020**, *6*, 175–180. [CrossRef]
5. Shah, N.; Engineer, S.; Bhagat, N.; Chauhan, H.; Shah, M. Research trends on the usage of machine learning and artificial intelligence in advertising. *Augment. Hum. Res.* **2020**, *5*, 1–15. [CrossRef]
6. Nawrocka, A.; Kot, A.; Nawrocki, M. Application of machine learning in recommendation systems. In Proceedings of the 2018 19th International Carpathian Control Conference (ICCC), Szilvasvarad, Hungary, 28–31 May 2018 ; IEEE: Piscataway, NJ, USA, 2018; pp. 328–331.
7. Aye, G.A.; Kim, S.; Li, H. Learning autocompletion from real-world datasets. In Proceedings of the 2021 IEEE/ACM 43rd International Conference on Software Engineering: Software Engineering in Practice (ICSE-SEIP), Madrid, Spain, 25–28 May 2021; IEEE: Piscataway, NJ, USA, 2021; pp. 131–139.
8. Kolb, S.; Teso, S.; Dries, A.; De Raedt, L. Predictive spreadsheet autocompletion with constraints. *Mach. Learn.* **2020**, *109*, 307–325. [CrossRef]
9. Ramesh, A.; Dhariwal, P.; Nichol, A.; Chu, C.; Chen, M. Hierarchical text-conditional image generation with clip latents. *arXiv* **2022**, arXiv:2204.06125 .
10. Roose, K. An AI-Generated Picture Won an Art Prize. Artists Aren't Happy. *New York Times*, 2 September 2022.
11. Das, S.; Dey, A.; Pal, A.; Roy, N. Applications of artificial intelligence in machine learning: Review and prospect. *Int. J. Comput. Appl.* **2015**, *115*, 31–41. [CrossRef]
12. Alpaydin, E. *Machine Learning: The New AI*; MIT Press: Cambridge, MA, USA, 2016.
13. Dehouche, N.; Dehouche, K. What is in a Text-to-Image Prompt: The Potential of Stable Diffusion in Visual Arts Education. *arXiv* **2023**, arXiv:2301.01902.
14. Rando, J.; Paleka, D.; Lindner, D.; Heim, L.; Tramèr, F. Red-Teaming the Stable Diffusion Safety Filter. *arXiv* **2022**, arXiv:2210.04610.
15. Munson, M.A. A study on the importance of and time spent on different modeling steps. *ACM SIGKDD Explor. Newsl.* **2012**, *13*, 65–71. [CrossRef]
16. Wani, M.A.; Bhat, F.A.; Afzal, S.; Khan, A.I. *Advances in Deep Learning*; Springer: Berlin/Heidelberg, Germany, 2020.
17. Zhong, B.; Xing, X.; Love, P.; Wang, X.; Luo, H. Convolutional neural network: Deep learning-based classification of building quality problems. *Adv. Eng. Inform.* **2019**, *40*, 46–57. [CrossRef]
18. Barbedo, J.G.A. Impact of dataset size and variety on the effectiveness of deep learning and transfer learning for plant disease classification. *Comput. Electron. Agric.* **2018**, *153*, 46–53. [CrossRef]
19. Soekhoe, D.; Van Der Putten, P.; Plaat, A. On the impact of data set size in transfer learning using deep neural networks. In Proceedings of the Advances in Intelligent Data Analysis XV: 15th International Symposium, IDA 2016, Stockholm, Sweden, 13–15 October 2016; Proceedings 15; Springer: Berlin/Heidelberg, Germany, 2016; pp. 50–60.

20. Ng, H.W.; Nguyen, V.D.; Vonikakis, V.; Winkler, S. Deep learning for emotion recognition on small datasets using transfer learning. In Proceedings of the 2015 ACM on International Conference on Multimodal Interaction, Seattle, WA, USA, 9–13 November 2015; pp. 443–449.
21. Pang, B.; Nijkamp, E.; Wu, Y.N. Deep learning with tensorflow: A review. *J. Educ. Behav. Stat.* **2020**, *45*, 227–248. [CrossRef]
22. Stevens, E.; Antiga, L.; Viehmann, T. *Deep Learning with PyTorch*; Manning Publications: Shelter Island, NY, USA , 2020.
23. Kuleto, V.; Ilić, M.; Dumangiu, M.; Ranković, M.; Martins, O.M.; Păun, D.; Mihoreanu, L. Exploring opportunities and challenges of artificial intelligence and machine learning in higher education institutions. *Sustainability* **2021**, *13*, 10424. [CrossRef]
24. Sanusi, I.T.; Oyelere, S.S.; Vartiainen, H.; Suhonen, J.; Tukiainen, M. A systematic review of teaching and learning machine learning in K-12 education. In *Education and Information Technologies*; Springer: Berlin/Heidelberg, Germany, 2022; pp. 1–31.
25. Wang, Y.E.; Wei, G.Y.; Brooks, D. Benchmarking TPU, GPU, and CPU platforms for deep learning. *arXiv* **2019**, arXiv:1907.10701.
26. Saha, S.S.; Sandha, S.S.; Srivastava, M. Machine learning for microcontroller-class hardware—A review. *IEEE Sens. J.* **2022**, *22*, 21362–21390 . [CrossRef] [PubMed]
27. Sze, V.; Chen, Y.H.; Emer, J.; Suleiman, A.; Zhang, Z. Hardware for machine learning: Challenges and opportunities. In Proceedings of the 2017 IEEE Custom Integrated Circuits Conference (CICC), Austin, TX, USA, 30 April–3 May 2017; IEEE: Piscataway, NJ, USA, 2017; pp. 1–8.
28. Berggren, K.; Xia, Q.; Likharev, K.K.; Strukov, D.B.; Jiang, H.; Mikolajick, T.; Querlioz, D.; Salinga, M.; Erickson, J.R.; Pi, S.; et al. Roadmap on emerging hardware and technology for machine learning. *Nanotechnology* **2020**, *32*, 012002. [CrossRef] [PubMed]
29. An, L.; Peng, K.; Yang, X.; Huang, P.; Luo, Y.; Feng, P.; Wei, B. E-TBNet: Light Deep Neural Network for automatic detection of tuberculosis with X-ray DR Imaging. *Sensors* **2022**, *22*, 821. [CrossRef] [PubMed]
30. Hadjis, S.; Abuzaid, F.; Zhang, C.; Ré, C. Caffe con troll: Shallow ideas to speed up Deep Learning. In Proceedings of the Fourth Workshop on Data analytics in the Cloud (DanaC'15), Melbourne, VIC, Australia, 31 May–4 June 2015; pp. 1–4.
31. Courville, V.; Nia, V.P. Deep learning inference frameworks for ARM CPU. *J. Comput. Vis. Imaging Syst.* **2019**, *5*, 3.
32. Sinha, I.; Foscht, T.; Sinha, I.; Foscht, T. The era of anti-marketing. In *Reverse Psychology Marketing: The Death of Traditional Marketing and the Rise of the New "Pull" Game*; Springer: Berlin/Heidelberg, Germany, 2007; pp. 143–167.
33. O'Gorman, D.E. The Fog of (marketing) Wars: The Need for Assumption-Based Decision Making Processes. In Proceedings of the Marketing Management Association 2007 Educators' Conference Proceedings, San Diego, CA, USA, 16–19 February 2007; p. 44.
34. Apple Inc. The 'Apple Event'—10 November 2020. Apple Event. 2020. Available online: https://www.youtube.com/watch?v=5AwdkGKmZ0I (accessed on 1 June 2023).
35. Apple Inc. Macbook Pro 13—Official Source Link with the Specification of the Macbook Pro 13 Laptop. 2023. Available online: https://support.apple.com/kb/SP870?locale=en_GB (accessed on 1 June 2023).
36. Apple Inc. Macbook Pro 14—Official Source Link with the Specification of the Macbook Pro 14 Laptop. 2023. Available online: https://support.apple.com/kb/SP889?locale=en_GB (accessed on 1 June 2023).
37. Kasperek, D.; Podpora, M.; Kawala-Sterniuk, A. Comparison of the Usability of Apple M1 Processors for Various Machine Learning Tasks. *Sensors* **2022**, *22*, 8005. [CrossRef] [PubMed]
38. Jana, A. Animals Detection Images Dataset. Collection of Wild Animal Species with Annotations. Available online: https://www.kaggle.com/datasets/antoreepjana/animals-detection-images-dataset (accessed on 1 June 2023).
39. Roy, R. Online Payments Fraud Detection Dataset. Online payment fraud big dataset for testing and practice purpose. Available online: https://www.kaggle.com/datasets/rupakroy/online-payments-fraud-detection-dataset (accessed on 1 June 2023).
40. Marko M. Steam Reviews Dataset 2021. Large Collection of Reviews of Steam Games. Available online: https://www.kaggle.com/datasets/najzeko/steam-reviews-2021 (accessed on 1 June 2023).
41. Apple Inc. Create ML Overview—Machine Learning—Apple Developer. 2022. Available online: https://developer.apple.com/machine-learning/create-ml/ (accessed on 1 June 2023).
42. Goodwill, J.; Matlock, W. The swift programming language. In *Beginning Swift Games Development for iOS*; Apress: Berkeley, CA, USA, 2015; pp. 219–244.
43. Kelly, M.; Nozzi, J. *Mastering Xcode: Develop and Design*; Peachpit Press: Berkeley, CA, USA, 2013.
44. Feiler, J. Introducing Swift Playgrounds. *Exploring Swift Playgrounds: The Fastest and Most Effective Way to Learn to Code and to Teach Others to Use Your Code*; Apress: Berkeley, CA, USA, 2017; pp. 1–11.
45. Apple Inc. Core ML Framework Documentation. 2023. Available online: https://developer.apple.com/documentation/coreml, (accessed on 1 June 2023).
46. Kasperek, D. The 'Benchmarker.playground' Project and Apple Devices Image Dataset. Available online: https://github.com/dKasperek/Benchmarker (accessed on 1 June 2023).
47. Apple Inc. Creating an Image Classifier Model | Apple Developer Documentation. 2022. Available online: https://developer.apple.com/documentation/createml/creating_an_image_classifier_model (accessed on 1 June 2023).
48. Apple Inc. "Getting a Core ML Model". 2023. Available online: https://developer.apple.com/documentation/coreml/getting_a_core_ml_model (accessed on 1 June 2023).
49. Apple Inc. MLModel | Apple Developer Documentation. 2023. Available online: https://developer.apple.com/documentation/coreml/mlmodel (accessed on 1 June 2023).

50. Apple Inc. WWDC 2018, What's New in Core ML, Part 1. In Proceedings of the Apple Worldwide Developers Conference, Online, 4–8 June 2018, Available online: https://developer.apple.com/videos/wwdc2018 (accessed on 1 June 2023).
51. Apple Inc. WWDC 2021, Tune your Core ML models. In Proceedings of the Apple Worldwide Developers Conference, Cupertino, CA, USA, 8–11 June 2021, Available online: https://developer.apple.com/videos/wwdc2021 (accessed on 1 June 2023).
52. Apple Inc. MLModelDescription | Apple Developer Documentation. 2023. Available online: https://developer.apple.com/documentation/coreml/mlmodeldescription (accessed on 1 June 2023).
53. Apple Inc. Metadata | Apple Developer Documentation. 2023. Available online: https://developer.apple.com/documentation/coreml/mlmodeldescription/2879386-metadata (accessed on 1 June 2023).
54. Apple Inc. MLModel Overview. 2023. Available online: https://coremltools.readme.io/docs/mlmodel (accessed on 1 June 2023).
55. Apple Inc. WWDC 2019, Core ML 3 Framework. In Proceedings of the Apple Worldwide Developers Conference, Online, 3–7 June 2019. Available online: https://developer.apple.com/videos/wwdc2019 (accessed on 1 June 2023).
56. Apple Inc. MLParameterKey. Apple Developer Documentation. 2023. Available online: https://developer.apple.com/documentation/coreml/mlparameterkey (accessed on 1 June 2023).
57. Apple Inc. Core ML Format Specification—Core ML Format Reference Documentation. 2023. Available online: https://apple.github.io/coremltools/mlmodel/index.html (accessed on 1 June 2023).
58. Apple Inc. Core ML Format Reference Documentation. 2023. Available online: https://apple.github.io/coremltools/mlmodel/Format/Model.html (accessed on 1 June 2023).
59. Kaggle. Kaggle Competitions Webpage. Available online: https://www.kaggle.com/competitions (accessed on 1 June 2023).
60. Apple Inc. macOS Ventura—Official Source Link from Mac App Store. 2023. Available online: https://apps.apple.com/us/app/macos-ventura/id1638787999?mt=12 (accessed on 1 June 2023).
61. Apple Inc. Xcode—Official Source Link from Mac App Store. 2023. Available online: https://apps.apple.com/us/app/xcode/id497799835?mt=12 (accessed on 1 June 2023).

sensors

MDPI

Article

Student Learning Behavior Recognition Incorporating Data Augmentation with Learning Feature Representation in Smart Classrooms

Zhifeng Wang [1,*], Longlong Li [1], Chunyan Zeng [2,*] and Jialong Yao [1]

1 Faculty of Artificial Intelligence in Education, Central China Normal University, Wuhan 430079, China
2 Hubei Key Laboratory for High-Efficiency Utilization of Solar Energy and Operation Control of Energy Storage System, Hubei University of Technology, Wuhan 430068, China
* Correspondence: zfwang@ccnu.edu.cn (Z.W.); cyzeng@hbut.edu.cn (C.Z.)

Abstract: A robust and scientifically grounded teaching evaluation system holds significant importance in modern education, serving as a crucial metric that reflects the quality of classroom instruction. However, current methodologies within smart classroom environments have distinct limitations. These include accommodating a substantial student population, grappling with object detection challenges due to obstructions, and encountering accuracy issues in recognition stemming from varying observation angles. To address these limitations, this paper proposes an innovative data augmentation approach designed to detect distinct student behaviors by leveraging focused behavioral attributes. The primary objective is to alleviate the pedagogical workload. The process begins with assembling a concise dataset tailored for discerning student learning behaviors, followed by the application of data augmentation techniques to significantly expand its size. Additionally, the architectural prowess of the Extended-efficient Layer Aggregation Networks (E-ELAN) is harnessed to effectively extract a diverse array of learning behavior features. Of particular note is the integration of the Channel-wise Attention Module (CBAM) focal mechanism into the feature detection network. This integration plays a pivotal role, enhancing the network's ability to detect key cues relevant to student learning behaviors and thereby heightening feature identification precision. The culmination of this methodological journey involves the classification of the extracted features through a dual-pronged conduit: the Feature Pyramid Network (FPN) and the Path Aggregation Network (PAN). Empirical evidence vividly demonstrates the potency of the proposed methodology, yielding a mean average precision (mAP) of 96.7%. This achievement surpasses comparable methodologies by a substantial margin of at least 11.9%, conclusively highlighting the method's superior recognition capabilities. This research has an important impact on the field of teaching evaluation system, which helps to reduce the burden of educators on the one hand, and makes teaching evaluation more objective and accurate on the other hand.

Keywords: teaching evaluation system; student learning behavior; data augmentation; smart classrooms

Citation: Wang, Z.; Li, L.; Zeng, C.; Yao, J. Student Learning Behavior Recognition Incorporating Data Augmentation with Learning Feature Representation in Smart Classrooms. *Sensors* **2023**, *23*, 8190. https://doi.org/10.3390/s23198190

Academic Editors: Peter Hockicko, Patrik Kamencay and Róbert Hudec

Received: 18 August 2023
Revised: 28 September 2023
Accepted: 29 September 2023
Published: 30 September 2023

1. Introduction

The rapid advancements in computer technologies, such as artificial intelligence, big data, and cloud computing, have led to the pervasive integration of smart classrooms in learning and teaching [1–3]. To bolster the development of these smart classrooms, the establishment of a robust and multifaceted teaching evaluation system is imperative [4,5], with a primary focus on comprehensive teaching evaluations. Among the array of evaluation tools, the recognition of student learning behaviors emerges as a particularly potent method for assessing teaching approaches [6,7]. By scrutinizing and identifying student learning behaviors, educators gain valuable insights into their students' progress and learning efficacy, enabling them to fine-tune teaching strategies and methods accordingly. This ultimately cultivates a more conducive learning environment [8].

In traditional classroom settings, educators and teaching personnel primarily rely on direct observation of students' learning behaviors to comprehend their ongoing learning trajectory [9–11]. Additionally, indirect evaluation methods, including grade records, homework completion, and class attendance tracking, are harnessed to gauge students' learning journey. With the integration of multimedia teaching in classrooms, sensor-based technologies have emerged, enabling the nuanced recognition of students' learning behaviors. For instance, Mohammed et al. [12] harnessed the Internet of Things (IoT) to devise an automated classroom behavior classification system. This system assesses class effectiveness through result recognition, furnishing administrators with insightful data to enhance overall classroom performance. In the field of psychology, researchers have explored the benefits of MOOC learning for students with dependent speech cognitive style and dependent image cognitive style based on students' brain waves, so as to provide targeted guidance for students' learning styles. However, traditional direct observation methods rely too much on the subjectivity of the observer and are time-consuming, while sensor technology and brainwave observation rely on expensive equipment and are difficult to popularize [13].

With the maturity of computer vision technology, it has become a trend to apply it to the intelligent classroom environment as an auxiliary tool for teaching evaluation. These technologies are characterized by continuous monitoring, objective evaluation and real-time interaction. Wang [14] proposed a deep residual network that leverages residual structures to identify students' engagement levels during class. This technological innovation serves as a pivotal cornerstone for the development of intelligent classrooms. It is worth noting that researchers in the field of education and teaching have recognized that factors such as different lighting conditions, different viewing angles, and the quality of image acquisition equipment in dense teaching environments pose challenges for the accurate identification of student learning behaviors [15–17]. Therefore, in addition to improving lighting facilities and optimizing the layout of camera equipment, the researchers are also working to develop and optimize more universal detection algorithms to address these challenges.

Our research question can be summarized as follows: how to accurately detect and recognize students' learning behaviors in complex smart classrooms (including a large number of students, occlusions, and different observation angles)? Building upon the foundation of previous research, this paper adopts the YOLOv7 framework as the principal network architecture, thereby achieving swift recognition while preserving accuracy. Moreover, we seamlessly integrate the attention recognition mechanism of CBAM to heighten recognition precision within complex background scenarios. Concretely, we employ the proposed algorithm to detect seven distinctive learning behaviors typically exhibited by elementary school students, encompassing actions like writing, reading, raising hands, and participating in discussions. We undertake a comprehensive performance comparison against state-of-the-art (SOTA) learning behavior recognition frameworks. This paper makes several significant contributions that advance the field of student learning behavior recognition:

- In response to the absence of specialized datasets for student learning behavior recognition in the field of education, this study introduces the Student Learning Behavior (SLB) dataset. This dataset is meticulously annotated to include comprehensive information about classroom learning behaviors among primary and secondary school students. Researchers and practitioners can access this valuable resource on GitHub through the following link: https://github.com/houhou34/Teaching-evaluation-dataset (accessed on 6 March 2023). The availability of this dataset addresses a critical gap in the domain, facilitating the development and evaluation of learning behavior recognition models.
- This paper proposes a novel learning behavior recognition method that builds upon the YOLOv7 architecture. This method has been designed to accurately detect multiple students' learning behaviors. Furthermore, we conduct an in-depth investigation into the impact of incorporating the attention detection mechanism into our proposed

method. This integration of the attention mechanism is shown to significantly enhance the recognition performance of our model.

- Our developed student learning behavior recognition network demonstrates a notable improvement over state-of-the-art (SOTA) methods. Specifically, it achieves an enhanced mean average precision (mAP) by 11.9%. Our model exhibits remarkable advantages in detecting students, even in scenarios with significant occlusion. It excels in identifying common classroom behaviors such as students turning their heads, raising their hands, and looking up. These achievements underscore the effectiveness of our proposed approach in real-world classroom environments.
- We tackle the challenge posed by limited training samples by employing two crucial techniques: data augmentation and transfer learning. These methods allow us to maximize the utility of the available data, enhancing the robustness and generalization capacity of our model.

The ensuing sections of this paper are structured as follows: Section 2 offers an in-depth review of pertinent research in student classroom behavior recognition and the YOLO algorithm. In Section 3, we provide a problem definition of this research and summarize the notations that appear in this paper. Section 4 elucidates the procedural approach and network framework adopted for detecting and identifying student behaviors. In Section 5, we present a detailed comparison and conduct ablation experiments to rigorously assess the efficacy of our proposed methodology. Lastly, Section 6 encapsulates our conclusions and outlines potential avenues for future research exploration.

2. Related Work

In this section, we will take a closer look at student learning behavior and current methods for detecting student behavior in the classroom, as well as provide an overview of the YOLO object recognition model. In addition, the application of attention mechanisms in computer vision is explored.

2.1. Student Learning Behavior

The manifestations, characteristics, and phenomena apparent in students' learning behaviors have been extensively studied in the fields of educational research and educational psychology [18]. Typically, these characteristics and phenomena are interrelated with various aspects of students' cognitive, affective, and social interactions. The impact of students' learning behaviors on their academic performance and learning experiences is significant. Learning analytics is a rapidly developing academic field that evaluates the status of students from various perspectives. In classroom instruction, attendance is vital, as it serves as the initial step towards participation in classroom activities and academic seminars [19]. Furthermore, attention plays a crucial role in gauging a student's dedication to learning. Students must maintain a high level of focus to comprehend and absorb course content effectively. Effective attention and focus contribute significantly to academic performance. Collaboration with peers further underscores the pivotal role of cooperative learning proficiency when working in teams [20,21]. Cultivating positive study behaviors and habits facilitates meeting academic challenges, enhancing academic performance, and improving lifelong learning outcomes. Learning behaviors and habits are instrumental in the success of learners. Therefore, educators must foster these behaviors during the teaching and learning process to assist students in achieving better academic outcomes and preparing for future careers.

2.2. Student Learning Behavior Recognition Based on Deep Learning

Student behavior within the classroom significantly impacts the quality of education [7]. Active participation, attentive listening, note-taking, and adherence to classroom rules all contribute to effective teaching and a conducive learning environment [22,23]. To address the significance of student behaviors, researchers have harnessed deep learning techniques to construct frameworks for detecting and recognizing these behaviors.

An enhanced Faster R-CNN model for student behavior recognition was proposed by Zheng et al. [24], which incorporated a novel scale-aware recognition head and a fresh feature fusion strategy for detecting low-resolution behaviors. Lv et al. [25] integrated the ResNet and FPN modules into the SSD model, addressing the challenge of recognizing small targets at the back of the classroom and significantly boosting image recognition efficiency. Mindoro et al. [26] introduced a method to predict student behavior based on facial expressions and real-time behavior recognition, implementing it with the YOLOv3 network. Tang et al. [27] employed a weighted bidirectional feature pyramid network (BIFPN) along with YOLOv5's feature pyramid structure, effectively transforming the target recognition issue into a fine-grained representation challenge. Furthermore, they enhanced the non-maximal value suppression algorithm to improve the differentiation of highly occluded objects. Yang et al. [28] developed an analytical system for assessing students' learning status, harnessing the YOLOv5 network and the CBAM attention module to extract robust features from student behavior. Mo et al. [29] proposed a multitask learning method to identify students' classroom behaviors, which used MTHN module to extract intermediate heat maps and combined key points and object locations to simulate students' behaviors. It is noteworthy that researchers in education and teaching face the challenge of detecting behaviors within densely populated classrooms, calling for a comprehensive solution to tackle missed and erroneous recognitions due to occlusion and observation angles during classroom behavior recognition.

In addition to computer vision-based techniques for identifying student classroom behaviors, Zhao et al. [30] leveraged deep learning and developed models grounded in Deep Belief Networks (DBNs) to assess teaching speech standards. Lu et al. [31] applied feature data mining methods to detect learning behaviors in online English classrooms. In Table 1, comparisons of deep learning-based student learning behavior recognition methods in terms of methods, implementation techniques, and results are shown.

Table 1. Comparison of student learning behavior recognition methods based on deep learning.

Method	Technique	Performance
Faster R-CNN [24]	1. Proposal of an improved Faster R-CNN model; 2. Introduction of a scale-aware detection head to handle scale variations; 3. Feature fusion strategy to detect low-resolution behaviors; 4. Use of Online Hard Example Mining (OHEM) to address class imbalances.	Experimental results on real corpus show that the performance of the proposed method is improved compared with the baseline method.
SSD [25]	1. Data enhancement techniques were applied to expand the dataset; 2. The improved model showed better feature extraction ability and small target recognition accuracy compared to the native SSD model.	The improved SSD model achieves high recognition accuracy and provides technical support for intelligent management and teaching in universities.
YOLOv3 [26]	The researchers used the YOLOv3 algorithm for face recognition and prediction of student behavior.	The proposed method offers a reasonable pace of identification and positive outcomes for measuring student interest based on observable actions in the classroom.
YOLOv5 [27]	1. Proposal of a classroom behavior detection algorithm using an improved YOLOv5 model; 2. Combination of feature pyramid structure and weighted bidirectional feature pyramid network; 3. Addition of spatial and channel convolutional attention mechanism; 4. Improvement of non-maximum suppression using distance-based intersection ratio.	The average precision of the algorithm on the self-built dataset is 89.8%, and the recall rate is 90.4%.

2.3. Target Detection Based on YOLO

This subsection provides an extensive overview of the YOLO networks' evolution, highlighting their multifarious applications in student learning behavior detection.

The You Only Look Once (YOLO) algorithm, introduced by Redmon et al. [32], represents a single-stage target detection approach. It reimagines target detection as a regression task, negating the need for candidate region extraction inherent in traditional two-stage target detection algorithms. The YOLO algorithm simultaneously ascertains target categories and regresses their positions using a singular network. Progressing iterations have yielded enhanced YOLO algorithm performance. YOLOv2 [33] introduced anchor boxes and batch normalization techniques to amplify detection accuracy. YOLOv3 [34] integrated the Darknet-53 architecture and Feature Pyramid Network (FPN) to elevate target detection efficacy. YOLOv4 [35] streamlined the target detection model by refining the training threshold. YOLOv5 [36] unveiled five distinct models of varying sizes, attaining performance amelioration through channel scaling and model size adjustments. Subsequently, YOLOv6 and YOLOv7 emerged. YOLOv6 embraced the RepVGG architecture, augmenting GPU device adaptability and engineering adaptations [37]. YOLOv7 incorporated module re-referencing and dynamic tag assignment strategies, bolstering both speed and accuracy, effectively outpacing existing target detectors in the 5 FPS to 160 FPS range [38].

Researchers have harnessed the YOLO family of algorithms across diverse applications within the realm of student behavior detection. Chen et al. [39] introduced an enhanced YOLOv4 behavior detection algorithm infused with Repulsion loss functions, thus amplifying detection capabilities for varying behaviors in the classroom. Mindoro et al. [26] utilized the YOLOv3 algorithm to decipher students' facial expressions and predict their behaviors. Wei and Ding [40] harnessed the OpenPose algorithm to extract global features of the human body and joint angle information, effectively distinguishing head-up from head to down states. Additionally, they employed the YOLO algorithm to discern hand-related information, determining whether a student was using a cellphone.

2.4. Attention Mechanism in Computer Vision

The attention mechanism, often referred to as selective attention, constitutes a cognitive ability observed in both humans and animals [41]. This ability enables the selective focusing on specific information while effectively ignoring irrelevant data when encountering intricate stimuli [42]. Such a mechanism serves to enhance cognitive efficiency and accuracy, thereby facilitating more streamlined information processing. The study of attention mechanisms finds application across a diverse array of fields, spanning cognitive psychology, neuroscience, and computer science. In the domain of clinical neuroscience, attention mechanisms are explored for their potential in diagnosing and addressing attention-related disorders, such as ADHD [43]. Within the realm of speech recognition, attention mechanisms find employment in critical tasks such as speech recognition and synthesis. Moreover, attention mechanisms have garnered substantial acceptance and utilization in recommendation systems, computer vision, and speech recognition, leading to tangible enhancements in model performance. In recommendation systems, these mechanisms prove invaluable in deciphering user interests and preferences, subsequently refining the accuracy of recommendations. Within the scope of computer vision, attention mechanisms are strategically deployed for tasks encompassing image classification, target recognition, and image generation, culminating in augmented model performance by virtue of their capacity to concentrate on pertinent image regions [44].

Specifically within the realm of computer vision, attention mechanisms can be broadly categorized into three distinctive types: channel attention mechanisms, spatial attention mechanisms, and region attention mechanisms. Channel attention mechanisms allocate distinct weights to individual channels within the feature map, thus amplifying recognition precision and efficiency. A salient illustration of channel attention mechanisms is the Squeeze-and-Excitation module as seen in SENet, which autonomously discerns the significance of each channel in the feature map, invariably heightening recognition accuracy [45]. Spatial attention mechanisms, on the other hand, endow diverse spatial locations within the feature map with varying weights, optimizing recognition precision and efficiency. The Spatial Attention module, a prime example of a spatial attention mechanism within

CBAM, effectively captures the significance of each spatial location in the feature map, resulting in an elevated level of recognition accuracy [46]. Region attention mechanisms further contribute by assigning distinct weights to discrete regions within an image, thereby enhancing recognition accuracy and efficiency. Notably, RoI Pooling as implemented in Faster R-CNN exemplifies a region attention mechanism, effectuating the extraction of feature maps of fixed dimensions through pooling operations applied to regions of interest [47]. These various attention mechanisms collectively bolster performance and precision across a spectrum of computer vision tasks.

Overall, the integration of attention mechanisms across diverse domains, including the realm of computer vision, holds substantial potential for elevating model performance and refining the concentration on pertinent information. By harnessing the capabilities of attention mechanisms, researchers have made substantial strides in areas such as student behavior recognition, target recognition, and related tasks. These developments pave the way for further exploration and refinement of attention-based models within the dynamic landscape of computer vision.

Summary: YOLO networks have attained widespread utilization and notable advancements in the arena of student learning behavior recognition. They have demonstrated remarkable efficacy in capturing localized visual features, such as body pose and facial expressions, that are integral for behavior recognition. Nonetheless, YOLO networks might encounter challenges in effectively modeling intricate relationships existing between diverse body parts and in comprehensively capturing overarching contextual information. Hence, attention mechanisms emerge as a prospective avenue to enhance the capabilities of student learning behavior recognition systems. These mechanisms possess the inherent capacity to apprehend global dependencies and contextual nuances. By adeptly modeling relationships between distinct body parts and judiciously considering the holistic context of students' classroom behaviors, we stand poised to amplify both the accuracy and the resilience of behavior recognition models.

3. Preliminaries

This section aims to elucidate the task of identifying student learning behaviors and the behavior recognition framework adopted in this study. To facilitate understanding, Table 2 presents crucial symbols alongside their corresponding interpretations.

Table 2. Symbols and notations.

Number	Notation	Description
1	(x, y, w, h, o)	The position coordinates and background of the marker box
2	$\mathbf{MLP}$	Multi-Layer Perceptron Calculator
3	$\mathbf{W_x}$	MLP layer x parameters
4	$\mathbf{F}$	Feature Map
5	$\otimes$	Pixel-level multiplication
6	$\mathbf{M_c(F)}$	Channel Attention maps
7	$\mathbf{M_s(F)}$	Spatial Attention Maps
8	$f^{7\times7}$	7×7 convolution

3.1. Problem Definition

The central challenge tackled pertains to the detection and recognition of student learning behaviors within classroom settings. Figure 1 provides an overview of the student classroom learning behavior testing system's workflow adopted in this research. The red and blue arrows symbolize the training and testing processes, correspondingly. The system starts with continuous video frame input from a classroom camera. It unfolds through three principal stages: the collection of student classroom learning behavior data, feature extraction, and behavior detection and recognition.

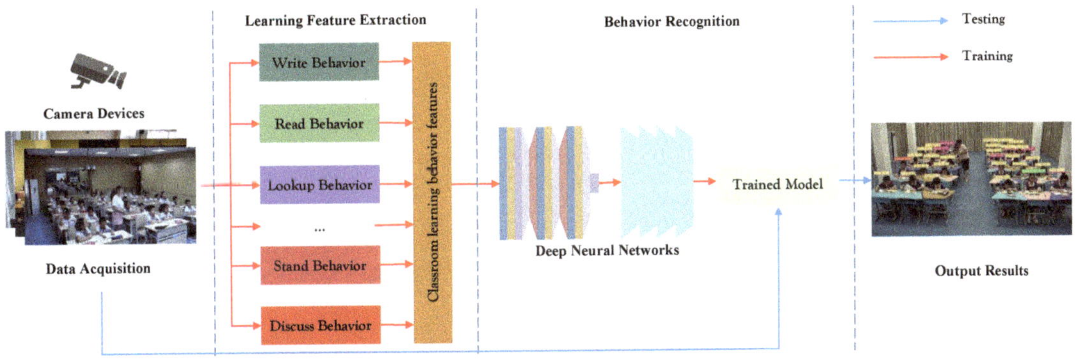

Figure 1. Process of student learning behavior recognition.

3.2. Definitions

Definition 1. *(Data collection of student learning behavior): Continuous video frames are captured by classroom cameras.*

Definition 2. *(Feature extraction of student learning behavior): The feature vectors corresponding to diverse student learning behaviors are extracted via deep neural network. These feature vectors are subsequently employed as inputs for the feature detection network during both the training and testing phases of the proposed network.*

Definition 3. *(Detection of student learning behavior characteristics): The proposed network amalgamates the CBAM attention mechanism with the YOLOv7 feature detection network to construct a model geared towards detecting and recognizing student classroom learning behaviors.*

4. Methods

This section provides details of the framework for analyzing student learning behavior. The task of detecting and recognizing student behavior within an smart classroom teaching environment is complex due to challenges such as diverse illumination sources, lighting variations, background interference, occlusion, and image noise. The You Only Look Once (YOLO) algorithm, known for its real-time performance, global sensing capability, multi-target recognition, and end-to-end training, has gained prominence in target recognition. Leveraging its versatility, this study presents a learning behavior recognition framework centered around YOLOv7.

4.1. Network Design of Student Behavior Recognition Task

The network structure adopted for this research is divided into four main components: Input, Backbone, Head, and Output. The input image is resized to 640 $\times$ 640 pixels and enters the backbone network. The head network generates three tiers of feature maps with varying dimensions. The RepConv module's recognition output yields both the target object's position and category information. The location information is typically represented as bounding box coordinates [33], where (t_x, t_y) denote the center coordinates, and (t_w, t_h) represent the bounding box's width and height. The category information is deduced using a multi-category classifier, obtaining confidence levels. This indicates the presence of the target object within the bounding box and provides a measure of accuracy. The method employs three anchor frames, resulting in 36 outputs per layer, computed as $(7+5) \times 3$. These outputs are then concatenated with the feature map size to produce the final output. Figure 2 visually outlines the network structure, with blue boxes signifying enhancements made.

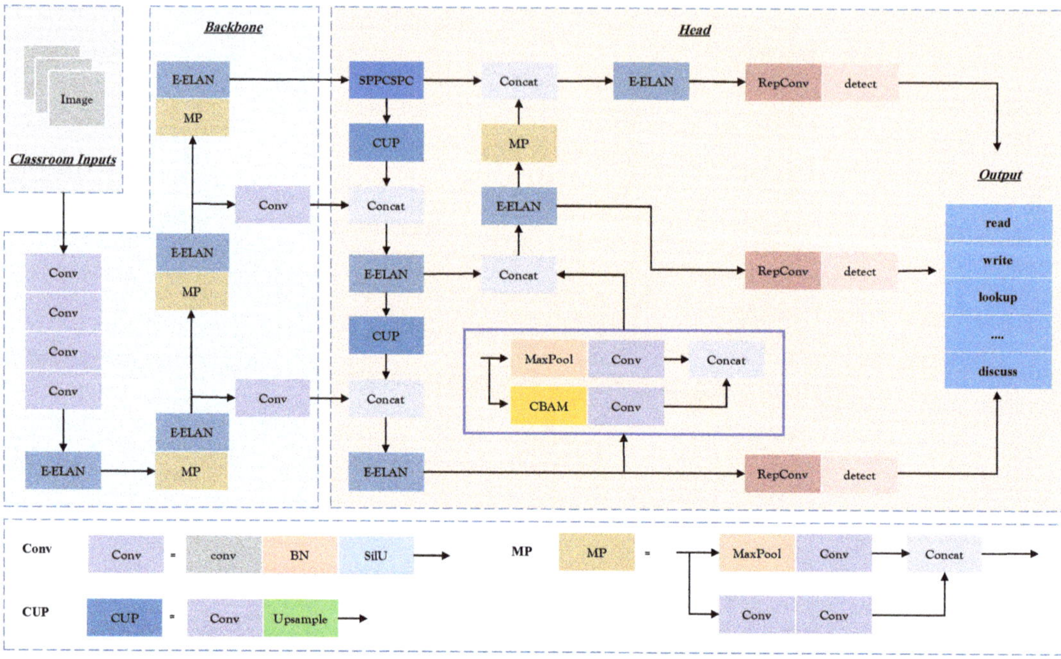

Figure 2. Student learning behavior recognition framework.

The backbone network, as depicted in Figure 2, incorporates the Conv module for input image normalization and nonlinear transformation. The MaxPool (MP) module and the E-ELAN module work together to learn image features with varying perceptual fields. The outcomes from three distinct E-ELANs are combined into the YOLOv7 feature detection network. The integration of the E-ELAN module enhances YOLOv7's learning capabilities, parameter utilization, and computational efficiency. To cater to targets of different scales, the head network augments feature map output from the backbone network using the Spatial Pyramid Pooling Combined with Spatial Context Prediction (SPPCSPC) module [38]. Furthermore, the E-ELAN module is employed to further heighten the network's computational efficiency.

This paper introduces the Convolutional Block Attention Module (CBAM) attention mechanism to underscore critical information concerning students' learning behaviors. CBAM effectively captures contextual features and bolsters the network's feature detection capabilities. Figure 2 illustrates the network architecture integrating the CBAM attention mechanism. The network's prediction results, post the RepConv module, comprise the multi-scale output from the head network.

4.2. Extended-Efficient Layer Aggregation Networks Module

This model introduces the Extended-efficient layer aggregation networks (E-ELAN) module, elevating network learning potential through the Expand, Shuffle, Merge Cardinality Network (EALN) approach. The E-ELAN module modifies both the backbone network and the head network's structure [38]. Group convolution is employed to expand the feature base count, and features from different groups are fused using shuffling and merging cardinality operations. This strategy improves parameter utilization, computational efficiency, and features learned from various feature maps. Figure 3 illustrates the architecture of the E-ELAN module.

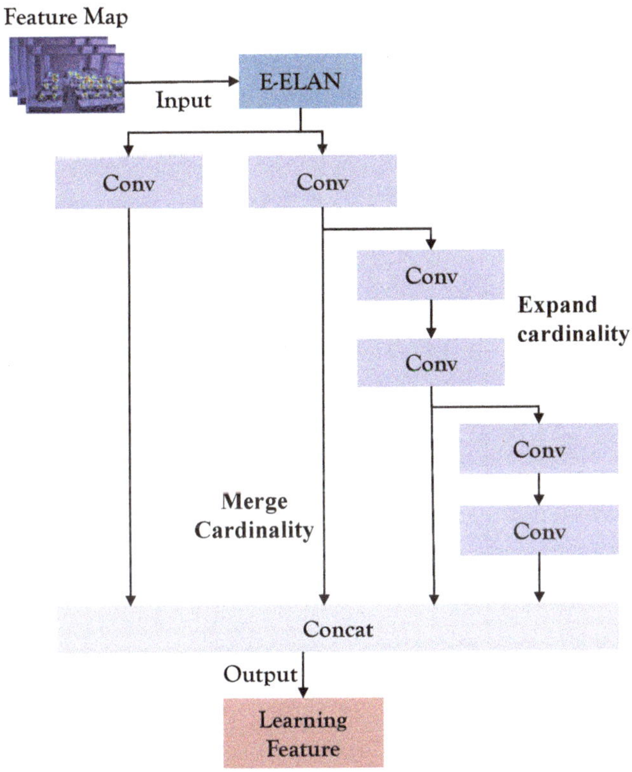

Figure 3. E-ELAN module's architectural depiction.

4.3. Convolutional Block Attention Module

The Convolutional Block Attention Module (CBAM) is an attention mechanism commonly employed in convolutional neural networks to enhance model performance in tasks such as recognition and classification, as shown in Algorithm 1. Its primary function is to adapt the input feature map by selectively highlighting crucial features through attention tuning across both channel and spatial dimensions. By doing so, the CBAM mechanism effectively improves the network's ability to understand and model complex images. The structure of the CBAM module is visually represented in Figure 4.

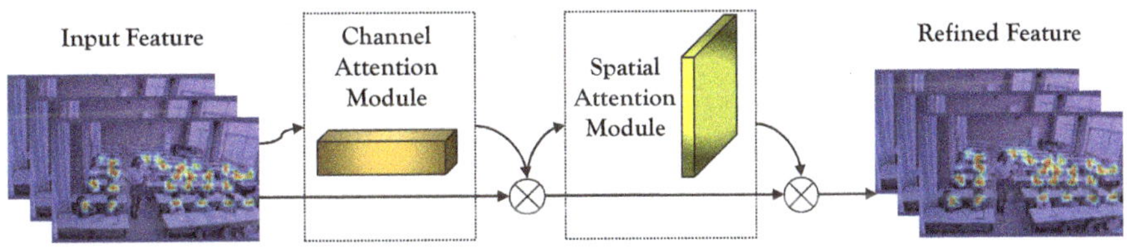

Figure 4. Schematic representation of the CBAM module.

The CBAM module is composed of two distinct sub-modules: the Channel Attention Module (CAM) and the Spatial Attention Module (SAM). The CAM's role is to perform spatially informative aggregation on the feature map, utilizing global max pooling and global average pooling techniques. Subsequently, the resulting feature maps undergo a two-layer Multi-Layer Perceptron (MLP) neural network. The output features are then element-wise summed and passed through a sigmoid activation function, ultimately producing the final channel attention feature.

Algorithm 1 Algorithm for CBAM attention mechanism

Input: Network intermediate volume characteristics map
Output: Attention maps
1: CAM performs a spatially informative aggregation operation on the feature map:

$$\begin{aligned}\mathbf{M_c}(\mathbf{F}) &= \sigma(\mathrm{MLP}(\mathrm{AvgPool}(\mathbf{F})) + (\mathrm{MLP}(\mathrm{MaxPool}(\mathbf{F}))) \\ &= \sigma\left(\mathbf{W_1}\left(\mathbf{W}_0\left(\mathbf{F^c_{avg}}\right)\right) + \mathbf{W_1}(\mathbf{W}_0(\mathbf{F^c_{max}}))\right)\end{aligned}$$

2: The channel attention map is multiplied at the pixel level with the original image:

$$\mathbf{F}' = \mathbf{M_c}(\mathbf{F}) \otimes \mathbf{F},$$

3: SAM performs feature focus and dimensionality reduction operations on feature maps:

$$\begin{aligned}\mathbf{M_s}(\mathbf{F}) &= \sigma\left(f^{7\times 7}([\mathrm{AvgPool}(\mathbf{F}'); \mathrm{Max\,Pool}(\mathbf{F}')])\right) \\ &= \sigma\left(f^{7\times 7}\left(\left[\mathbf{F^s_{avg}}; \mathbf{F^s_{max}}\right]\right)\right)\end{aligned}$$

4: Spatial Attention Maps are multiplied by the original map at the pixel level:

$$\mathbf{F}'' = \mathbf{M_s}(\mathbf{F}') \otimes \mathbf{F}'$$

5: Output CBAM processed feature maps.

SAM, on the other hand, operates on the feature map obtained from the CAM. It undertakes global max pooling and global average pooling along the channel dimension to yield two distinct feature maps. These feature maps are used to calculate the spatial attention feature using a sigmoid activation function. Finally, the spatial attention feature is element-wise multiplied with the initial feature map to generate the ultimate feature output.

The combined utilization of these modules, namely E-ELAN and CBAM, significantly contributes to the enhancement of object recognition models by bolstering the network's learning capacity and selectively highlighting important features. CBAM attention mechanism can enhance CNN's attention to important student behavior features, focus attention on features related to student behavior, and resist redundant information in the data. In addition, the CBAM attention mechanism can reduce unnecessary computational burden, filter out irrelevant feature information.

4.4. Data Augmentation

The efficacy of deep learning in the realm of computer vision hinges on access to extensive, meticulously annotated datasets [48]. However, constructing a high-quality dataset for target recognition introduces a host of challenges:

(1) Labeling complexity: Target recognition mandates the precise identification and classification of each object, rendering the annotation process more time-intensive. Ensuring accuracy and consistency necessitates skilled annotators. Moreover, complexities like occlusion, rotation, and pose variations further compound the labeling process.

(2) Dataset imbalance: Target recognition datasets often exhibit significant disparities in the number of samples across different categories, leading to suboptimal recognition for certain categories. It is crucial to achieve a balanced distribution of samples to alleviate category imbalance during dataset creation.

(3) Dataset diversity: The target recognition dataset must encompass a wide spectrum of characteristics—varied target objects, scenes, lighting conditions, poses, and viewpoints—to bolster generalization capabilities. Yet, curating such diverse datasets entails substantial investments of time and labor.

(4) Dataset size: Successful target recognition models typically demand a substantial volume of training data for optimal performance. However, constructing a sufficiently large dataset poses challenges due to the heightened complexity and time required for annotation.

(5) Dataset quality: Producing a high-quality target recognition dataset mandates annotators with professional acumen to ensure precision and uniformity. Moreover, datasets may inadvertently contain errors or noise, necessitating thorough screening and cleaning to preserve data quality.

To address these challenges, this paper turns to data augmentation, an essential technique to bolster the training of deep learning models using limited effective training data. Data augmentation diversifies the dataset, mitigates overfitting, and enhances the model's capacity for generalization. In this study, data augmentation is performed on the self-built Student Learning Behavior (SLB) dataset through a series of techniques.

4.4.1. Random Rotation Enhancement

To enhance the variability of the SLB dataset and mitigate the risk of overfitting, random rotation enhancement is introduced. This technique introduces diversity by applying random rotations to images. Figure 5 visually illustrates this process, depicting the original image, a 15° clockwise rotation, and a 15° counterclockwise rotation. Algorithm 2 provides the underlying principle behind random rotation enhancement.

Algorithm 2 Image random rotation enhancement algorithm

Input: image max_angle
Output: enhanced_image

function RANDOM_ROTATION(image, max_angle)
  height, width = image.shape
  Randomly generate the rotation angle. Randomly generated rotation angle θ can be represented by the random number function rand:

$$\theta = \text{rand} \times 2 \times \text{max\_angle} - \text{max\_angle}$$

  Calculate the center of rotation. Assume that the coordinates of the center of rotation are (x_center, y_center).

$$center\_x = \frac{width}{2}, \quad center\_y = \frac{height}{2}$$

  Define the rotation matrix.

$$\text{rotation\_matrix} = \begin{bmatrix} \cos(\theta) & -\sin(\theta) & (1-\cos(\theta)) \cdot \text{center\_x} + \sin(\theta) \cdot \text{cente} \\ \sin(\theta) & \cos(\theta) & -\sin(\theta) \cdot \text{center\_x} + (1-\cos(\theta)) \cdot \text{cent} \\ 0 & 0 & 1 \end{bmatrix}$$

  Performs image rotation. The position of each pixel after rotation is (x',y').

$$(x', y') = \begin{cases} x' = (x - \text{center\_x}) \cdot \cos(\theta) - (y - \text{center\_y}) \cdot \sin(\theta) + \text{center\_x} \\ y' = (x - \text{center\_x}) \cdot \sin(\theta) + (y - \text{center\_y}) \cdot \cos(\theta) + \text{center\_y} \end{cases}$$

  return rotated_image
end function

(a) The original image from a smart classroom

(b) Rotate clockwise to enhance the image

(c) Rotate counterclockwise to enhance the image

Figure 5. Illustration of random rotation enhancement.

4.4.2. Grayscale Enhancement

In the context of smart classroom environments, certain image details and features often pose challenges for human observation and automated recognition, particularly when influenced by low light conditions or shadows. Enhancing the grayscale level of an image serves to accentuate these intricate aspects, thereby bolstering the differentiation of features within the image. Moreover, grayscale enhancement transcends the realm of object color and encompasses other crucial cues like shape and edges.

Within this paper, we employ grayscale enhancement to the Single-Label Behavior (SLB) dataset. The technique involves extending the grayscale dynamic range of the original image to a predefined interval, facilitated through a linear relationship equation. The equation reads as follows, where $f(x,y)$ and $g(x,y)$ denote the grayscale values of pixels at positions (x,y) before and after enhancement. The parameters a and b stand for the minimum and maximum values of grayscale levels in the original image, while c and d pertain to the minimum and maximum values in the enhanced image:

$$g(x,y) = \begin{cases} c, & [0,a) \\ \dfrac{d-c}{b-a} \times f(x,y) + c, & [a,b] \\ d, & (b,255] \end{cases} \quad (1)$$

This study embraces a grayscale transformation employing $k = 25\%$ on the SLB dataset. An exemplar of the grayscale enhancement process is portrayed in Figure 6.

(a) The original image

(b) Grayscale enhanced image

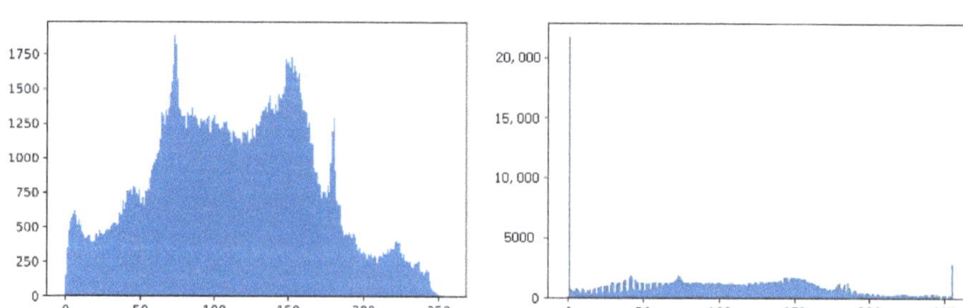

(c) Histogram of original image

(d) Histogram of Grayscale image

Figure 6. Illustration of grayscale enhancement.

4.4.3. Noise Enhancement

Within real classroom scenarios, images often encounter inevitable disturbances and noise, encompassing phenomena like lighting variations, image blurring, and image distortion. To bolster the model's aptitude for discerning robust features and enhancing its resilience against disturbances and noise, this paper introduces noise to the images through a noise enhancement algorithm grounded in the mean filter approach.

Figure 7 offers a visual demonstration of this transformation. In this context, the mean filter operates by replacing the value of the central pixel with the average of the pixel values within a window centered on the pixel. Typically, the size of the filter's window corresponds to an odd number, determining its dimensions. The specifics of the image noise enhancement algorithm, utilizing mean filtering, are elucidated in Algorithm 3:

Algorithm 3 Image noise enhancement algorithm based on mean filtering

Input: image kernel_size
Output: enhanced_image

function ENHANCE_NOISE_WITH_AVERAGE_FILTER(image, kernel_size)
    height, width = image.shape
    Creates a blank image of the same size as the original image.
    Gets the height and width of the image.
    Performs noise enhancement on the image.
    for y = 0 to height − 1 **do**
        for x = 0 to width − 1 **do**
            Calculate the boundary coordinates of the filter. Assume that the coordinates of the computational boundary are (x, y).

$$(x.y) = \begin{cases} \text{top} = \max(y - \lfloor N/2 \rfloor, 0) \\ \text{bottom} = \min(y + \lfloor N/2 \rfloor, \text{height} - 1) \\ \text{left} = \max(x - \lfloor N/2 \rfloor, 0) \\ \text{right} = \min(x + \lfloor N/2 \rfloor, \text{width} - 1) \end{cases}$$

            The average value is calculated for the pixels within the filter.

$$\text{sum\_value} = \sum_{i=\text{top}}^{\text{bottom}} \sum_{j=\text{left}}^{\text{right}} \text{image}(x + i - \lfloor N/2 \rfloor, y + j - \lfloor N/2 \rfloor)$$

            The average value is used as the enhanced pixel value.

$$\text{average} = \frac{\text{sum\_value}}{N \times N}$$

        end for
    end for
    return enhanced_image
end function

(a) The original image (b) Noise enhanced image

Figure 7. Noise enhancement example.

In this study, the SLB dataset is enriched through the noise enhancement algorithm based on mean filtering, thereby simulating an array of disturbances and noise scenarios.

5. Experimental Results and Analysis

Building upon the aforementioned framework for learning behavior recognition, we conducted empirical research using real smart classroom data.

5.1. Dataset for Experiments

In the realm of computer vision, a plethora of datasets have emerged to cater to diverse visual tasks. Examples include the MNIST dataset for digit recognition, the KITTI dataset for autonomous driving research, and the ADE20K dataset for scene understanding. These datasets have furnished researchers and developers with extensive image samples and annotated information, thereby enabling comprehensive investigations and algorithmic advancements tailored to specific vision tasks. At the intersection of computer vision technology and pedagogy lies the significant challenge of detecting and recognizing diverse learning behaviors exhibited by students.

In response to this challenge, we have crafted a dataset called Student Learning Behavior (SLB) to facilitate researchers and educators in comprehending students' behavioral patterns and learning states. This dataset is constructed from classroom videos acquired from NPS, and comprises 600 high-resolution RGB color images, with one image extracted every 150 frames. Each image boasts dimensions of 2048 × 1152 pixels. The vott [49] open-source software was employed for annotating learning behaviors and bounding boxes of students in each image. The dataset encompasses seven categories of student classroom learning behaviors: write, read, lookup, turn_head, raise_hand, stand, and discuss. Furthermore, the dataset covers four smart classroom scenarios, each boasting distinct layouts. Within each scenario, approximately 30 individual students are present, culminating in around 120 distinct student objects. An illustrative instance of each behavior type is depicted in Figure 8. The dataset is divided into three segments: 420 images for training, 120 images for validation, and 60 images for testing. The distribution of different labels, conforming to the VOC2012 dataset format [38], is provided in Table 3.

Figure 8. Example images from the SLB dataset.

Table 3. Details of the SLB dataset.

Number	Classes	Num of Labels	Train	Val	Test
1	write	1025	452	491	82
2	read	1075	810	139	126
3	lookup	5725	3620	1656	449
4	turn_head	1025	748	117	160
5	raise_hand	725	561	82	82
6	stand	94	50	30	14
7	discuss	242	172	50	20

5.2. Evaluation Metrics

To evaluate the effectiveness of the proposed approach, this study employs the PASCAL VOC metric [27]. This metric assesses the accuracy of object identification based on four key parameters: true positive (TP), true negative (TN), false positive (FP), and false negative (FN). TP signifies correct positive predictions, where both the predicted and true values are positive samples. TN denotes correct negative predictions, where both the predicted and true values are negative samples. FP represents incorrect positive predictions, where the predicted value is positive but the true value is negative. FN indicates incorrect negative predictions, where the predicted value is negative but the true value is positive. Precision and Recall are calculated based on these parameters.

$$Precision = \frac{TP}{TP + FP} \tag{2}$$

$$Recall = \frac{TP}{TP + FN} \tag{3}$$

Furthermore, the concepts of Average Precision (AP) and mean Average Precision (mAP) are introduced to jointly evaluate Precision and Recall. AP serves as a metric to assess the performance of detecting individual targets within specific categories. It quantifies the model's accuracy by calculating the area under the precision-recall curve, as defined in Equation (4). AP values range from 0 to 1, with higher values indicating superior model performance. In contrast, mAP represents the mean Average Precision for detecting multiple targets across various categories. In multi-category target recognition tasks, an AP value can be computed for each category. These AP values are then averaged to derive mAP, as shown in Equation (5) [27].

In addition, this study uses Frames Per Second (FPS) to evaluate the real-time performance and efficiency of the model [27]. Its calculation method is given in Equation (6), where inference time refers to the time from the preprocessed image input model to the model output result, and NMS is the post-processing time.

$$AP_i = \int_0^1 P(r)dr \tag{4}$$

$$mAP = \frac{1}{n}\sum_i^n (AP_i) \tag{5}$$

$$FPS = 1/inferencetime + NMS \tag{6}$$

5.3. Baselines

In order to comprehensively evaluate the efficacy of our proposed method, we conducted a comparative analysis against several established models within the domain of student learning behavior recognition systems. Furthermore, we included YOLOv6, a recent addition to the YOLO family, as well as the lightweight YOLOv7-Tiny model from YOLOv7, for comparative purposes.

SSD [25]: This method employs the ResNet network for feature extraction and integrates the Region Proposal Network (RPN) for generating bounding boxes. Subsequently, a k-means algorithm is employed for post-processing and filtering.

Faster R-CNN [24]: Faster R-CNN, a classic two-stage target detection approach, is widely applied for various behavior detection tasks. It leverages Region Proposal Network (RPN) networks to streamline model computation and enhance detection efficiency.

YOLOv3 [26]: YOLOv3-SPP employs a feature extraction network to capture features, followed by the utilization of the Spatial Pyramid Pooling (SPP) module for multi-scale feature extraction. The detection network is then utilized for behavior classification and positional regression to derive final behavior detection outcomes.

YOLOv5 [27]: The YOLOv5 model adapts anchor frames during computation. It employs k-means clustering to determine n anchors and employs a genetic algorithm to randomize anchor width and height (wh). An anchor fitness approach is employed for evaluating obtained fitness.

YOLOv6 [50]: YOLOv6 introduces varied backbone networks based on model scale and employs distinct activation functions for different scenarios to balance between field deployment and model accuracy. The model training incorporates the ATSS label assignment strategy during the initial stages. In sum, YOLOv6 is particularly well-suited as a behavior detection method for industrial applications.

YOLOv7-Tiny [38]: YOLOv7-Tiny, a lightweight network introduced by the YOLOv7 system, features fewer layers and parameters, making it more compatible with GPU devices in specific deployment contexts. Consequently, YOLOv7-Tiny holds promise for application in industrial environments.

5.4. *Training*

The experiments were executed using an Intel(R) Xeon(R) Platinum 8358P CPU boasting 24 GB of RAM, alongside an NVIDIA Tesla 3090 GPU. The software stack utilized PyTorch 1.8.1, Python 3.8, and CUDA 11.1.

For the training phase, we employed the pre-training weights (yolov7.pt) provided by YOLOv7. The stochastic gradient descent (SGD) [51] algorithm was adopted as the optimizer for updating and refining the network model weights. To mitigate model oscillations due to a high initial learning rate, a warm-up strategy was incorporated during training. Within this warm-up phase, the model's learning rate was gradually increased to reach 0.01. Following the warm-up, the network's learning rate was dynamically adjusted using the cosine annealing algorithm. Specific experimental parameters were set as follows: batch size of 12, learning rate of 0.01, and weight update factor of 0.0005.

The progress of the detector's loss was tracked during training, as depicted in Figure 9. Notably, the training and validation losses for the student learning behavior detector converged satisfactorily after 75 rounds.

The observed training progress conclusively establishes that the learned model successfully avoids both overfitting and underfitting, attesting to its optimal suitability for subsequent experiments.

5.5. *Comparison with YOLOv7 on Individual Learning Behavior Recognition*

We conducted empirical research from two perspectives: a comparison with the baseline YOLOv7 in single-class learning behavior recognition and a comparison with six benchmark methods in overall learning behavior recognition. For the sake of convenience in describing, we have named our method the Student Learning Behavior Recognition Framework (SLBRF).

The results of the single-class comparison between our proposed method and the baseline models on the SLB test dataset are provided in Table 4. These results clearly illustrate the superior single-class average precisions (APs) achieved by our proposed method in comparison to the benchmark models. Notably, our method surpasses the YOLOv7 network model on the SLB dataset, showcasing its effectiveness in the domain of student learning behavior recognition.

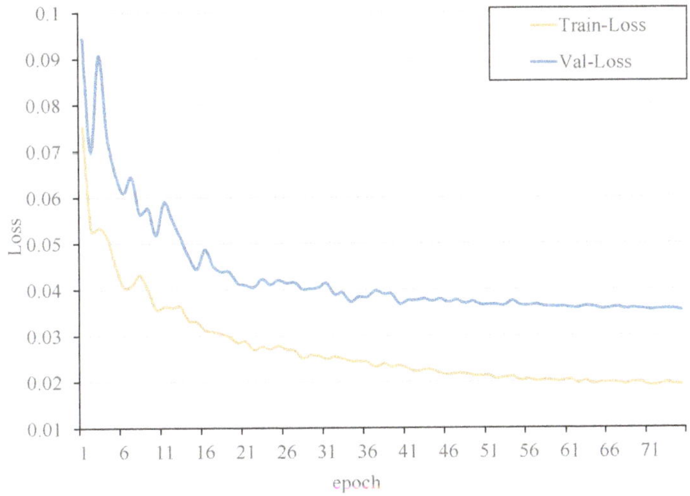

Figure 9. Training and validation losses for students' learning behavior recognition.

Furthermore, Figure 10 displays the heat map generated using the GradCAM [52] visualization model on the original SLB test set images, highlighting key behavioral features. The prominent orange areas in the heat map indicate the successful localization of relevant image features by the student behavior recognition network. This visualization solidifies the efficacy of our proposed method in accurately identifying and highlighting learned behaviors.

(a) Input Images (b) Category Heat Maps (c) Heat Maps

Figure 10. Heat map generated by the GradCAM visualization model on SLB test dataset.

These compelling results underscore the superiority of our proposed method in detecting student learning behaviors, substantiating its efficacy and potential applications in educational contexts.

Table 4. Single-class AP results on the test set. The bold means the best performance.

Classes	YOLOv7 (%)	Proposed Method (%)
write	87.2	**98.7**
read	71.3	**93.0**
lookup	94.4	**98.6**
turn_head	84.7	**96.4**
raise_hand	95.0	**99.1**
stand	78.2	**93.0**
discuss	96.6	**98.1**
Total	84.8	**96.7**

5.6. Comparison with Six Baseline Methods

To further substantiate the performance of our proposed method, a comparison is conducted with YOLOv7-Tiny, the SSD lightweight network, and other prominent classical networks. The comparison experiments employ the same dataset and data configuration, with the results summarized in Table 5.

Table 5. Comparison of performance and speed of different networks. The bold means the best performance. The underline means the second best performance.

Methods	mAP@0.5 (%)	AP50 (%)	FPS (f/s)		
			2080Ti 11G*1	3090 24G*1	12v 8255C (CPU)
SSD	65.7	43.0	10.5	18.0	0.1
Faster R-CNN	68.9	49.6	30.6	54.6	0.1
YOLOv3-SPP	63.1	43.0	80.6	**95.2**	3.9
YOLOv5	76.9	<u>54.9</u>	**97.9**	92.5	1.8
YOLOv6	<u>77.7</u>	52.9	62.8	89.1	4.7
YOLOv7-Tiny	54.3	34.8	86.2	81.3	**12.8**
SLBRF(Our Method)	**96.7**	**75.8**	80	84.8	2.7

As evident in Table 5, our proposed method consistently outperforms YOLOv7-Tiny, SSD, YOLOv3-SPP, Faster R-CNN, YOLOv5, and YOLOv6, with a noteworthy minimum improvement of 19% in terms of mAP. These results unequivocally establish the superiority of our proposed method over other mainstream networks. In addition to this, in order to analyze the time cost, inference is performed on RTX 2080TI (11G), 3090 (24G) and 12v 8255 cpu. The results show that our method achieves a good balance between precision and speed.

5.7. Ablation Experiments

Ablation experiments play a pivotal role in dissecting the impact of network structure modifications on performance. In this section, we present and analyze the outcomes of four essential experiments: YOLOv7, SLBRF with Data Enhancement (SLBRF_DE), SLBRF with the CBAM attention mechanism (SLBRF_CB), and SLBRF with both Data Enhancement and CBAM (SLBRF_DE_CB). The visual training process of these methods on the SLB dataset is depicted in Figure 11, with corresponding training results summarized in Table 6.

In Figure 12, we provide a visual comparison between the proposed method in this paper and the baseline YOLOv7 network's recognition results in different scenarios. Specifically, the left figure in Figure 12 illustrates the recognition outcomes of the proposed method, while the right figure presents the recognition results of the YOLOv7 baseline.

Table 6. Performance comparison of different models on the SLB dataset. The bold means the best performance.

Model	mAP@0.5 (%)	mAP@0.5:0.95 (%)	Params (10^4)	FLOPs (G)	Weight (M)
YOLOv7	84.8	62.7	3722.3	105.2	71.4
SLBRF_CB	**85.5**	-	**3721.2**	105.2	71.4
SLBRF_DE	95.5	73.6	3722.7	105.2	71.4
SLBRF_DE_CB(ours)	**96.7**	**75.9**	**3721.2**	**105.0**	71.4

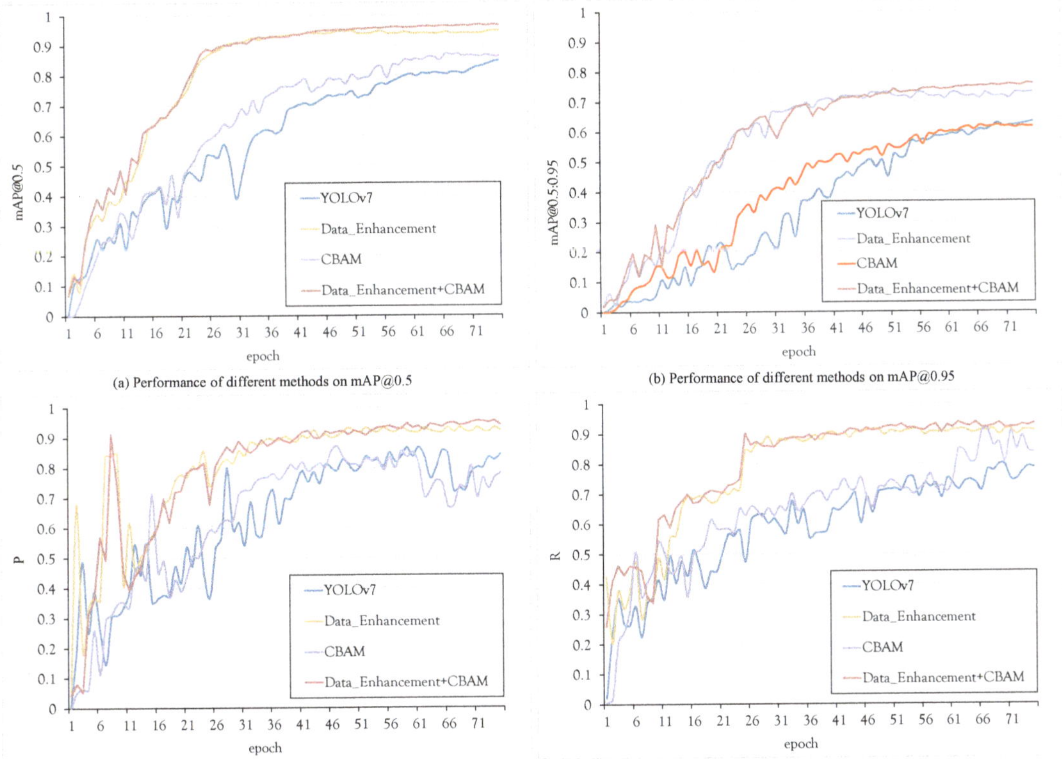

(a) Performance of different methods on mAP@0.5

(b) Performance of different methods on mAP@0.95

(c) Performance of different methods on Precision

(d) Performance of different methods on Recall

Figure 11. Experimental results of each model on the SLB dataset.

5.7.1. The Effectiveness of CBAM

The integration of the CBAM attention mechanism exhibited a tangible enhancement in the performance of the student classroom learning behavior recognition network. It led to a 0.7% increase in mAP on the SLB dataset, while concurrently maintaining the number of parameters and FLOPs at the baseline network level. This observation underlines the constructive influence of the CBAM module on enhancing the accuracy of student classroom learning behavior recognition.

5.7.2. The Effectiveness of Data Enhancement

Data augmentation, a crucial technique for mitigating small dataset limitations, not only expanded the dataset to boost the model's generalization capacity but also increased its adaptability to diverse input data, thereby reinforcing the model's robustness. The results presented in Table 6 clearly demonstrate an improvement of up to 10.7% in the network's mAP on the SLB dataset, all without introducing additional computational overhead, as indicated by the steady parameters count and FLOPs.

(a) The write behavior detection comparison.

(b) The read, stand behavior detection comparison.

(c) The lookup behavior detection comparison.

(d) The turn_head behavior detection comparison.

Figure 12. Results of student learning behavior recognition.

5.7.3. The Effectiveness of the Model Ensemble

The ultimate recognition outcomes of the integrated model, as presented in Table 4, substantiate its superiority over the baseline model for each category. Improved Average Precision (AP) values were observed across all categories, attesting to the model's exceptional overall recognition performance. Moreover, Figure 12 provides visual evidence of diverse behaviors within different scenarios. A stark enhancement in recognition accuracy over the baseline model is evident in Figure 12d, where the head-turn behavior, denoted by the black box, is correctly detected, rectifying the previous erroneous outcomes and markedly improving accuracy.

Lastly, the visual representation in Figure 12 provides vivid confirmation of the proposed method's efficacy in boosting recognition accuracy, especially in challenging scenarios where the YOLOv7 model may encounter difficulties. These findings provide further validation of the proposed method's superior performance and its capability to accurately detect targets across a variety of real-world scenarios.

5.7.4. The Effectiveness of Attention Mechanism in Learning Behavior Recognition

Given the proven widespread efficacy and performance of attention mechanisms in target recognition, this study extended its exploration to include other attention mechanism modules within the YOLOv7 network. As demonstrated in Table 7, the inclusion of the CBAM module notably amplified the recognition accuracy of the network compared to alternative attention mechanism modules, including CA, SE, and SimAM modules. This outcome emphasizes the CBAM module's potential in boosting recognition performance.

Table 7. Comparison of networks using different attention mechanism modules.

Model	mAP (%)	Precision (%)	Recall (%)	FLOPs (G)	Params (10^4)	Weight (M)
YOLOv7	84.8	86.24	77.8	105.2	3722.3	71.4
SLBRF_SE	84.2 ↓	83.4	78.4	105.2	3723.5	113
SLBRF_CA	82.0 ↓	82.81	77.83	105.0	3716.7	71.3
SLBRF_SimAM	84.6 ↓	81.7	84.8	105.0	3721.0	71.3
SLBRF_CBAM(ours)	85.5	79.73	88.02	105.0	3721.2	71.4

In Table 7, a comprehensive comparison is presented between networks utilizing different attention mechanism modules. The metrics evaluated include mAP, precision, recall, FLOPs, parameters (Params), and model weight size. The results demonstrate the following:

- YOLOv7 achieves an mAP of 84.8%, with a precision of 86.24% and a recall of 77.8%. It has 105.2 billion FLOPs, 3722.3×10^4 parameters, and a model weight size of 71.4 million;
- Incorporating the SE module in YOLOv7 leads to a slight decrease in mAP (84.2%), precision (83.4%), and recall (78.4%), while maintaining the same FLOPs and increasing the parameters to 3723.5×10^4 and model weight size to 113 million;
- Utilizing the CA module in YOLOv7 results in a further decrease in mAP (82.0%), precision (82.81%), and recall (77.83%). The FLOPs remain the same, while the parameters decrease to 3716.7×10^4 and the model weight size remains at 71.3 million;
- Applying the SimAM module in YOLOv7 leads to a slight improvement in mAP (84.6%), but precision (81.7%) and recall (84.8%) show mixed changes. The FLOPs and parameters remain consistent, and the model weight size remains at 71.3 million;
- Incorporating the CBAM module (proposed in this paper) in YOLOv7 results in the highest mAP of 85.5% (an increase from the baseline). The precision is 79.73%, and the recall reaches 88.02%. The FLOPs and parameters remain consistent with YOLOv7, while the model weight size remains at 71.4 million.

In summary, the inclusion of the CBAM attention mechanism in YOLOv7 proves to be the most effective in terms of improving recognition accuracy, surpassing other attention mechanism modules such as SE, CA, and SimAM. It achieves the highest mAP, precision, and recall while maintaining a comparable number of parameters, FLOPs, and model weight size. These findings highlight the superiority of the proposed method in this paper in terms of attention mechanisms for target recognition.

6. Conclusions

In this study, we have presented an effective student learning behavior detector based on the YOLOv7 network, enabling accurate recognition of classroom learning behaviors. We have addressed the challenge of limited data samples through the strategic application of transfer learning and data augmentation techniques. Additionally, by integrating the CBAM attention mechanism into the YOLOv7 feature detection network, we have amplified its ability to extract vital information about students' learning behaviors, thereby enhancing feature recognition accuracy in classroom settings. The experimental results substantiate the superiority of our approach over YOLOv7-Tiny, SSD lightweight networks, and other prominent classical networks in the context of student learning behavior recognition. We have achieved noteworthy advancements in the precise identification and categorization of various learning behaviors exhibited by students in the classroom.

However, it is essential to acknowledge the limitations of our study. Notably, the dataset employed in our experiments includes behaviors that should have been more rigorously defined and labeled. This limitation may have introduced some ambiguity and noise during both training and evaluation. Moreover, we recognize that the current network model we have proposed, while powerful, is relatively large, which may present performance challenges when deployed on embedded devices with limited GPU resources. To overcome this limitation, we acknowledge the necessity of developing lightweight yet high-performing models tailored for real-time recognition in resource-constrained settings. Our upcoming research efforts will be directed towards substantial improvements in these areas.

In future studies, we are committed to addressing this issue by refining the dataset and providing more precise behavior definitions, ensuring higher-quality training data. Additionally, due to objective factors such as data acquisition equipment and acquisition angles, the dataset we constructed is not sufficiently large, resulting in lower data resolution. Therefore, we intend to expand a high-quality dataset for future research. In order to enhance users' personal privacy and data security, we plan to introduce federated learning in future model deployments. This initiative will ensure the adequate protection of students' personal data. Simultaneously, we will further optimize the model by moving the data acquisition and processing phases to the client side, avoiding the transfer of students' image data to the server side, thereby reducing the potential risk of data transmission. This approach not only helps protect users' personal privacy, but also reduces the need for server-side data storage, improving the overall system security.

Introducing computer vision technology into modern education evaluation system can provide many useful tools and methods for education and teaching staff to improve the quality, efficiency, and fairness of education evaluation. At the same time, our results show that the use of visual techniques has great potential for classroom analysis.

Author Contributions: Conceptualization, Z.W. and C.Z.; methodology, Z.W.; software, Z.W. and C.Z.; validation, Z.W., J.Y., L.L. and C.Z.; formal analysis, Z.W.; investigation, Z.W., J.Y., L.L. and C.Z.; resources, Z.W.; data curation, Z.W.; writing—original draft preparation, Z.W. and L.L.; writing—review and editing, Z.W. and J.Y.; visualization, Z.W. and J.Y.; supervision, Z.W.; project administration, Z.W.; funding acquisition, Z.W. All authors have read and agreed to the published version of the manuscript.

Funding: The research work in this paper was supported by the National Natural Science Foundation of China (No. 62177022, 61901165, 61501199), AI and Faculty Empowerment Pilot Project (No. CCNUAI&FE2022-03-01), Collaborative Innovation Center for Informatization and Balanced Development of K-12 Education by MOE and Hubei Province (No. xtzd2021-005), and Natural Science Foundation of Hubei Province (No. 2022CFA007).

Institutional Review Board Statement: Not applicable.

Informed Consent Statement: Not applicable.

Data Availability Statement: Data will be made available on reasonable request.

Conflicts of Interest: The authors declare no conflict of interest.

Abbreviations

The following abbreviations are used in this manuscript:

FPN	Feature Pyramid Network
PAN	Path Aggregation Network
IoT	Internet of Things
SOTA	State-of-the-Art
CNNs	Convolutional Neural Networks
CBAM	Convolutional Block Attention Module
CAM	Channel Attention Module
SAM	Spatial Attention Module
SSD	Single Shot MultiBox Detector
DBN	Deep Belief Network
YOLO	You Only Look Once
ViT	Vision Transformer
DETR	Detection Transformer
W-MSA	Windowed Multihead Self-Attention
SW-MSA	Sliding-Window Multihead Self-Attention
MLP	Multilayer Perceptron
LN	Layer Normalization
CARAFE	Content-Aware Reassembly of Features
mAP	Mean Average Precision
RPN	Region Proposal Network
CBL	Convolutional Block Layer
PAN	Path Aggregation Network
ELAN	Efficient Local Attention Network
CAT	Category-aware Transformation

References

1. Alfoudari, A.M.; Durugbo, C.M.; Aldhmour, F.M. Understanding Socio-Technological Challenges of Smart Classrooms Using a Systematic Review. *Comput. Educ.* **2021**, *173*, 104282. [CrossRef]
2. Kaur, A.; Bhatia, M.; Stea, G. A Survey of Smart Classroom Literature. *Educ. Sci.* **2022**, *12*, 86. [CrossRef]
3. Saini, M.K.; Goel, N. How Smart Are Smart Classrooms? A Review of Smart Classroom Technologies. *ACM Comput. Surv.* **2019**, *52*, 130:1–130:28. [CrossRef]
4. Harlen, W. Criteria for Evaluating Systems for Student Assessment. *Stud. Educ. Eval.* **2007**, *33*, 15–28. [CrossRef]
5. Wang, Z.; Yan, W.; Zeng, C.; Tian, Y.; Dong, S. A Unified Interpretable Intelligent Learning Diagnosis Framework for Learning Performance Prediction in Intelligent Tutoring Systems. *Int. J. Intell. Syst.* **2023**, *2023*, e4468025. [CrossRef]
6. Xia, X. Interaction Recognition and Intervention Based on Context Feature Fusion of Learning Behaviors in Interactive Learning Environments. *Interact. Learn. Environ.* **2023**, *31*, 2033–2050. [CrossRef]
7. Wang, Z.; Yao, J.; Zeng, C.; Li, L.; Tan, C. Students' Classroom Behavior Detection System Incorporating Deformable DETR with Swin Transformer and Light-Weight Feature Pyramid Network. *Systems* **2023**, *11*, 372. [CrossRef]
8. Han, L.; Yao, X.; Yu, J. Application of Deep Learning in Medical English Teaching Evaluation. *Wirel. Commun. Mob. Comput.* **2022**, *2022*, e8671806. [CrossRef]
9. Zahner, W.; Wynn, L. Rethinking Learning Trajectories in Light of Student Linguistic Diversity. *Math. Think. Learn.* **2023**, *25*, 100–114. [CrossRef]
10. Liu, Y.; Culpepper, S.A.; Chen, Y. Identifiability of Hidden Markov Models for Learning Trajectories in Cognitive Diagnosis. *Psychometrika* **2023**, *88*, 361–386. [CrossRef] [PubMed]
11. McManus, S.M.; Gettinger, M. Teacher and Student Evaluations of Cooperative Learning and Observed Interactive Behaviors. *J. Educ. Res.* **1996**, *90*, 13–22. [CrossRef]
12. Mohammed, A.; Hiny, L.E.; Asal, S.; Zualkernan, I. Evaluating Classroom Activities Using Low-Cost Sensors and IOT Technologies. In Proceedings of the EDULEARN16 Proceedings, Barcelona, Spain, 4–6 July 2016; pp. 5547–5557. [CrossRef]
13. Chang, J.J.; Lin, W.S.; Chen, H.R. How attention level and cognitive style affect learning in a MOOC environment? Based on the perspective of brainwave analysis. *Comput. Hum. Behav.* **2019**, *100*, 209–217. [CrossRef]
14. Ling, W. Automatic Recognition of Students' Classroom Behavior Based on Computer Vision. *Acad. J. Comput. Inf. Sci.* **2022**, *5*, 31–34. [CrossRef]
15. Tham, J.C.K.; Verhulsdonck, G. Smart Education in Smart Cities: Layered Implications for Networked and Ubiquitous Learning. *IEEE Trans. Technol. Soc.* **2023**, *4*, 87–95. [CrossRef]

16. Sá, M.J.; Serpa, S.; Ferreira, C.M. Citizen Science in the Promotion of Sustainability: The Importance of Smart Education for Smart Societies. *Sustainability* **2022**, *14*, 9356. [CrossRef]
17. Singh, H.; Miah, S.J. Smart Education Literature: A Theoretical Analysis. *Educ. Inf. Technol.* **2020**, *25*, 3299–3328. [CrossRef]
18. Snowman, J. Educational psychology: What do we teach, what should we teach? *Educ. Psychol. Rev.* **1997**, *9*, 151–170. [CrossRef]
19. Casalino, G.; Grilli, L.; Guarino, A.; Schicchi, D.; Taibi, D. Intelligent Knowledge Understanding from Students Questionnaires: A Case Study. In Proceedings of the Higher Education Learning Methodologies and Technologies Online, Palermo, Italy, 21–23 September 2022; Casalino, G., Cimitile, M., Ducange, P., Padilla Zea, N., Pecori, R., Picerno, P., Raviolo, P., Eds.; Springer: Cham, Switzerland, 2022; pp. 74–86.
20. Lodge, J.M.; Harrison, W.J. Focus: Attention science: The role of attention in learning in the digital age. *Yale J. Biol. Med.* **2019**, *92*, 21.
21. Arnold, J. Attention to affect in language learning. *Online Submiss.* **2011**, *22*, 11–22.
22. Zaletelj, J.; Košir, A. Predicting Students' Attention in the Classroom from Kinect Facial and Body Features. *EURASIP J. Image Video Process.* **2017**, *2017*, 80. [CrossRef]
23. Wang, Z.; Yao, J.; Zeng, C.; Wu, W.; Xu, H.; Yang, Y. YOLOv5 Enhanced Learning Behavior Recognition and Analysis in Smart Classroom with Multiple Students. In Proceedings of the 2022 International Conference on Intelligent Education and Intelligent Research (IEIR), Wuhan, China, 18–20 December 2022; IEEE: New York, NY, USA, 2022; pp. 23–29. [CrossRef]
24. Zheng, R.; Jiang, F.; Shen, R. Intelligent Student Behavior Analysis System for Real Classrooms. In Proceedings of the ICASSP 2020—2020 IEEE International Conference on Acoustics, Speech and Signal Processing (ICASSP), Barcelona, Spain, 4–8 May 2020; pp. 9244–9248. [CrossRef]
25. Lv, W.; Huang, M.; Zhang, Y.; Liu, S. Research on Intelligent Recognition Algorithm of College Students' Classroom Behavior Based on Improved SSD. In Proceedings of the 2022 IEEE 2nd International Conference on Computer Communication and Artificial Intelligence (CCAI), Beijing, China, 6–8 May 2022; pp. 160–164. [CrossRef]
26. Mindoro, J.N.; Pilueta, N.U.; Austria, Y.D.; Lolong Lacatan, L.; Dellosa, R.M. Capturing Students' Attention Through Visible Behavior: A Prediction Utilizing YOLOv3 Approach. In Proceedings of the 2020 11th IEEE Control and System Graduate Research Colloquium (ICSGRC), Shah Alam, Malaysia, 8 August 2020; pp. 328–333. [CrossRef]
27. Tang, L.; Xie, T.; Yang, Y.; Wang, H. Classroom Behavior Detection Based on Improved YOLOv5 Algorithm Combining Multi-Scale Feature Fusion and Attention Mechanism. *Appl. Sci.* **2022**, *12*, 6790. [CrossRef]
28. Yang, S.Q.; Chen, Y.H.; Zhang, Z.Y.; Chen, J.H. Student In-Class Behaviors Detection and Analysis System Based on CBAM-YOLOv5. In Proceedings of the 2022 7th International Conference on Intelligent Computing and Signal Processing (ICSP), Xi'an, China, 15–17 April 2022; pp. 440–443. [CrossRef]
29. Mo, J.; Zhu, R.; Yuan, H.; Shou, Z.; Chen, L. Student behavior recognition based on multitask learning. *Multimed. Tools Appl.* **2023**, *82*, 19091–19108. [CrossRef]
30. Zhao, X.; Jin, X. Standardized Evaluation Method of Pronunciation Teaching Based on Deep Learning. *Secur. Commun. Netw.* **2022**, *2022*, e8961836. [CrossRef]
31. Shi, L.; Di, X. A Recognition Method of Learning Behaviour in English Online Classroom Based on Feature Data Mining. *Int. J. Reason.-Based Intell. Syst.* **2023**, *15*, 8–14. [CrossRef]
32. Redmon, J.; Divvala, S.; Girshick, R.; Farhadi, A. You Only Look Once: Unified, Real-Time Object Detection. In Proceedings of the IEEE Conference on Computer Vision and Pattern Recognition, Las Vegas, NV, USA, 26 June–1 July 2016; pp. 779–788.
33. Redmon, J.; Farhadi, A. YOLO9000: Better, Faster, Stronger. In Proceedings of the IEEE Conference on Computer Vision and Pattern Recognition, Honolulu, HI, USA, 21–26 July 2017; pp. 7263–7271.
34. Redmon, J.; Farhadi, A. YOLOv3: An Incremental Improvement. *arXiv* **2018**, arXiv:1804.02767. https://doi.org/10.48550/arXiv.1804.02767.
35. Bochkovskiy, A.; Wang, C.Y.; Liao, H.Y.M. YOLOv4: Optimal Speed and Accuracy of Object Detection. *arXiv* **2020**, arXiv:2004.10934. https://doi.org/10.48550/arXiv.2004.10934.
36. Zhu, X.; Lyu, S.; Wang, X.; Zhao, Q. TPH-YOLOv5: Improved YOLOv5 Based on Transformer Prediction Head for Object Detection on Drone-Captured Scenarios. In Proceedings of the IEEE/CVF International Conference on Computer Vision, Montreal, BC, Canada, 11–17 October 2021; pp. 2778–2788.
37. Li, C.; Li, L.; Jiang, H.; Weng, K.; Geng, Y.; Li, L.; Ke, Z.; Li, Q.; Cheng, M.; Nie, W.; et al. YOLOv6: A Single-Stage Object Detection Framework for Industrial Applications. *arXiv* **2022**, arXiv:2209.02976. https://doi.org/10.48550/arXiv.2209.02976.
38. Wang, C.Y.; Bochkovskiy, A.; Liao, H.Y.M. YOLOv7: Trainable Bag-of-Freebies Sets New State-of-the-Art for Real-Time Object Detectors. In Proceedings of the IEEE/CVF Conference on Computer Vision and Pattern Recognition, Vancouver, BC, Canada, 18–22 June 2023; pp. 7464–7475.
39. Chen, H.; Guan, J. Teacher–Student Behavior Recognition in Classroom Teaching Based on Improved YOLO-v4 and Internet of Things Technology. *Electronics* **2022**, *11*, 3998. [CrossRef]
40. Wei, L.; Ding, J. Behavior Detection Method of OpenPose Combined with Yolo Network. In Proceedings of the 2020 International Conference on Communications, Information System and Computer Engineering (CISCE), Kuala Lumpur, Malaysia, 3–5 July 2020; pp. 326–330. [CrossRef]
41. Wang, Z.; Yang, Y.; Zeng, C.; Kong, S.; Feng, S.; Zhao, N. Shallow and Deep Feature Fusion for Digital Audio Tampering Detection. *EURASIP J. Adv. Signal Process.* **2022**, *2022*, 69. [CrossRef]

42. Li, L.; Wang, Z. Calibrated Q-Matrix-Enhanced Deep Knowledge Tracing with Relational Attention Mechanism. *Appl. Sci.* **2023**, *13*, 2541. [CrossRef]
43. Cortese, S.; Peñalver, C.M. Comorbidity between ADHD and Obesity: Exploring Shared Mechanisms and Clinical Implications. *Postgrad. Med.* **2010**, *122*, 88–96. [CrossRef]
44. Guo, M.H.; Xu, T.X.; Liu, J.J.; Liu, Z.N.; Jiang, P.T.; Mu, T.J.; Zhang, S.H.; Martin, R.R.; Cheng, M.M.; Hu, S.M. Attention Mechanisms in Computer Vision: A Survey. *Comput. Vis. Media* **2022**, *8*, 331–368. [CrossRef]
45. Hu, J.; Shen, L.; Sun, G. Squeeze-and-Excitation Networks. In Proceedings of the IEEE Conference on Computer Vision and Pattern Recognition, Salt Lake City, UT, USA, 18–23 June 2018; pp. 7132–7141.
46. Woo, S.; Park, J.; Lee, J.Y.; Kweon, I.S. CBAM: Convolutional Block Attention Module. In Proceedings of the European Conference on Computer Vision (ECCV), Munich, Germany, 8–14 September 2018; pp. 3–19.
47. Girshick, R. Fast R-CNN. In Proceedings of the IEEE International Conference on Computer Vision, Santiago, Chile, 7–13 December 2015; pp. 1440–1448.
48. Hao, X.; Liu, L.; Yang, R.; Yin, L.; Zhang, L.; Li, X. A Review of Data Augmentation Methods of Remote Sensing Image Target Recognition. *Remote Sens.* **2023**, *15*, 827. [CrossRef]
49. VOTT, 2019. Available online: https://github.com/Microsoft/VoTT/releases (accessed on 10 June 2022).
50. Bist, R.B.; Subedi, S.; Yang, X.; Chai, L. A Novel YOLOv6 Object Detector for Monitoring Piling Behavior of Cage-Free Laying Hens. *AgriEngineering* **2023**, *5*, 905–923. [CrossRef]
51. Gower, R.M.; Loizou, N.; Qian, X.; Sailanbayev, A.; Shulgin, E.; Richtárik, P. SGD: General analysis and improved rates. In Proceedings of the International Conference on Machine Learning, Long Beach, CA, USA, 9–15 June 2019; pp. 5200–5209.
52. Selvaraju, R.R.; Das, A.; Vedantam, R.; Cogswell, M.; Parikh, D.; Batra, D. Grad-CAM: Why did you say that? *arXiv* **2016**, arXiv:1611.07450.

sensors

MDPI

Article

A Self-Supervised Model Based on CutPaste-Mix for Ductile Cast Iron Pipe Surface Defect Classification

Hanxin Zhang, Qian Sun and Ke Xu *

Collaborative Innovation Center of Steel Technology, University of Science and Technology Beijing, Beijing 100083, China; m202121256@xs.ustb.edu.cn (H.Z.); d202210616@xs.ustb.edu.cn (Q.S.)
* Correspondence: xuke@ustb.edu.cn

Abstract: Online surface inspection systems have gradually found applications in industrial settings. However, the manual effort required to sift through a vast amount of data to identify defect images remains costly. This study delves into a self-supervised binary classification algorithm for addressing the task of defect image classification within ductile cast iron pipe (DCIP) images. Leveraging the CutPaste-Mix data augmentation strategy, we combine defect-free data with enhanced data to input into a deep convolutional neural network. Through Gaussian Density Estimation, we compute anomaly scores to achieve the classification of abnormal regions. Our approach has been implemented in real-world scenarios, involving equipment installation, data collection, and experimentation. The results demonstrate the robust performance of our method, in both the DCIP image dataset and practical field application, achieving an impressive 99.5 AUC (Area Under Curve). This presents a cost-effective means of providing data support for subsequent DCIP surface inspection model training.

Keywords: ductile cast iron pipe; defect classification; self-supervised; CutPaste-Mix

Citation: Zhang, H.; Sun, Q.; Xu, K. A Self-Supervised Model Based on CutPaste-Mix for Ductile Cast Iron Pipe Surface Defect Classification. *Sensors* **2023**, *23*, 8243. https://doi.org/10.3390/s23198243

Academic Editors: Peter Hockicko, Róbert Hudec and Patrik Kamencay

Received: 24 August 2023
Revised: 27 September 2023
Accepted: 29 September 2023
Published: 4 October 2023

1. Introduction

Defect detection plays a crucial role in industrial production as the timely identification of surface defects is essential for enhancing production quality [1]. A ductile cast iron pipe [2] is a type of pipeline material that can withstand higher pressure and loads and has a longer lifespan and lower maintenance costs. During the production process of Ductile Cast Iron Pipes (DCIPs), various defects [3,4] such as cracks, heavy skin, and pore voids can inevitably arise, impacting product quality and safety. Employing effective defect detection techniques for DCIPs is pivotal in ensuring product longevity and quality. With the advancement of machine vision [5,6] and deep learning [7–9], there has been growing interest among numerous research teams in developing online surface defect detection methods [10–12]. These systems primarily utilize methods based on two-dimensional feature information. They involve capturing two-dimensional images of the target surface using cameras, analyzing defect information within these images, and subsequently conducting defect classification and detection.

Making an excellent online surface detection system involves many steps. With a sufficient amount of data, existing detection models [13,14] have the potential to reach the state-of-the-art (SOTA) level in particular scenarios. Usually, the accuracy of these methods is positively correlated with the number of defect samples. Thanks to the mature casting process of DCIPs, the production often results in a high yield rate. However, the occurrence of defects is unpredictable both in time and space, making the collection of defect samples a challenging task. We made a simple estimate of the amount of defect data in a production line. Unfortunately, the number of defect images is less than 0.2% of the total collected images. The cost and inefficiency of manually filtering these samples are clearly prohibitive. The production scenario mentioned above poses significant challenges to the collection of defect data. Therefore, our research focuses on efficiently filtering negative samples [15] rather than on proposing a new detection model. This distinction is crucial for understanding the significance of our study.

Constructing an efficient way of collecting defect images is a key factor in the implementation of surface detection. Essentially, the task is defined as anomaly classification, a prerequisite task that supports surface detection. Due to the limited availability of anomaly data, constructing effective anomaly classification models is often achieved through semi-supervised or self-supervised methods. Given the unknown distribution of anomaly regions, training is typically conducted solely on defect-free samples. Thus, the creation of a meaningful pre-training task becomes crucial. Deep One-class [16] has demonstrated an effective end-to-end training model, which is a parameterized deep neural network-based one-class classifier. Notably, its superiority lies in its deeper network architecture, outperforming shallower classification models such as one-class support vector machines or autoencoders [17], as demonstrated through comparative experiments with deep neural networks. In self-supervised feature learning methods, techniques like geometric transformations [18] and contrastive learning [19] have shown success in successfully discriminating normal and defective images.

In this work, we approximately define defect regions as a particular case of image anomalies, wherein a binary classification model should focus on all regions divergent from the normal surface. We adopt a two-stage framework [20], wherein we initially create anomaly regions on normal DCIP surface images through the construction of a self-supervised learning pre-training task. Specifically, we propose a strategy called CutPaste-Mix. This is an improved method based on CutPaste [21], creating abnormal regions by cutting patches from images. CutPaste-Mix includes three strategies, as follows: 1. Not reattaching the patch to the image, thus preserving the incomplete area; 2. Enlarging the patch by a certain factor and then randomly pasting it back onto the image; 3. Randomly rotating the patch and subsequently pasting it back into a random position within the image. Our intention is to introduce irregularities approximating defects in images. Moreover, we utilize ResNet-18 for feature learning on both types of images and employ defect-free feature vectors to compute a Gaussian Density Estimator (GDE) [22,23]. The GDE is leveraged to compute abnormal scores for each image, thereby achieving the classification of anomalous images.

We deployed our equipment on-site and collected actual DCIP images, creating a dataset. We tested our proposed method on the DCIP dataset and compared it with other models. Remarkably, without utilizing any defective data for training, our method achieved an impressive 99.5 AUC (Area Under Curve) [24]. Additionally, we conducted ablation experiments demonstrating the efficacy of the three methods contained within CutPaste-Mix when used individually and in combination. The outcomes demonstrated the effectiveness of the self-supervised classification model based on CutPaste-Mix in anomaly classification. The model we propose proves valuable for dataset creation and holds promise for practical engineering deployment.

2. Methods

2.1. Definition of Self-Supervised Binary Classifier

A self-supervised binary classifier [25] is a machine learning model designed to learn meaningful representations from unlabeled data by formulating a binary classification task within the data itself. The core idea behind self-supervised binary classification is to create surrogate positive and negative samples from the input data and train the model to differentiate between them. This allows the model to learn useful features from the data without the need for explicit human labeling. Common techniques used to generate positive and negative samples include data augmentation, transformations, or utilizing context information from the same data instance.

In the context of image data, techniques such as data augmentation can be employed to train a self-supervised binary classifier. In this approach, parts of an image are cut and pasted to create positive and negative pairs. The model is then trained to distinguish between the original images and the manipulated ones. This encourages the model to learn relevant features for the given classification task. Mathematically, let x represent an input image, x^+ represent a positive sample (e.g., an image with a manipulated patch), and x^- represent a negative sample (e.g., the original image or a different image). The self-supervised binary classification loss function can be formulated as follows:

$$L(x, x^+, x^-) = -log\left(\frac{exp(f(x) \cdot f(x^+))}{exp(f(x) \cdot f(x^+)) + exp(f(x) \cdot f(x^-))}\right)$$

where $L(x, x^+, x^-)$ represents the loss for a triplet of samples: the input image x, positive sample x^+, and negative sample x^-. $f(x)$ represents the feature embedding of sample x. The formula calculates the binary classification loss using logistic loss (cross-entropy) for positive (x and x^+) and negative (x and x^-) pairs.

In summary, a self-supervised binary classifier needs to enable the model to learn informative features from unlabeled data by creating its own binary classification task. This approach taps into the inherent structure of the data to generate supervision signals, making it a powerful technique for learning meaningful representations without the need for extensive manual labeling.

2.2. Self-Supervised Learning with CutPaste-Mix

A well-crafted pretext task [26,27] is a cornerstone for achieving successful self-supervised learning. The essence of such a task lies in setting up a specific objective within unlabeled data, enabling the model to glean meaningful representations from it. This task is meticulously designed to generate supervision signals intrinsically from the data itself, thus obviating the need for manual annotations. By tackling this pretext task, the model becomes adept at unraveling the inherent structures and patterns latent within the data in an unsupervised manner. Consequently, it provides a more robust foundation for initializing subsequent supervised learning tasks or refining feature representations. Commonly encountered pretext tasks in the realm of self-supervised learning encompass predicting image rotation angles, completing missing segments, colorizing images, and capturing contextual relationships within images. These tasks harness the intrinsic information embedded within the data to generate supervision signals, thereby prompting the model to assimilate valuable features and elevate its capacity for generalization.

Due to the requirement of training the model exclusively on defect-free data while also formulating an effective pretext task, the utilization of data augmentation techniques to generate images with anomalous regions emerges as the most suitable strategy. This methodology ensures the model's robust acquisition of distinctive features. While techniques such as rotation, translation, and contrastive learning methods enhance accuracy in single-classification tasks and augment the model's generalization and resilience against interference, they prove to be less effective when directly applied to images of Ductile Cast Iron Pipes (DCIPs), which are characterized by their high resolution and small defective areas. In essence, augmentation strategies like rotation and translation entail fundamental geometric transformations that are applied uniformly across the entire image. While these strategies are adept at capturing objective and conceptual features (often referred to as semantic features) within images, they fall short in addressing the learning of continuous local features within images. Within the context of DCIP images, defects manifest as irregular and disjointed anomalous regions. Consequently, it becomes imperative to design a strategy that effectively emulates these distinct abnormal regions. In summary, the optimal approach necessitates an augmentation strategy that mimics the irregular and discontinuous nature of DCIP defects, rather than relying solely on generic geometric transformations. This approach will significantly enhance the model's ability to effectively capture localized anomalies, thereby refining its capability to discern subtle yet crucial features.

We present a self-supervised methodology called CutPaste-Mix, a data augmentation strategy. CutPaste-Mix constitutes a comprehensive augmentation technique encompassing three specific variants: Erase, Enlarge, and Rotate. Although the task may seem straightforward to implement through conventional programming, it is gratifying to note that it indeed empowers deep neural networks to glean distinctive features from normal regions. Fundamentally, CutPaste-Mix involves the random cropping of a patch from a defect-free image and its subsequent overlay onto the original image to simulate an anomalous region.

CutPaste-Mix. A patch is randomly cropped from an image of the dataset; then, one or more of the following three strategies (Erase, Enlarge, or Rotate) are randomly used. This process creates anomalous regions on a normal surface, and the processed images serve as positive samples only for training the backbone of the classification network. We also set a consistent random seed to ensure the reproducibility of random processes.

1. **Erase.** Analogous to the Cutout technique, this strategy directly excises a portion from the image, discarding the patch area. Given that DCIP images are grayscale, the residual area's color adheres to the original grayscale values, with the mean of the original pixels serving as a substitute color. However, this approach engenders appreciable information loss in images and can be harmonized with other methods.
2. **Enlarge.** Commencing with the random cropping of a rectangular region, this method subsequently enlarges and reattaches it to a random location on the original image. We confine the cropped region's dimensions to not exceed 20% of the image area, while the enlargement factor remains within the image's boundaries. This approach often yields significant abnormal regions while mitigating the substantial information loss entailed by erase.
3. **Rotate.** Analogous to Enlarge, this approach involves initial patch cropping, followed by rotation by a random angle before reintegration into the original image. This tactic imposes minimal information loss while concurrently generating abnormal regions.

We chose ResNet-18 as the backbone of the classifier. As shown in Figure 1, the input images serve as negative samples for the training set. Simultaneously, all the output images generated through CutPaste-Mix are positive samples for the training set. It is important to note that the training of the network backbone requires both positive and negative samples. However, when fitting the parameters of the Gaussian density estimator, we exclusively used negative samples. To test whether a new image has anomalies or not, ResNet-18 first computes the feature vector. Then, the Gaussian density estimator calculates the final anomaly score for this feature vector to achieve the classification goal.

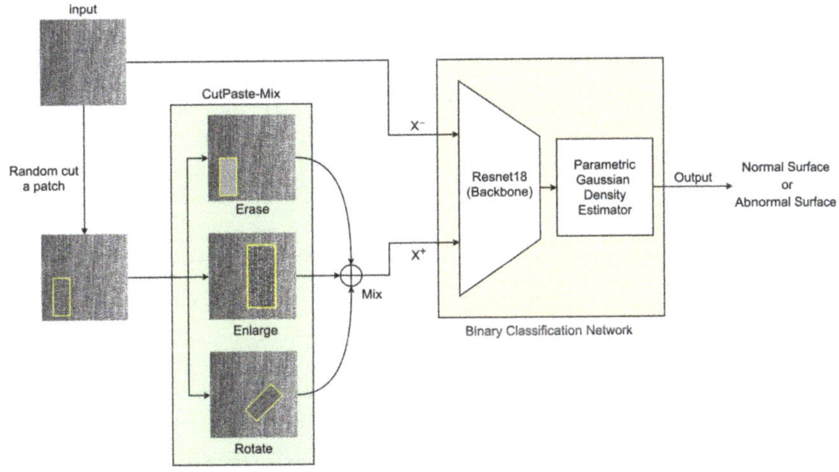

Figure 1. CutPaste-Mix includes three augmentation methods (Erase, Enlarge, Rotate) for creating abnormal regions on a normal surface. The binary classification network consists of ResNet-18 (backbone) and a parametric Gaussian density evaluator. "Normal Surface" refers to a surface that is within the expected or acceptable condition, and "Abnormal Surface" is the opposite.

This paper introduces the training loss function for the proposed supervised representation learning, as follows:

$$L_{CPM} = E_{x \in \chi}\{CE(g(x), 0) + CE(g(CP(x), 1))$$

where χ represents the defect-free dataset, $L_{CPM}(\cdot)$ stands for the loss function of the CutPaste-Mix, and $g(\cdot)$ is the singular classifier composed of a deep neural network. $CE(\cdot, \cdot)$ denotes the cross-entropy loss. During the implementation of the code, all the data augmentation strategies mentioned in CutPaste-Mix are completed before inputting the sample x into the singular classifier $g(\cdot)$.

2.3. *Deep Residual Learning with ResNet-18*

We recognize that there are many outstanding Convolutional Neural Networks (CNNs) available for us to choose from as the backbone of our classification model. These CNNs exhibit exceptional feature extraction capabilities. In the production pipeline of DCIPs, the speed of data collection is high. For our proposed classification model to achieve rapid classification on the production line, it must have a lower number of layers and lower computational complexity. Therefore, we excluded models like Inception [28] and Xception [29] due to their higher computational complexity. Traditional backbone networks like VGG [30] and AlexNet [31] sometimes suffer from the vanishing gradient problem because of less effective gradient propagation strategies. We found that ResNet-18's [32] network complexity and residual strategy align well with our requirements, which is why we chose it as the backbone for constructing the classification model.

Deep Residual Learning, commonly referred to as ResNet, has emerged as a significant advancement in the realm of deep neural networks. One prominent architecture within the ResNet family is ResNet-18 [32,33], which exhibits remarkable capabilities in overcoming the challenges posed by training extremely deep networks. ResNet-18 utilizes skip connections, also known as residual connections, to enable the effective training of very deep networks by mitigating the vanishing gradient problem.

The core innovation of ResNet-18 lies in its residual blocks, where the input to a layer is added to the output of a subsequent layer, allowing for the preservation of features across different layers. Figure 2 showcases the architecture of a residual block. The whole network architecture is composed of several stacked residual blocks, each consisting of multiple convolutional layers, batch normalization, and ReLU activation functions. This enables the network to learn complex hierarchical features while still maintaining a manageable overall network depth. Mathematically, this residual connection can be represented as follows:

$$x_{l+1} = x_l + F(x_l, W_l)$$

where x_l represents the input to the l-th layer, x_{l+1} is the output of the l+1-th layer, $F(x_l, W_l)$ is the residual mapping, and W_l denotes the weights of the l-th layer.

The ResNet-18 architecture serves as the foundational convolutional backbone in this investigation. Notably, an appended global average pooling layer precedes the terminal fully connected layer within the convolutional hierarchy. The resultant feature representation extracted from this pooling operation serves as the direct input to the ensuing fully connected stratum, which is seamlessly linked to a multilayer perceptron. To facilitate the accommodation of diverse image dimensions and sustain training efficiency invariant to input image size variations, a global average pooling layer is seamlessly integrated prior to the ultimate fully connected stratum of the ResNet-18 framework. Noteworthy is the scenario when images of dimensions $256 \times 256 \times 3$ are channeled into the model, thereby engendering an $8 \times 8 \times 512$ nodal feature map. Notably, the terminal fully connected layer of ResNet-18 encompasses 1000 nodes, necessitating the propagation of $32{,}768 \times 1000$ weights, thereby demanding substantial memory resources. The strategic inclusion of the global average pooling layer streamlines the transition from the feature map to the ultimate classification outcome, thereby manifesting substantial empirical efficacy. Concomitantly, this architectural augmentation yields a streamlined parameter space, thus fortifying model resilience and affording a pronounced mitigation of the proclivities toward overfitting.

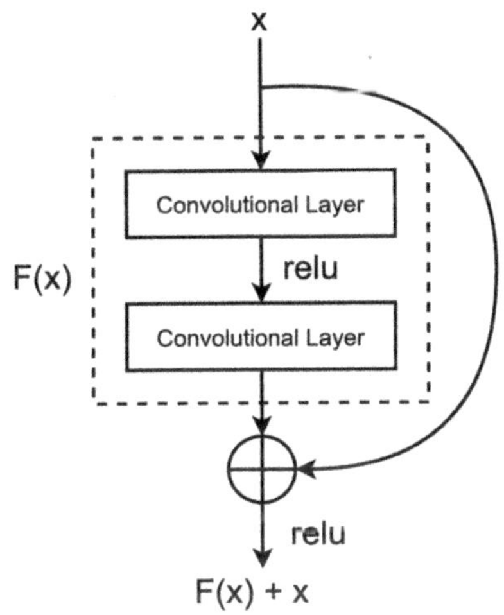

Figure 2. In a residual structure, the input features undergo initial processing through a sequence of convolutional layers and activation functions. These processed features are combined with the output through a skip connection, achieved by element-wise addition.

2.4. Computing Anomaly Score for Classification

The realm of binary classifiers offers a diverse array of methodologies for computing anomaly scores. The proposed approach involves the direct computation of anomaly scores based on the features extracted by the convolutional backbone. This is executed through techniques like kernel density estimation [34] or Gaussian density estimation [23]. Over the course of the past several decades, both kernel density estimation and Gaussian density estimation methods have undergone thorough investigation and widespread application. While these methods may exhibit certain limitations in specific contexts, such as parameter selection and computational intricacy, they stand as robust mechanisms for estimating probability density functions. These approaches offer invaluable tools and insights for tasks such as anomaly region classification. In our study, we harness the Gaussian density estimation technique to calculate anomaly scores for anomalies present in DCIPs, thereby facilitating an effective method for classifying surface defects in railway tracks.

Kernel Density Estimation offers several advantages, including the absence of prior assumptions about data distribution and the avoidance of the necessity for pre-estimation of parameters. We construct a simple parametric Gaussian density estimator, the basic principle of which is expressed mathematically as follows:

$$\log p_{\text{gde}}(x) \propto \left\{-\frac{1}{2}(f(x)-\mu)^{\top}\Sigma^{-1}(f(x)-\mu)\right\}$$

where μ and Σ represent the Gaussian parameters learned from defect-free negative samples.

3. Experiments

3.1. Description of DCIPs

The experiment focuses on inspecting T-type centrifugal ductile cast iron pipes (DCIPs), as illustrated in Figure 3. These specific pipes play crucial roles in municipal, industrial, and mining water supply and drainage systems, as well as contributing to rural water supply networks. Their significance extends to enhancing both urban and rural water supply conditions and augmenting regional sources of rural drinking water. Hence, it becomes imperative to ensure the quality and performance of cast pipes through thorough and dependable testing before they are dispatched from the factory.

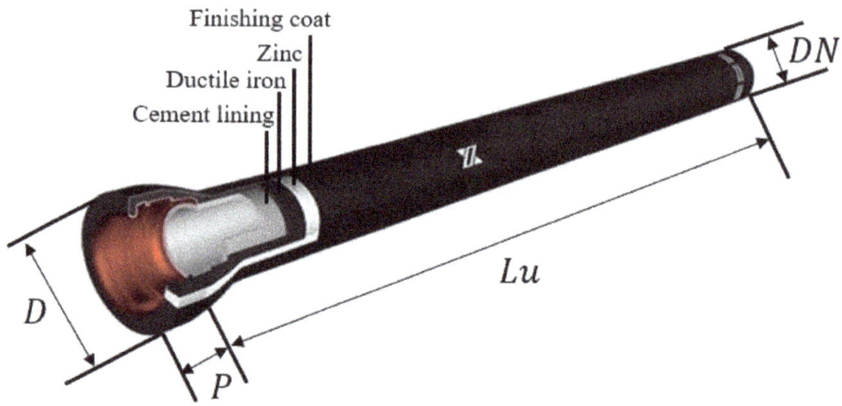

Figure 3. The T-type centrifugal DCIP.

The practical application scenarios vary, giving rise to diverse prerequisites concerning DCIP models, sizes, and various parameters. The specifications of the T-type centrifugal ductile cast iron pipes intended for assessment on the production line are elucidated in Table 1. The dimensions of DCIPs on the production line exhibit variations, encompassing a range from DN350 to DN1000 mm, with a maximum nominal diameter deviation of 650 mm. Given the cylindrical nature of the DCIP, rotational motion is indispensable for effectively detecting the circumferential surface. Moreover, these cast pipes extend up to a length of 6 m, necessitating the deployment of 6 cameras to cover the entire span, even though a single camera can encompass a physical field of view of 1 m.

Table 1. The specifications of the T-type centrifugal DCIPs.

DN/mm	D/mm	P/mm	Lu/m
350	448	110	6
400	500	110	6
450	540	120	6
500	604	120	6
600	713	120	6
700	824	150	6
800	943	160	6
900	1052	175	6
1000	1158	185	6

3.2. Experiment Setup

To address the challenge posed by variations in the nominal diameters of Ductile Cast Iron Pipes (DCIPs), we introduce an automated lifting adjustment system predicated on Programmable Logic Controllers (PLCs) and servo motor technologies. This system operates by discerning the specific diameter of the targeted DCIP for inspection and subsequently orchestrating the repositioning of the vision system to an optimized focal length for precise image acquisition. This real-time dynamic adjustment ensures the preservation of high-fidelity image capture, which in turn serves as a foundational cornerstone for subsequent algorithmic processing. As delineated in Figure 4, the multistep defect detection procedure for DCIPs within the production environment involves the following pivotal phases:

Step 1: Upon the DCIP's spatial alignment with the designated detection position, the automated lifting and adjustment system conducts a precise dimensional assessment of the DCIP, subsequently dictating the precise positioning of the imaging module for optimal acquisition.

Step 2: Following the precise alignment of the imaging module with the designated acquisition position, a rotational mechanism is actuated to induce the DCIP's rotation. This rotational mechanism maintains a consistent angular velocity despite varying linear speeds resultant from inherent DCIP size discrepancies. The synchronization of the rotational velocity with the light source system and camera acquisition is accomplished through encoder feedback, enabling real-time modulation of the light source's luminance duration and the camera's exposure time. Each imaging module seamlessly transfers acquired image data to a centralized server.

Figure 5a,b provides an illustrative depiction of the prototype system's configuration seamlessly integrated within the production line, emanating from the foundational prototype framework aforementioned. The illumination infrastructure within this context is meticulously orchestrated by two distinct lighting controllers, both intricately interfaced with a computational unit. Employing high-resolution BASLER Mono CMOS line scan cameras (Ahrensburg, Germany) boasting 4096 pixels, accompanied by ZEISS Planar T* 1.4/50 mm lenses, the image capture process is orchestrated. These cameras are intricately synchronized with the LED lighting infrastructure through a Field-Programmable Gate Array (FPGA) control board, ensuring temporal harmony. The captured DCIP images are promptly relayed to the computational unit for further comprehensive processing. The computational infrastructure boasts a 64-bit Windows 10 operating system, a 2.4 GHz central processing unit, and a substantial 32 GB of Random Access Memory (RAM).

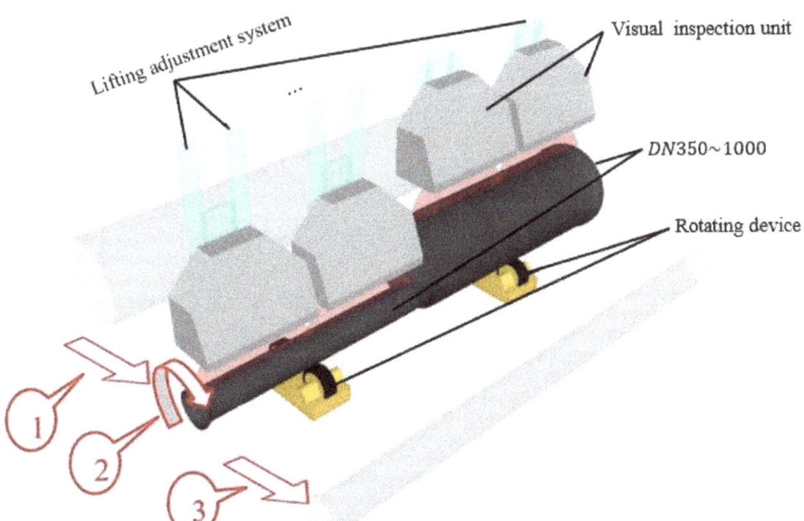

Figure 4. The prototype system of capture DCIP data.

3.3. Description of Image in Dataset

The imaging assessment apparatus deployed at the production site was employed for image testing on each machined DCIP. The system utilized a line scan camera with a resolution of 4096 pixels. By configuring the image acquisition parameters through the Pylon software, 4096 rows of data were simultaneously captured, resulting in a DCIP image with dimensions of 4096 × 2048 pixels. An exemplar image of the DCIP sample acquired on the production line is depicted in Figure 6.

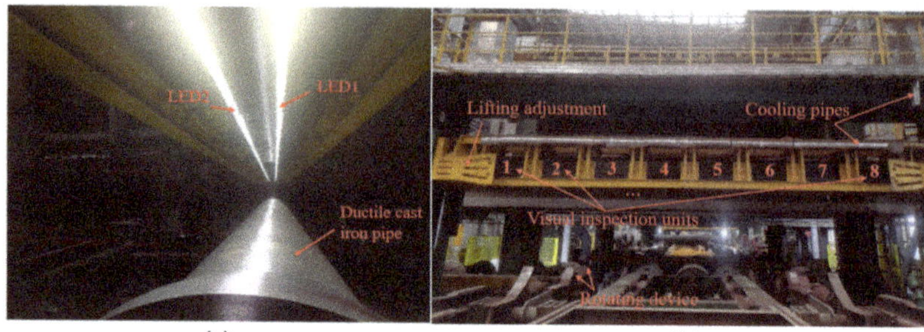

(**a**) (**b**)

Figure 5. The experimental prototype system at the production site. (**a**) Local scene; (**b**) global scene.

Figure 6. Sample images of DCIP (uncut).

There are six common types of surface defects found on DCIP: crazing, cracks, heavy skin, iron bean, mold powder, and pores. The visual characteristics of these defects are presented in Figure 7. After segmenting the captured 4096 × 2048-pixel images into consecutive 512 × 512-pixel sections and applying subsequent filtering, the obtained surface images devoid of any anomalies are shown in Figure 8. The captured images with dimensions of 4096 × 2048 pixels were segmented into consecutive images of size 512 × 512 pixels.

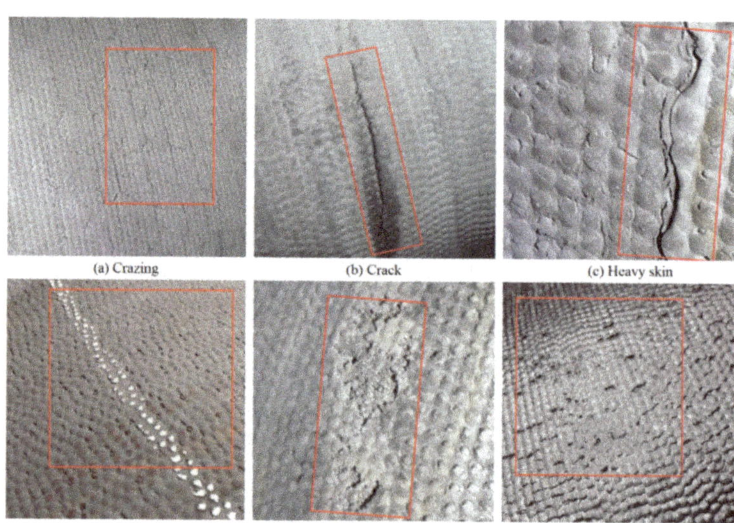

Figure 7. Images of DCIP with real surface defects, (**a**–**f**) in turn expressed as crazing, crack, heavy skin, iron bean, mold powder, pore.

Figure 8. Normal defect-free surface of DCIP.

3.4. *Training and Testing Process of Classifier*

Training Details: Our experiments were conducted on an NVIDIA RTX 3090. The input images were of size 512 × 512 pixels. The model typically converges with 150 iterations. We also utilized a list-type hyperparameter, ['erase', 'enlarge', 'rotate'], to control the involvement of the three strategies in CutPaste-Mix. The optimizer used was SGD (Stochastic Gradient Descent) with a learning rate of 0.03, a momentum parameter of 0.9, and a weight decay parameter set to 3×10^{-5}.

Dataset Preparation: Utilizing the aforementioned acquisition equipment, images of DCIPs were collected. Around 2000 defect-free images were meticulously selected to serve as the negative samples for the training set. These images were subsequently cropped into a 512 × 512-pixel format, as illustrated in Figure 8. The dataset was then subjected to CutPaste-Mix to generate images containing anomalous regions, thereby constituting the positive samples for the training set. Within the employed training dataset, an equal number of positive and negative samples were maintained.

The training steps of combining a pre-trained deep convolutional network with Gaussian density estimation to create a binary classifier can be represented as follows:

Computation of Feature Vectors: Employing a pre-trained ResNet-18, the training dataset undergoes feature extraction. Each datum is fed through the initial layers of the convolutional neural network, thus obtaining the image's feature representation. These representations can be conceived as high-dimensional feature vectors of the images.

Computing the parameters of GDE: The mean and covariance matrix of the Gaussian density estimator are computed using defect-free DCIP data. For each positive and negative sample, we aim to set an appropriate threshold to distinguish between them. Thus, we use the testing set to iteratively adjust suitable thresholds. This threshold is continuously optimized as new defect data are gathered in subsequent iterations.

New Samples classification: The images are first passed through the pre-trained convolutional neural network to obtain the feature representation of new samples. Next, these feature representations are input into the Gaussian Density Estimator (GDE) model to calculate the likelihood probabilities of the sample belonging to the positive and negative classes. Finally, the class with the higher probability is chosen as the prediction result.

Accuracy evaluation: We used 200 real samples with defects as the testing set, all of which were collected using the device proposed in this paper. The AUC (Area Under Curve) is the evaluation metric used to test the performance of our classifier and other models. The classifier calculates the True Positive Rate (TPR) and False Positive Rate (FPR) using various thresholds on the test dataset and plots these values to create the ROC curve. The ROC curve typically has TPR on the y-axis and FPR on the x-axis. Finally, the area under ROC curve, denoted as AUC (Area Under the Curve), is calculated to evaluate the classifier's performance.

3.5. Main Results

Table 2 showcases the performance of the novel self-supervised classification approach introduced in this study. Through 10 distinct experiments employing diverse random seeds, we present the averaged AUC accompanied by its standard error across the testing set. We not only evaluated the overall performance of CutPaste-Mix but also separately evaluated the three augmentation strategies it includes: Erase, Enlarge, and Rotate. Additionally, we conducted comparative assessments with three alternative techniques: Deep One-Class (DOCC) [16], Uninformed-Student [35], and Patch-SVDD [36].

Table 2. The ablation experiments evaluated the performance of CutPaste-Mix and its sub-methods separately on the DCIP dataset. To reduce randomness, each experimental group was tested 10 times using different random seeds. We report the AUC and standard deviation for each experimental group.

Erase	Enlarge	Rotate	AUC
*			84.3 ± 2.3
	*		95.8 ± 1.1
		*	92.5 ± 1.8
*	*		88.6 ± 0.5
*		*	97.2 ± 0.7
	*	*	98.9 ± 0.2
*	*		94.8 ± 1.5
*	*	*	$\mathbf{99.4 \pm 0.1}$

We refer to the three strategies included in CutPaste-Mix as Enlarge, Erase, and Rotate just for clarity. Enlarge excelled as a standalone data augmentation technique, even surpassing Rotate, achieving an impressive 95.8 AUC. The performance was poorest while only using Erase, with 84.3 AUC. When combining two data augmentation methods, Enlarge and Rotate together achieved an even higher 98.9 AUC. Furthermore, the results of CutPaste-Mix reached a remarkable 99.4 AUC. We also compared the top-performing CutPaste-Mix approach with other models, and it consistently yielded the best results, as demonstrated in Table 3.

Table 3. The performance of various anomaly detection models on the DCIP dataset evaluated using the AUC metric, serving as a comparative experiment against the CutPaste-Mix.

DOCC [16]	U-Student [35]	P-SVDD [36]	CutPaste-Mix (Best)
83.4	92.8	95.6	**99.5**

We applied our most effective classifier to the field-collected dataset using the acquisition equipment. The results were gathered from 10,000 consecutive screening operations. Our proposed self-supervised classifier ultimately identified 43 images as anomalous. Following on-site comparisons, we validated the presence of 20 genuine defect images, as illustrated in Figure 9. However, it is important to note that the DCIP's surface is inevitably subject to disturbances like water stains and oil smudges, as depicted in Figure 10. These noise interferences also contribute to the anomalous regions. While it may not be feasible to entirely distinguish whether the detected anomalous regions represent defects or noise, our approach substantially reduced the time required for manual defect screening. Moreover, its on-site applicability offers significant convenience for subsequent model optimization endeavors.

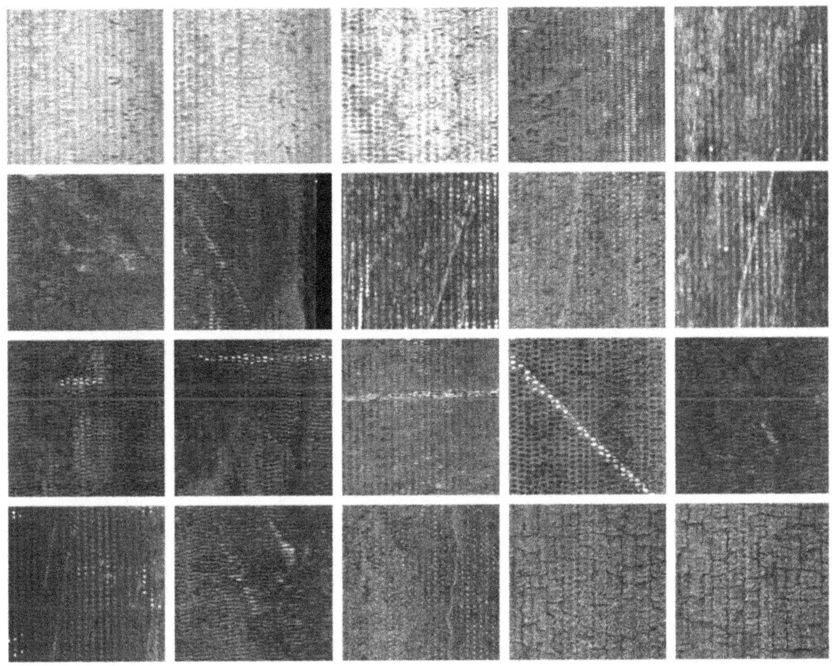

Figure 9. The true positives in the model's classification results encompass defect types such as iron bean, crack, crazing, and pore.

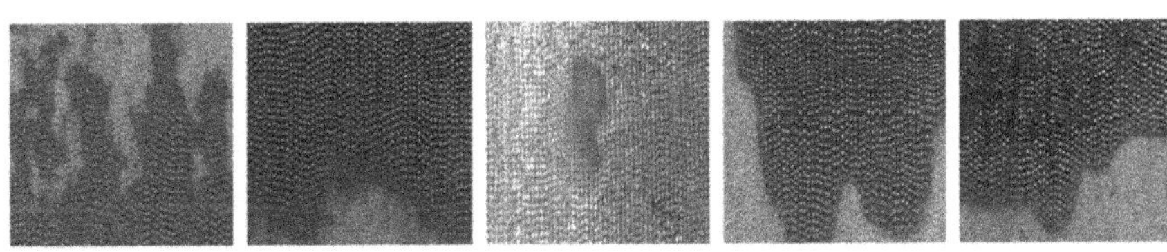

Figure 10. The false positives in the model's classification results are primarily caused by oil stains, water marks, and iron oxide patches on the surface of the DCIPs. Although these anomalies are not defects, they still represent abnormal areas distinct from the normal surface.

4. Conclusions

The aforementioned experimental results confirm the feasibility and superiority of the self-supervised classification algorithm based on CutPaste-Mix on the DCIP dataset. The use of data augmentation to generate abnormal regions in images effectively yields positive samples for classifier training. The ResNet-18 backbone network efficiently captures image features and computes feature vectors. The Gaussian Density Estimation (GDE) is utilized to compute anomaly scores for the extracted features, thus achieving anomaly classification. Our experimental outcomes demonstrate the advantages of this approach compared to other methods. Through ablation experiments, we discussed the results of using different augmentation strategies individually and in combination. For each method, we conducted 10 trials and calculated the average AUC and standard deviation to showcase the superiority of CutPaste-Mix.

This study also has a limitation. In the classification results, we encountered non-defective surface images, such as water stains or oil stains. This suggests that the proposed classification model struggles to distinguish between genuine defects and noise interference, both of which fall into the category of anomalies. The reason for this challenge may be that both genuine defects and noise interference have a large Euclidean distance from normal surfaces in the feature space, but their feature similarity is high. Therefore, we believe that distinguishing between noise interference and real defects could be a valuable direction for future research.

Author Contributions: Conceptualization, H.Z. and Q.S.; methodology, H.Z. and Q.S.; software, H.Z. and Q.S.; validation, H.Z. and Q.S.; formal analysis, H.Z., Q.S. and K.X.; investigation, H.Z., Q.S. and K.X.; resources, H.Z., Q.S. and K.X.; data curation, H.Z., Q.S. and K.X.; writing—original draft preparation, H.Z. and Q.S.; writing—review and editing, H.Z. and Q.S.; visualization, H.Z. and Q.S.; supervision, K.X.; project administration, K.X.; funding acquisition, K.X.; All authors have read and agreed to the published version of the manuscript.

Funding: This work is sponsored by Beijing Science and Technology Planning Project (No. Z221100005822012), and Key Technologies Research and Development Program of China (No. 2021YFB3202403).

Institutional Review Board Statement: Not applicable.

Informed Consent Statement: Informed consent was obtained from all subjects involved in the study.

Data Availability Statement: The data that support the findings of this study are available from the corresponding author upon reasonable request.

Conflicts of Interest: The authors declare no conflict of interest.

References

1. Zhou, D.; Xu, K.; Lv, Z.; Yang, J.; Li, M.; He, F.; Xu, G. Intelligent Manufacturing Technology in the Steel Industry of China: A Review. *Sensors* **2022**, *22*, 8194. [CrossRef]
2. Latif, J.; Shakir, M.Z.; Edwards, N.; Jaszczykowski, M.; Ramzan, N.; Edwards, V. Review on Condition Monitoring Techniques for Water Pipelines. *Measurement* **2022**, *193*, 110895. [CrossRef]
3. Makar, J.M.; Desnoyers, R.; McDonald, S.E. Failure Modes and Mechanisms in Gray Cast Iron Pipes. In *Underground Infrastructure Research*; Knight, M., Thomson, N., Eds.; CRC Press: Boca Raton, FL, USA, 2020; pp. 303–312. ISBN 978-1-00-307748-0.
4. Ji, J.; Hong Lai, J.; Fu, G.; Zhang, C.; Kodikara, J. Probabilistic Failure Investigation of Small Diameter Cast Iron Pipelines for Water Distribution. *Eng. Fail. Anal.* **2020**, *108*, 104239. [CrossRef]
5. Javaid, M.; Haleem, A.; Singh, R.P.; Rab, S.; Suman, R. Exploring Impact and Features of Machine Vision for Progressive Industry 4.0 Culture. *Sens. Int.* **2022**, 3, 100132. [CrossRef]
6. Smith, M.L.; Smith, L.N.; Hansen, M.F. The Quiet Revolution in Machine Vision—A State-of-the-Art Survey Paper, Including Historical Review, Perspectives, and Future Directions. *Comput. Ind.* **2021**, *130*, 103472. [CrossRef]
7. Gupta, C.; Farahat, A. Deep Learning for Industrial AI: Challenges, New Methods and Best Practices. In *Proceedings of the 26th ACM SIGKDD International Conference on Knowledge Discovery & Data Mining, Online, 6–10 July 2020*; Association for Computing Machinery: New York, NY, USA, 2020; pp. 3571–3572.
8. Khalil, R.A.; Saeed, N.; Masood, M.; Fard, Y.M.; Alouini, M.-S.; Al-Naffouri, T.Y. Deep Learning in the Industrial Internet of Things: Potentials, Challenges, and Emerging Applications. *IEEE Internet Things J.* **2021**, *8*, 11016–11040. [CrossRef]
9. Wang, D.; Wang, J.-G.; Xu, K. Deep Learning for Object Detection, Classification and Tracking in Industry Applications. *Sensors* **2021**, *21*, 7349. [CrossRef] [PubMed]
10. Sika, R.; Szajewski, D.; Hajkowski, J.; Popielarski, P. Application of instance-based learning for cast iron casting defects prediction. *Manag. Prod. Eng. Rev.* **2019**, *10*, 101–107.
11. Di, H.; Ke, X.; Peng, Z.; Dongdong, Z. Surface Defect Classification of Steels with a New Semi-Supervised Learning Method. *Opt. Lasers Eng.* **2019**, *117*, 40–48. [CrossRef]
12. Pourkaramdel, Z.; Fekri-Ershad, S.; Nanni, L. Fabric Defect Detection Based on Completed Local Quartet Patterns and Majority Decision Algorithm. *Expert Syst. Appl.* **2022**, *198*, 116827. [CrossRef]
13. Wang, C.-Y.; Bochkovskiy, A.; Liao, H.-Y.M. YOLOv7: Trainable Bag-of-Freebies Sets New State-of-the-Art for Real-Time Object Detectors. In Proceedings of the IEEE/CVF Conference on Computer Vision and Pattern Recognition (CVPR), Vancouver, BC, Canada, 18–22 June 2023.
14. Liu, Z.; Hu, H.; Lin, Y.; Yao, Z.; Xie, Z.; Wei, Y.; Ning, J.; Cao, Y.; Zhang, Z.; Dong, L.; et al. Swin Transformer V2: Scaling Up Capacity and Resolution. In Proceedings of the IEEE/CVF Conference on Computer Vision and Pattern Recognition (CVPR), New Orleans, LA, USA, 19–24 June 2022.

15. Zhao, P.; Lin, Q. RLNF: Reinforcement Learning Based Noise Filtering for Click-Through Rate Prediction. In Proceedings of the 44th International ACM SIGIR Conference on Research and Development in Information Retrieval, Online, 11–15 June 2021.
16. Ruff, L.; Vandermeulen, R.; Goernitz, N.; Deecke, L.; Siddiqui, S.A.; Binder, A.; Müller, E.; Kloft, M. Deep One-Class Classification. *PMLR* **2018**, *80*, 4393–4402.
17. Masci, J.; Meier, U.; Cireşan, D.; Schmidhuber, J. Stacked Convolutional Auto-Encoders for Hierarchical Feature Extraction. In *Proceedings of the Artificial Neural Networks and Machine Learning—ICANN, Espoo, Finland, 14–17 June 2011*; Honkela, T., Duch, W., Girolami, M., Kaski, S., Eds.; Springer: Berlin, Heidelberg, 2011; pp. 52–59.
18. Bergman, L.; Hoshen, Y. Classification-Based Anomaly Detection for General Data. *arXiv* **2020**, arXiv:2005.02359.
19. Schlegl, T.; Seeböck, P.; Waldstein, S.M.; Schmidt-Erfurth, U.; Langs, G. Unsupervised Anomaly Detection with Generative Adversarial Networks to Guide Marker Discovery. In Proceedings of the Information Processing in Medical Imaging, Boone, NC, USA, 25–30 June 2017; Niethammer, M., Styner, M., Aylward, S., Zhu, H., Oguz, I., Yap, P.-T., Shen, D., Eds.; Springer International Publishing: Cham, Switzerland, 2017; pp. 146–157.
20. Sohn, K.; Li, C.-L.; Yoon, J.; Jin, M.; Pfister, T. Learning and Evaluating Representations for Deep One-Class Classification. *arXiv* **2021**, arXiv:2011.02578.
21. Li, C.-L.; Sohn, K.; Yoon, J.; Pfister, T. CutPaste: Self-Supervised Learning for Anomaly Detection and Localization. In Proceedings of the IEEE/CVF Conference on Computer Vision and Pattern Recognition (CVPR), Online, 19–25 June 2021; pp. 9664–9674.
22. Otneim, H.; Tjøstheim, D. The Locally Gaussian Density Estimator for Multivariate Data. *Stat. Comput.* **2017**, *27*, 1595–1616. [CrossRef]
23. Varanasi, M.K.; Aazhang, B. Parametric Generalized Gaussian Density Estimation. *J. Acoust. Soc. Am.* **1989**, *86*, 1404–1415. [CrossRef]
24. Lobo, J.M.; Jiménez-Valverde, A.; Real, R. AUC: A Misleading Measure of the Performance of Predictive Distribution Models. *Glob. Ecol. Biogeogr.* **2008**, *17*, 145–151. [CrossRef]
25. Sazzed, S.; Jayarathna, S. SSentiA: A Self-Supervised Sentiment Analyzer for Classification from Unlabeled Data. *Mach. Learn. Appl.* **2021**, *4*, 100026. [CrossRef]
26. Kinakh, V.; Taran, O.; Voloshynovskiy, S. ScatSimCLR: Self-Supervised Contrastive Learning with Pretext Task Regularization for Small-Scale Datasets. In Proceedings of the IEEE/CVF International Conference on Computer Vision (ICCV) Workshops, Online, 11–17 October 2021; pp. 1098–1106.
27. Albelwi, S. Survey on Self-Supervised Learning: Auxiliary Pretext Tasks and Contrastive Learning Methods in Imaging. *Entropy* **2022**, *24*, 551. [CrossRef]
28. Szegedy, C.; Ioffe, S.; Vanhoucke, V.; Alemi, A. Inception-v4, Inception-ResNet and the Impact of Residual Connections on Learning. *Proc. AAAI Conf. Artif. Intell.* **2016**, *31*, 4278–4284. [CrossRef]
29. Chollet, F. Xception: Deep Learning with Depthwise Separable Convolutions. In Proceedings of the 2017 IEEE Conference on Computer Vision and Pattern Recognition (CVPR), Honolulu, HI, USA, 21–26 July 2017; pp. 1800–1807.
30. Simonyan, K.; Zisserman, A. Very Deep Convolutional Networks for Large-Scale Image Recognition. *arXiv* **2015**, arXiv:1409.1556.
31. Krizhevsky, A.; Sutskever, I.; Hinton, G.E. ImageNet Classification with Deep Convolutional Neural Networks. *Commun. ACM* **2017**, *60*, 84–90. [CrossRef]
32. He, K.; Zhang, X.; Ren, S.; Sun, J. Deep Residual Learning for Image Recognition. In Proceedings of the IEEE Conference on Computer Vision and Pattern Recognition (CVPR), Las Vegas, NV, USA, 26 June–1 July 2016; pp. 770–778.
33. He, F.; Liu, T.; Tao, D. Why ResNet Works? Residuals Generalize. *IEEE Trans. Neural Netw. Learn. Syst.* **2020**, *31*, 5349–5362. [CrossRef] [PubMed]
34. Terrell, G.R.; Scott, D.W. Variable Kernel Density Estimation. *Ann. Stat.* **1992**, *20*, 1236–1265. [CrossRef]
35. Bergmann, P.; Fauser, M.; Sattlegger, D.; Steger, C. Uninformed Students: Student-Teacher Anomaly Detection with Discriminative Latent Embeddings. In Proceedings of the IEEE/CVF Conference on Computer Vision and Pattern Recognition (CVPR), Seattle, WA, USA, 13–19 June 2020; pp. 4182–4191.
36. Yi, J.; Yoon, S. Patch SVDD: Patch-Level SVDD for Anomaly Detection and Segmentation. In Proceedings of the Asian Conference on Computer Vision (ACCV), Kyoto, Japan, 30 November–4 December 2020; pp. 375–390.

sensors

MDPI

Article

Comparing Three Methods of Selecting Training Samples in Supervised Classification of Multispectral Remote Sensing Images

Hongying Zhang [1], Jinxin He [1,*], Shengbo Chen [2], Ye Zhan [3], Yanyan Bai [1] and Yujia Qin [1]

[1] College of Earth Sciences, Jilin University, Changchun 130061, China; hongyingz21@mails.jlu.edu.cn (H.Z.); yanyanbai21@mails.jlu.edu.cn (Y.B.); qinyj2220@mails.jlu.edu.cn (Y.Q.)
[2] College of Geoexploration Science and Technology, Jilin University, Changchun 130061, China; chensb@jlu.edu.cn
[3] Aviation Operations Service College, Aviation University Air Force, Changchun 130021, China; xuzj@jlu.edu.cn
* Correspondence: hejx@jlu.edu.cn; Tel.: +86-431-88502327

Abstract: Selecting training samples is crucial in remote sensing image classification. In this paper, we selected three images—Sentinel-2, GF-1, and Landsat 8—and employed three methods for selecting training samples: grouping selection, entropy-based selection, and direct selection. We then used the selected training samples to train three supervised classification models—random forest (RF), support-vector machine (SVM), and k-nearest neighbor (KNN)—and evaluated the classification results of the three images. According to the experimental results, the three classification models performed similarly. Compared with the entropy-based method, the grouping selection method achieved higher classification accuracy using fewer samples. In addition, the grouping selection method outperformed the direct selection method with the same number of samples. Therefore, the grouping selection method performed the best. When using the grouping selection method, the image classification accuracy increased with the increase in the number of samples within a certain sample size range.

Keywords: remote sensing classification; sample selection method; classification model; sample size

Citation: Zhang, H.; He, J.; Chen, S.; Zhan, Y.; Bai, Y.; Qin, Y. Comparing Three Methods of Selecting Training Samples in Supervised Classification of Multispectral Remote Sensing Images. *Sensors* **2023**, *23*, 8530. https://doi.org/10.3390/s23208530

Academic Editors: Róbert Hudec, Patrik Kamencay and Peter Hockicko

Received: 25 August 2023
Revised: 4 October 2023
Accepted: 10 October 2023
Published: 17 October 2023

1. Introduction

Multispectral remote sensing images contain a large amount of ground object information, which is encoded in various bands of the image. Ground object information can be quickly extracted from multiple bands. Different ground objects have different spectral characteristics, and similar ground objects have the same or similar spectral characteristics under the same conditions. The same ground object exhibits different radiation energy in different bands, resulting in differences between images obtained from different bands. With the rapid development of remote sensing technology, the information that we obtain from images is becoming increasingly rich and, correspondingly, image information extraction technology is constantly improving. In the field of remote sensing, image classification has always been the focus of research for professionals, who are committed to researching advanced classification methods to improve the accuracy of remote sensing image classification. As a fundamental image processing method, remote sensing image classification is the basis for environmental and socioeconomic applications. It has been widely used in various fields, such as environmental protection, land change monitoring, agricultural planning, water resource analysis, natural disaster detection, biodiversity monitoring, etc. [1].

Remote sensing image classification is a complex process, in which every step is crucial, from the selection of data sources and the design of sample selection schemes to the selection of classification methods and the evaluation of classifier performance. The complexity of the landscape in the study area, the scale of the study area, and economic conditions are also important factors that influence the selection of remote sensing data, the design of classification algorithms, and the quality of classification results [2]. However, many previous studies have focused specifically on advanced classification methods and improving the accuracy of remote sensing image classification [3–7] but paid little attention to the research on the selection methods for training samples in classification algorithms.

The structure of this paper is as follows: In this study, the background and motivation of the research are first introduced, outlining the objective of achieving better classification results in remote sensing imagery by selecting appropriate sample selection methods. The paper provides an overview of the current research status in this area. Subsequently, the study elaborates on the three sample selection methods employed. The data and models used in this research, including Sentinel-2, GF-1, and Landsat 8 remote sensing data, as well as SVM [8], RF [9], and KNN [10] classification models, are then introduced. The design and execution processes of the experiments, encompassing sample selection, model training, and the utilization of remote sensing imagery, are described. In the section presenting the experimental results, a comprehensive analysis and comparison of the classification accuracy achieved through different sample selection methods is presented. Additionally, this study explores the influence of varying sample sizes on classification accuracy under the optimal sample selection method. Finally, the Conclusions and Outlook section summarizes the main findings and contributions of this study, offering recommendations for future research directions.

2. Related Works

Samples in remote sensing image classification are mainly used as training data for classification models and as test data to evaluate the accuracy of map products. In supervised classification of multispectral remote sensing images, common classification methods include SVM, RF, and KNN. Training samples are required to train the classifier, and the method of selecting the samples determines the quality of the training samples and has a significant impact on remote sensing image classification and accuracy evaluation. With the development of remote sensing technology, the amount of information contained in remote sensing data is increasing. Many researchers have also begun to pay attention to the impact of samples on classification results. Koreen et al. conducted a study analyzing the effects of input data features on the RF classification algorithm. Their results showed that the RF classification algorithm was highly sensitive to the training dataset, and the selection strategy of specific input variables (i.e., image channels) and training data used in classification had a significant impact on the overall accuracy of image classification [11]. Zhen et al. delineated reference polygons to obtain training and validation data and studied the effects of four selection schemes on the classification accuracy and accuracy estimates obtained from validation data in object-based classification research [12]. J. Corcoran et al. selected training samples using three methods—namely, selecting points interpreted from the field and photographs, fixed windows around points, and image objects intersecting with points—and studied the effects of point- and polygon-based training data on the accuracy of wetland RF classification [13]. Shahriar et al. studied the effects of classifier selection, reference sample size, reference class distribution, and scene heterogeneity on the accuracy of pixel classification. They found that SVM and KNN have a significant advantage in accuracy for edge pixels, and that the class distribution in the training dataset has an impact on the accuracy calculation [14]. Ming et al. investigated the impact of training samples and classifiers on Landsat 8 image classification and found that SVM had the highest classification accuracy. The accuracy of the classifier increased with the increase in the training sample size, providing guidance for the selection of training samples and classifiers [15]. Li et al. researched the effect of sample size on remote sensing classification [16–18]. Christopher et al. evaluated four sample selection methods (simple random, proportionate stratified random, disproportionate stratified random, and judicious sampling) and three cross-validation adjustment methods (k-fold, leave-one-out, and the

Monte Carlo method) for regional-scale machine learning classification. They trained supervised machine learning algorithms using the selected samples to generate land cover maps of large geographic areas [4]. Additionally, Jin et al. studied the effects of four training sample selection schemes on urban and non-urban binary classification in the Denver metropolitan area of Colorado, including two stratification schemes (spatial and class-specific) and two sample allocation options (proportional area and equal allocation) [19]. Lv et al. proposed a grouping-based sample selection method that applies histogram analysis to select more distinctive samples [20].

These studies indicate that sample selection significantly influences the outcomes of remote sensing image classification. Researchers have explored various factors, such as input data features, strategies for selecting training datasets, acquisition of reference polygons for training and validation data, training data based on points and polygons, classifier selection, reference sample size, reference class distribution, scene heterogeneity, sample size, and sampling and cross-validation adjustment strategies. They have found that these factors have a notable impact on the overall accuracy of image classification. Additionally, researchers have proposed specific sample selection methods. In particular, researchers have employed various sample selection methods, including random sampling, stratified random sampling, disproportionate stratified random sampling, and judicious sampling. They have also investigated different cross-validation adjustment methods, such as k-fold, leave-one-out, and the Monte Carlo method. The choice of these methods produces varying effects in different scenarios. For instance, in the context of regional-scale machine learning classification, selecting appropriate sample selection methods and cross-validation adjustment methods can generate land cover maps for large geographic areas. Furthermore, researchers have focused on classifier selection and observed distinct advantages for specific classifiers such as SVM and KNN in certain situations. These studies provide valuable guidance for remote sensing image classification, enabling researchers to more accurately select samples and classifiers, thereby enhancing the accuracy and reliability of remote sensing image classification.

3. Materials and Methods

The process of identifying features in remote sensing images essentially involves transforming the identification problem into sample classification. The quality and representativeness of training samples directly impact the classification results of remote sensing image classifiers. Therefore, it is crucial to quickly and effectively select representative training samples. In this paper, we consider each pixel in the remote sensing images as a sample. Each pixel is composed of multiple bands, and these band data contain abundant information. We chose to use the band data of each pixel as sample data. We chose the grouping method and the "entropy"-based selection method for sample selection. Using Sentinel-2, GF-1, and Landsat 8 remote sensing images, as well as SVM, RF, and KNN classification models, we further analyzed the influence of these selection methods on the classification accuracy of remote sensing images.

3.1. Image Data

This experiment primarily utilized three remote sensing images: Sentinel-2, Landsat 8, and GF-1. Table 1 provides specific details about these images, all of which underwent preprocessing steps such as radiometric calibration and atmospheric correction.

Table 1. Characteristics of the three remote sensing images.

Satellite	Key Features
Sentinel-2	High-resolution multispectral imaging satellite equipped with 13 bands. Primarily used for monitoring terrestrial environments and providing information on vegetation, soil, and coastal conditions.
Landsat 8	Equipped with the Operational Land Imager (OLI) and Thermal Infrared Sensor (TIRS). The OLI has 9 bands with a resolution of 30 m. It is widely used in fields such as global change, agriculture, and water quality.
GF-1	Equipped with high-resolution and multispectral cameras. It can cover large areas with high spatial and temporal resolution.

3.2. Classification Models

SVM, RF, and KNN are widely used land cover classification models in the field of remote sensing. These models exhibit excellent performance and versatility when it comes to handling remote sensing data and addressing land cover classification tasks. They are chosen for land cover classification because of their ability to effectively process various types of remote sensing data, including multispectral, hyperspectral, and remote sensing imagery, and to provide accurate classification results across different land cover scenarios. These models play a crucial role in geographic information systems (GISs), environmental monitoring, land-use planning, resource management, disaster monitoring, and more, offering robust support for decision-making and spatial analysis. Therefore, selecting these models for remote sensing land cover classification is a logical choice, given their proven excellence and broad application prospects in this field.

SVM is a supervised learning algorithm used to find the optimal hyperplane for separating the feature space. It uses support vectors to determine the position of the hyperplane, which are the training samples closest to it. SVM transforms the feature space into a higher-dimensional space using kernel functions, enabling linearly inseparable data to become linearly separable in the new space. There are many types of kernels, and in this experiment, the radial basis function (RBF) kernel [21], commonly used in remote sensing, was used as a baseline for evaluating the performance of new SVM kernels.

RF is a machine learning classifier that leverages multiple decision trees for ensemble learning and improves accuracy by aggregating their classification results. Each decision tree acts as a classifier, and the final classification result is determined by the voting of all classifiers.

KNN is a lazy supervised learning classification algorithm that determines the class of a new sample by finding the k most similar samples. The value of k in the KNN algorithm affects the complexity of the decision boundary, with smaller k leading to complex decision boundaries and larger k improving the model's generalization ability.

3.3. Sample Selection Method

We conducted experiments using the group-based sampling method [20]. The group-based sampling method evaluates the feature distribution of each category by analyzing the histogram of the number of samples in each category, and then it selects training samples from different groups in the histogram. This method considers the heterogeneity of categories and is capable of selecting training samples with more prominent features while excluding mislabeled sample points. Additionally, a method based on "entropy" for selecting training samples was proposed. Through experimental analysis, the optimal sample selection method was determined, enabling supervised classification with a small number of representative samples, thus improving the classification efficiency and accuracy. The following is a detailed introduction to the sample selection methods used:

3.3.1. Group-Based Selection Method

Based on the existing method [20], considering that different land features in the image have varying areas and spectral ranges, we developed a specific calculation method to increase the number of groups. The number of groups for each land feature was calculated based on its spectral range and the number of sample labels. In addition, a variable P was introduced to control the number of selected samples. Below are the specific methods for selecting grouped samples.

Firstly, the remote sensing image to be classified needs to be labeled to select the initial samples. As different features have different proportions in the image, a proportional stratified random sampling method is adopted to ensure the representativeness of the training samples. This means that every pixel in the image has the chance to be selected, and the number of samples selected is determined based on the percentage of each feature's area in the image to ensure that the number of samples for each feature corresponds to its proportion in the image. At the same time, the number of labeled feature blocks should also match the proportion of each feature in the image.

After obtaining the labeled sample data, a feature distribution map for each feature class is generated based on the gray value of each pixel. The gray value of each pixel is calculated as gray = 0.299 × red + 0.587 × green + 0.114 × blue. Red, green, and blue represent the red, green, and blue bands used in the image, respectively, and the calculation of the gray value equalizes the values of the bands and facilitates data processing.

$$\mathrm{Bins_k} = \sqrt[2]{(\mathrm{band_{max}} - \mathrm{band_{min}})} * \frac{\mathrm{m_k}}{\mathrm{M}} \tag{1}$$

where $\mathrm{Bins_k}$ represents the number of groups of the kth land cover type, $\mathrm{band_{max}}$ represents the maximum band value in the labeled sample of the kth land cover type, $\mathrm{band_{min}}$ represents the minimum band value in the labeled sample of the kth land cover type, $\mathrm{m_k}$ represents the number of labeled samples of the kth land cover type, and M represents the total number of labeled samples of all land cover types. Using Formula (1), each land cover type can be divided into different numbers of groups according to its proportion in the image. Different groups represent different spectral ranges, and these spectral ranges correspond to different variation patterns within the same category. By selecting different numbers of samples from each group, it is possible to consider and capture the internal heterogeneity within the category, thus covering the intraclass heterogeneity.

Next, calculate the number of labeled samples within each group and determine the number of samples selected from each group using the following formula:

$$\mathrm{C_{b_i}} = \mathrm{P} * \mathrm{N_{b_i}^2} / \mathrm{T_K} \tag{2}$$

where $\mathrm{C_{b_i}}$ represents the number of samples obtained for the ith grouping, $\mathrm{N_{b_i}}$ represents the ratio of the total number of selected samples to the total number of labeled samples, $\mathrm{T_k}$ represents the total number of pixels in the ith grouping, and $\mathrm{b_i}$ represents the total number of pixels of the kth land cover type. Samples of pixels are selected for each group in each land cover type in turn. From this formula, it can be seen that the higher the frequency of $\mathrm{b_i}$, the more labeled samples are selected from this group. When $\mathrm{b_i}$ approaches 0, no samples are selected from this group. The samples within this range may be mislabeled compared to most of the samples of this land cover type. Therefore, they are not suitable as characteristic training samples.

Figure 1 shows the characteristic distribution map of the forestland. According to Formula (1), each column represents a group in the histogram. According to Formula (1), the labeled pixels of forestland are divided into eight groups. We need to select a certain number of representative samples from these groups. Formula (2) is used to calculate the number of samples that should be selected for each group. From the histogram, it can be observed that the number of pixels in groups 6–8 is very small, indicating a small N in Formula (2). Therefore, the calculated C approaches 0, which means that no sample points are selected from this interval. This is because pixels within this range may be mislabeled compared to most forestland sample points.

The method of group-based selection takes into account the differences in land features and spectral distributions in remote sensing images. It enables the more accurate selection of representative training samples and ensures that the sample quantity matches the proportion of land categories in the image [20]. This helps to improve the performance and accuracy of classification models.

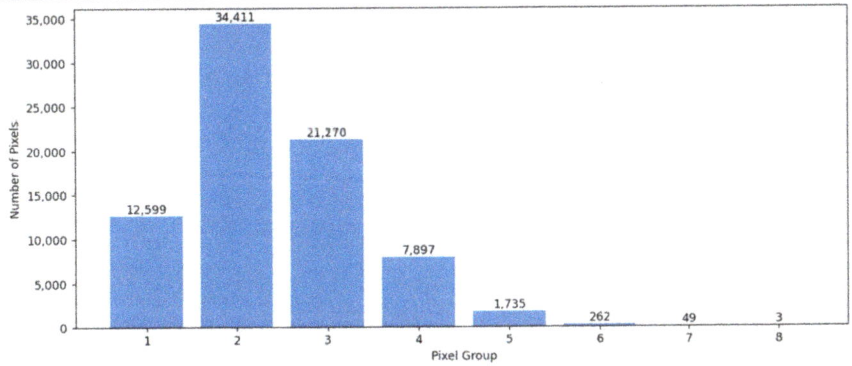

Figure 1. Distribution of woodland features.

3.3.2. The Selection Method Based on "Entropy"

In information theory and probability statistics, information entropy is a measure of the uncertainty of a random variable, and it also represents the expected value of information. It is commonly used as a quantifying indicator of information content and serves as a criterion for optimizing system equations or selecting parameters. As the information entropy increases, the number of unstable factors and the level of uncertainty within an object also increase. The magnitude of the information reflects the extent to which uncertainty events are reduced, while the magnitude of information entropy reflects the degree of uncertainty associated with the events.

Drawing upon the definition of information entropy, we propose a sample selection method based on "entropy". This method employs the "entropy" value to assess whether a sample should be selected. A higher entropy value signifies greater uncertainty and a more significant impact on the overall result, indicating that the sample contains more information. Consequently, we can choose samples with higher entropy values from a multitude of samples as training samples, as they possess more information and are suitable for training purposes. Assuming that the total area of the remote sensing sample to be classified is S and the total number of marked pixels is T, each pixel is calculated according to the following formula:

$$P(s) = -\sum_{k=1}^{n} p(x_k)\log_2 p(x_k) \quad (3)$$

where n represents the number of bands, x_k is the value of pixel x in the kth band, and $p(x_k)$ is the probability of x appearing in the kth band. Based on the total number of training samples required, the sample size of each type of land cover is calculated using stratified sampling. The proportion of stratified sampling is the percentage of the area of each type of land cover in the total image area. Samples with higher P(x) values for each type of land cover are selected as training samples, and the higher the P(x) value, the more information it contains.

3.3.3. Direct Sampling Method

In the direct sampling method, we randomly select a portion of samples from the remote sensing image and directly choose a specific number of samples as training samples through the relevant procedure. When compared to the group-based sampling method, we must ensure that the number of samples selected for each category is the same as the number of samples selected in the group-based sampling method.

4. Experiments

Three types of remote sensing data—namely, Sentinel-2, GF-1, and Landsat 8—were selected. The remote sensing images were chosen from the red, green, and blue bands.

Figure 2 displays both the original images and the classified images of these three datasets. Then, three supervised classification models were chosen, including RF, KNN, and SVM. Sample selection was carried out using the grouping-based sampling method, entropy-based sampling method, and direct sampling method. These three classification models were used to train and classify three images, and the results were compared and analyzed using a test set.

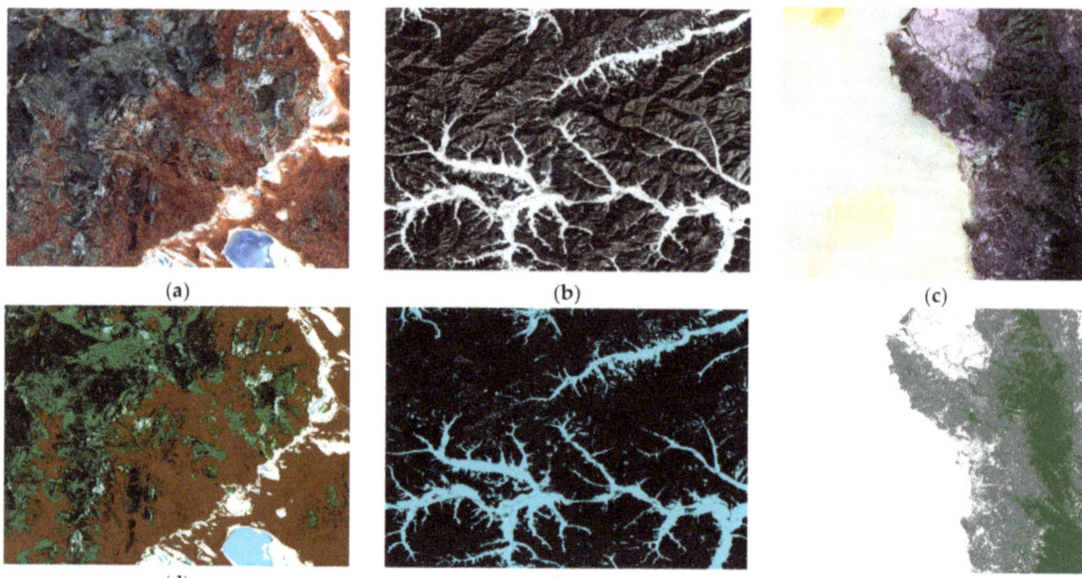

Figure 2. Original images of (**a**) Sentinel-2, (**b**) GF-1, and (**c**) Landsat 8. Classification images for (**d**) Sentinel-2, (**e**) GF-1, and (**f**) Landsat 8.

4.1. Experimental Steps

4.1.1. Selecting Training Samples

In this experiment, we selected three types of remote sensing data, including Sentinel-2, GF-1, and Landsat 8, as shown in Figure 2. We calculated the land area ratio for each image and collected samples using a stratified sampling method.

Taking Sentinel-2 as an example, the total area of the image was 500,000 square kilometers, of which forests accounted for 50%, grassland accounted for 18%, swamps accounted for 22%, saline–alkali land accounted for 6%, and water bodies accounted for 4%. In Table 2, A, B, C, D, and E represent different land cover categories. We labeled a total of 154,172 sample pixels in the image, including 78,225 forest pixels, 26,844 grassland pixels, 33,643 swamp pixels, 8962 saline–alkali land pixels, and 6498 water body pixels. Table 2 shows the number of pixels of each type of land selected for each image. It can be seen from the table that the number of pixels of each type maintains the area ratio of each land type.

Table 2. Number of total samples for three images.

Image Name	Object A	Object B	Object C	Object D	Object E	Total Sample Size
Sentinel-2	78,225	26,844	33,643	8962	6498	154,172
Landsat 8	118,247	93,162	93,053	24,295	/	328,757
GF-1	249,954	11,202	/	/	/	261,156

We took the sample sizes presented in Table 1 and divided them into groups for selection using the "entropy" method or directly selecting three methods for screening. When using the grouping method, objects within each region are divided into different groups. The method of selecting samples through grouping helps to identify more representative samples and reduce the overall sample size. The size of the sample can be controlled by adjusting the *p*-value. When the *p*-value was set to 0.5, we selected representative samples from each group and, finally, determined the number of samples for each category, as shown in Table 3. The number of forest samples decreased from 78,225 to 116, grassland from 26,844 to 76, marshland from 33,643 to 150, saline–alkali land from 8962 to 44, and water bodies from 6498 to 23. A total of 409 samples were selected as training samples for model training. It can be seen that the number of samples used was significantly reduced, and that a small number of training samples can reduce the training time of the model. When selecting training samples using the "entropy" method, we calculated the "entropy" value of each pixel and selected a certain number of samples with smaller "entropy" values from each type of object based on the proportion of object area. We trained these training samples using three different models and applied the trained models to the test set to determine their classification accuracy.

Table 3. Numbers of training samples for the three images.

Image Name	Object A	Object B	Object C	Object D	Object E	Total Sample Size
Sentinel-2	116	76	150	44	23	409
Landsat 8	364	192	193	89	/	828
GF-1	203	50	/	/	/	253

4.1.2. Selecting Model Parameters

Many supervised classification models require the selection of appropriate parameters to optimize their performance on specific objectives or datasets. Selecting classifier parameters is a crucial step in the classification process. The optimal parameters cannot be directly determined and require tuning through cross-validation methods. In this experiment, K-fold cross-validation was employed for parameter tuning, with K set to 5. Additionally, the kappa coefficient was used to assess the model parameters. The kappa coefficient is a statistical metric used to assess the performance of classification models, particularly suited for evaluating the accuracy of classification tasks, and is not influenced by class imbalance issues. It considers the consistency between the predicted results of the classification model and the actual observed results by quantifying the model's performance through comparing correct classifications with random classifications [22]. Table 4 presents the parameter ranges for each classifier, while Table 5 displays the optimal parameters selected through fivefold cross-validation and the model evaluation criterion (kappa coefficient).

Table 4. The range of model parameter selection.

Classifier	Parameter	Tested Parameter Ranges
SVM(RBF)	C	0.25, 0.50, 1, 2, 4, 8, 16, 32, 64, 128
	gamma	0.001, 0.01, 0.1, 1, 10, 100
RF	num.trees	10, 50, 100, 200
	mtry	1, 3, 5, 7, 9
KNN	K	1, 3, 5, 7, 9

We needed to analyze the classification accuracy achievable for each image using the three sample selection methods, evaluate the best sample selection method, and evaluate the influence of sample size under this sample selection method by controlling the number of samples with different labeled sample ratios (i.e., *p*-values) in the total samples.

Table 5. The optimal parameters of the model under different conditions.

Classification Model	Optimal Parameters Under Different Sample Selection Methods									
	Sample selection methods	Direct sampling method			Group-based selection method			The selection method based on "entropy"		
	Image name	GF-1	Landsat 8	Sentinel-2	GF-1	Landsat 8	Sentinel-2	GF-1	Landsat 8	Sentinel-2
SVM(RBF)	C	8	2	16	8	2	128	0.25	1	0.25
	gamma	0.01	0.001	0.001	0.01	0.001	0.001	0.001	0.001	0.001
	Kappa coefficient	0.7	0.8	0.85	0.76	0.7	0.913	0.6	0.7	0.73
RF	num.trees	200	50	100	100	200	100	10	10	10
	mtry	3	1	3	1	3	1	1	1	1
	Kappa coefficient	0.90	0.83	0.89	0.95	0.94	0.89	0.91	0.86	0.8
KNN	K	3	3	3	5	9	7	1	1	1

4.2. Analysis of Experimental Results

From Table 6, it can be observed that the different sample selection methods, including direct sampling, entropy-based sampling, and grouping selection, significantly influenced the classification results across different remote sensing images (GF-1, Landsat 8, and Sentinel-2).

Table 6. The accuracy of the three classification methods.

Sample Selection Methods	Image Name	Classification Model	Accuracy (%)
Direct sampling method	GF-1	SVM	86
		RF	88
		KNN	89
	Landsat 8	SVM	31
		RF	57
		KNN	70
	Sentinel-2	SVM	87
		RF	86
		KNN	88
Group-based selection method	GF-1	SVM	95
		RF	93
		KNN	95
	Landsat 8	SVM	81
		RF	94
		KNN	93
	Sentinel-2	SVM	90
		RF	93
		KNN	90
The selection method based on "entropy"	GF-1	SVM	69
		RF	88
		KNN	81
	Landsat 8	SVM	79
		RF	91
		KNN	91
	Sentinel-2	SVM	81
		RF	67
		KNN	78

Firstly, using the direct sampling method, the classification accuracy for GF-1 ranged from 86% to 89%, while for Landsat 8 it ranged from 31% to 70% and for Sentinel-2 it ranged from 86% to 88%. This method lacks clear selection criteria or strategies. Due to the randomness of the sample selection process and the uncertainty factors, such as different land features and sizes in different images, the classification performance on these three types of images was not satisfactory.

Under the entropy-based sampling method, the classification accuracy for GF-1 ranged from 69% to 88%, while for Landsat 8 it ranged from 79% to 91% and for Sentinel-2 it ranged from 67% to 81%. Compared to direct sampling, the entropy-based method improved the classification accuracy in some cases, but its performance was unstable across different images and models. It did not adapt well to all images and models.

However, the grouping selection method outperformed the others across all remote sensing images and classification models. In GF-1 images, the SVM and KNN classifiers achieved a classification accuracy of 95% using the grouping selection method—significantly higher than the direct sampling and entropy-based methods. In Landsat 8 images, the RF classifier achieved the highest classification accuracy of 94% using the grouping selection method. Similarly, in Sentinel-2 images, the grouping selection method also demonstrated significant improvement, with the SVM classifier's accuracy reaching over 90%.

The success of the grouping selection method lies in its ability to ensure sample representativeness, thereby enhancing the performance of classification models. It allows the models to adapt better to different images, increasing the accuracy of remote sensing image classification. In contrast, direct sampling and entropy-based methods may lead to non-representative samples, affecting the models' generalizability and accuracy.

In summary, the grouping selection method exhibits significant advantages in sample selection. It not only enhances the performance of classification models but also increases their generalizability. These models can be applied to different images, achieving high classification accuracy.

Next, let us explore the influence of sample size under the grouping selection method by controlling the number of samples with different labeled sample ratios (p-values) and investigate the impact of different sample sizes on the classification accuracy when using the grouping selection method.

For example, Sentinel-2 images were selected to test different p-values; nine sets of different numbers of labeled samples were selected, and the samples were used as training samples for three models of KNN, random forest, and SVM classifiers to examine the effects of different sample sizes on the classification accuracy. Table 7 shows the classification accuracies when different p-values are taken. From the experimental results in the table, it can be seen that the larger the p-value, the higher the classification accuracy. A classification accuracy of 90% can be achieved by all three classification models when the p-value is 0.005. The classification accuracy of the RF classifier is improved by 6.1%, that of the KNN classifier is improved by 17.9%, and that of the SVM classifier is improved by 11.8% when the p-value is increased from 0.001 to 0.1 and the number of selected labeled samples is increased from 78 to 8311. All of the classification accuracies can reach more than 93%. Therefore, the larger the value of p is in the range of 0.001 to 0.1, the larger the number of selected landmark samples and the higher the classification accuracy.

Table 7. The classification accuracy at different p-values.

The Value of p	RF Classification Accuracy (%)	KNN Classification Accuracy (%)	SVM Classification Accuracy (%)
0.001	87.3	76	82
0.003	92.7	89.1	88.4
0.005	93	91.4	90.8
0.007	92.2	90.2	90.6
0.01	92.3	90.4	91.2
0.03	93.2	92.6	92.7
0.05	93.1	93.4	93
0.07	93.3	93.6	93.6
0.1	93.4	93.9	93.8

5. Conclusions and Outlook

This paper used three methods for selecting training samples: grouping-based sampling, entropy-based sampling, and direct sampling. The KNN, SVM, and RF algorithms were used as classification models. Grouping-based sampling has significant advantages in improving classification accuracy and stability for different image classifications. Additionally, it requires only a small amount of training data, improving the classification efficiency.

When using the grouping-based sampling method, the classification accuracy improves as the sample size increases within a certain range. This has practical implications for the classification and identification of land features in remote sensing images. The research findings clearly indicate that incorporating a grouping strategy during the sample selection stage significantly enhances the performance of classification models. This provides effective support for sample selection methods in future remote sensing image processing and analysis.

However, this study also faces certain limitations that need to be considered. Firstly, the experimental results might have been influenced by the choice of datasets and the configuration of the model parameters. In this experiment, we primarily utilized three types of remote sensing images—Sentinel-2, GF-2, and Landsat 8—employing common classification models such as SVM, RF, and KNN. These choices will have had a certain impact on the research outcomes. Secondly, different remote sensing images may exhibit variations in land cover types and differences in land cover distribution and features, potentially affecting the model's performance. Therefore, it is necessary to further investigate the potential impact of image characteristics on classification results.

In addition, this study only considered three sample selection methods and three classification models. Future research could explore the integration of additional methods and incorporate a broader range of classification models, such as the BP neural network model, to further enhance classification performance. Furthermore, given the crucial role of the spatial distribution characteristics of training samples in determining the final classification accuracy, future studies should comprehensively investigate this aspect to further improve the accuracy of remote sensing image classification.

Supplementary Materials: The following supporting information can be downloaded at: https://www.mdpi.com/article/10.3390/s23208530/s1.

Author Contributions: Writing—original draft, H.Z.; writing—review and editing, J.H.; supervision and making improvements to the manuscript, S.C.; programming, Y.Z.; data collection and processing, Y.B.; data processing, Y.Q. All authors have read and agreed to the published version of the manuscript.

Funding: This research was funded by the National Key Research and Development Program of China grant number (No. 2020YFA0714103) and the Young Teachers and Students' Cutting-Edge Funding of Jilin University, China (No. 2022-JCXK-31).

Institutional Review Board Statement: Not applicable.

Informed Consent Statement: Not applicable.

Data Availability Statement: Details please see in Supplementary Materials.

Acknowledgments: This research work was supported by the National Key Research and Development Program of China (No.2020YFA0714103) and the Young Teachers and Students' Cutting-Edge Funding of Jilin University, China (No. 2022-JCXK-31). We also thank the anonymous reviewers for their helpful suggestions.

Conflicts of Interest: The authors declare no conflict of interest.

References

1. Xi, J.; Ersoy Okan, K.; Cong, M.; Zhao, C.; Qu, W.; Wu, T. Dynamic Wide and Deep Neural Network for Hyperspectral Image Classification. *Remote Sens.* **2021**, *13*, 2575. [CrossRef]
2. Lu, D.; Weng, Q. A survey of image classification methods and techniques for improving classification performance. *Int. J. Remote Sens.* **2007**, *28*, 823–870. [CrossRef]
3. Yang, F.; Zhou, G.Q.; Xiao, J.R.; Li, Q.; Jia, B.; Wang, H.Y.; Gao, J. Multispectral remote sensing image classification based on quantum entanglement. *Remote Sens. Spat. Inf. Sci.* **2020**, *42*, 667–670. [CrossRef]
4. Geiß, C.; Pelizari, P.A.; Blickensdörfer, L.; Taubenböck, H. Virtual Support Vector Machines with self-learning strategy for classification of multispectral remote sensing imagery. *ISPRS J. Photogramm. Remote Sens.* **2019**, *151*, 42–48. [CrossRef]
5. Li, H.; Zhang, S.; Zhang, C.; Li, P.; Cropp, R. A novel unsupervised Levy flight particle swarm optimization (ULPSO) method for multispectral remote-sensing image classification. *Int. J. Remote Sens.* **2017**, *38*, 6970–6992. [CrossRef]

6. Wieland, W.; Pittore, M. Performance Evaluation of Machine Learning Algorithms for Urban Pattern Recognition from Multi-spectral Satellite Images. *Remote Sens.* **2014**, *6*, 2912–2939. [CrossRef]
7. Shivam, P.; Biplab, B. Self-supervision assisted multimodal remote sensing image classification with coupled self-looping convolution networks. *Neural Netw.* **2023**, *164*, 1–20.
8. Tarabalka, Y.; Fauvel, M.; Chanussot, J.; Benediktsson, J.A. SVM- and MRF-Based Method for Accurate Classification of Hyperspectral Images. *IEEE Geosci. Remote Sens. Lett.* **2010**, *7*, 736–740. [CrossRef]
9. Li, J.; Bioucas-Dias, J.M.; Plaza, A. Semisupervised hyperspectral image classification using soft sparse multinomial logistic regression. *IEEE Geosci. Remote Sens. Lett.* **2012**, *10*, 318–322.
10. Tu, B.; Zhang, X.; Zhang, G.; Wang, J.; Zhou, Y. Hyperspectral image classification via recursive filtering and KNN. *Remote Sens. Land Resour.* **2019**, *31*, 22–23.
11. Millard, K.; Richardson, M. On the Importance of Training Data Sample Selection in Random Forest Image Classification: A Case Study in Peatland Ecosystem Mapping. *Remote Sens.* **2015**, *7*, 8489–8515. [CrossRef]
12. Zhen, Z.; Quakenbush, L.; Stehman, S.; Zhang, L. Impact of training and validation sample selection on classification accuracy assessment when using reference polygons in object-based classification. *Remote Sens.* **2013**, *34*, 6914–6930. [CrossRef]
13. Corcoran, J.; Knight, J.; Pelletier, K.; Rampi, L.; Wang, Y. The effects of point or polygon based training data on RandomForest classification accuracy of wetlands. *Remote Sens.* **2015**, *7*, 4002–4025. [CrossRef]
14. Heydari, S.S.; Mountrakis, G. Effect of classifier selection, reference sample size, reference class distribution and scene heterogeneity in per-pixel classification accuracy using 26 Landsat sites. *Remote Sens. Environ.* **2017**, *204*, 648–658. [CrossRef]
15. Shang, M.; Wang, S.; Zhou, Y.; Du, C. Effects of Training Samples and Classifiers on Classification of Landsat-8 Imagery. *J. Indian Soc. Remote Sens.* **2018**, *46*, 1333–1340. [CrossRef]
16. Li, C.; Wang, J.; Wang, L.; Hu, L.; Gong, P. Comparison of classification algorithms and training sample sizes in urban land classification with landsat thematic mapper imagery. *Remote Sens.* **2014**, *6*, 964–983. [CrossRef]
17. Fassnacht, F.E.; Hartig, F.; Latifi, H. Importance of sample size, data type and prediction method for remote sensing-based estimations of aboveground forest biomass. *Remote Sens. Environ.* **2014**, *154*, 102–114. [CrossRef]
18. Foody, G.M. Sample size determination for image classification accuracy assessment and comparison. *Int. J. Remote Sens.* **2009**, *30*, 5273–5291. [CrossRef]
19. Jin, H.; Stehman, S.V.; Mountrakis, G. Assessing the impact of training sample selection of accuracy of an urban classification: A case study in Denver, Colorado. *Remote Sens.* **2014**, *35*, 2067–2081. [CrossRef]
20. Zhiyong, L.; Li, G.; Yan, J. Training Samples Enriching Approach for Classification Improvement of VHR Remote Sensing Image. *IEEE Geosci. Remote Sens. Lett.* **2021**, *99*, 1–5.
21. Pal, M. Kernel Methods in Remote Sensing: A Review. *ISH J. Hydraul. Eng.* **2012**, *15*, 194–215. [CrossRef]
22. Ramezan, C.A.; Warner, T.A. Effects of Training Set Size on Supervised Machine-Learning Land-Cover Classification of Large-Area High-Resolution Remotely Sensed Data. *Remote Sens.* **2021**, *13*, 368. [CrossRef]

MDPI
St. Alban-Anlage 66
4052 Basel
Switzerland
www.mdpi.com

Sensors Editorial Office
E-mail: sensors@mdpi.com
www.mdpi.com/journal/sensors

www.ingramcontent.com/pod-product-compliance
Lightning Source LLC
Chambersburg PA
CBHW040814250926
51850CB00007BA/1168
9783725811717